VOLUME SIXTY SEVEN

ADVANCES IN QUANTUM CHEMISTRY

Proceedings of MEST 2012: Exponential Type Orbitals for Molecular Electronic Structure Theory

VOLUME SIXTY SEVEN

ADVANCES IN QUANTUM CHEMISTRY

Proceedings of MEST 2012: Exponential Type Orbitals for Molecular Electronic Structure Theory

Edited by

PHILIP E. HOGGAN

Series Editors
John R. Sabin
and Erkki Brändas

ELSEVIER

Amsterdam • Boston • Heidelberg • London
New York • Oxford • Paris • San Diego
San Francisco • Singapore • Sydney • Tokyo
Academic Press is an imprint of Elsevier

Academic Press is an imprint of Elsevier
The Boulevard, Langford Lane, Kidlington, Oxford, OX51GB, UK
32, Jamestown Road, London NW1 7BY, UK
Radarweg 29, PO Box 211, 1000 AE Amsterdam, The Netherlands
225 Wyman Street, Waltham, MA 02451, USA
525 B Street, Suite 1900, San Diego, CA 92101-4495, USA

First edition 2013

ISBN: 978-0-12-411544-6
ISSN: 0065-3276

For information on all Academic Press publications
visit our web site at *store.elsevier.com*

CONTENTS

PREFACE

The present volume of Advances in Quantum Chemistry collects proceedings of Molecular Electronic Structure Theory (MEST) 2012.

This international workshop drew about 100 participants from all five continents to Çanakkale, on the Dardanelles in Turkey.

This book deals with a scientific theme covered at MEST 2012; Exponential Type Orbitals (ETO) for molecular electronic structure calculations. It is an update on the conference held in Tallahassee, Florida in 1981 [1], where the term ETO was coined and two of the original participants contribute to this workshop proceedings (F. Harris and H. Silverstone). Major progress has been made in the field and we hope this snapshot is helpful to readers interested in the state of the art.

Other themes from MEST 2012 proceedings are to be presented in volume 68 of Advances in Quantum Chemistry.

The workshop was organized by a local committee, led by Israfil Guseinov and Murat Erturk and by an international committee led by Philip Hoggan. The venue was chosen after a long visit by Dr Hoggan in 2010 and scientific exchanges between us. The Turkish scientific research funding agency TUBITAK covered living expenses for speakers and travel in Turkey as well as bursaries for younger Turkish participants. A fund from Cambridge University in the UK enabled 18 younger participants from abroad to attend.

The hosting institution was Çanakkale Onsekiz Mart University, who provided the lecture-rooms and lunch free of charge and conference website

(http://physics.comu.edu.tr/MEST/).

All who attended or worked toward MEST 2012 are grateful to these generous organizations.

It should be stressed that the location was very beautiful, participant accommodation was a university residence on the beach with a view from the Asian shore of the Dardanelles into Europe. Many had a swim.

About 30 km west, the ruins of Troy, still surprisingly massive and imposing after much warfare, were the destination of the conference outing (a guided tour).

The Trojan horse was large as life, although 'troops' in it seemed very friendly!

The conference also awarded three poster prizes.

First prize: to Mr Juan Martin Randazzo, for

The Complete Model Of The Hydrogen Atom Confined In An Impenetrable Spherical Box.

by J. M. Randazzo (1) and K.V. Rodriguez (2)

(1) Division de Colisiones Atomicas, CAB-CONICET, (8400) S. C. de Bariloche, Rio Negro, Argentina

(2) Departamento de Fisica, UNS-CONICET, Av. Alem 1253, 8000 Bahia Blanca, Buenos Aires, Argentina

The second poster prize provided two awards, one to: Ms Lafifi Ismahan, for Theoretical Investigation Of The Intramolecular H-bonding On Tautomerism by Lafifi Ismahan and Khatmi Djameleddine

Laboratory of computational chemistry, University of Guelma, Algeria and one to: Mr Niyazi Yukcu for Reducing and Solving of Electric Multipole Moment Integrals by Niyazi Yukcu (1) and Emin Oztekin (2)

(1) Department of Technical Programs, Vocational School, Sinop University, Sinop, Turkey

(2) Department of Physics, Faculty of Arts and Sciences, Ondokuz Mayis University, Samsun, Turkey

To conclude, this volume contains 11 contributions that are more or less expanded from talks and some poster presentations at MEST 2012 as a lasting record of some of the rich scientific exchanges at this workshop.

As chair of MEST 2012 it is my pleasure to thank all who played a role in its success and as editor of this volume of Advances in Quantum Chemistry, seeing it grow and develop to cover one of the main themes of MEST 2012 has been, I hope, as satisfying as to you who read it.

Philip E. Hoggan, Cricqueboeuf, France, 18 October 2013.

REFERENCE

[1] Proceedings edited by Weatherford and Jones, D. Reidel Publishing Company, Dordrecht, 1982; pp 208.

CONTRIBUTORS

S. Aksoy
Department of Physics, Faculty of Arts and Sciences, Canakkale Onsekiz Mart University, Canakkale, 17100, Turkey.

V. V. Albert
Department of Physics, Yale University, P.O. Box 208120, New Haven, CT 06520, USA.

C.J. Alturria Lanzardo
Departamento de Matemática, FCEFQyN, Universidad Nacional de Río Cuarto, Ruta Nac. 36 Km 601, CP X5804BYA Río Cuarto, Argentina.

L.U. Ancarani
Théorie, Modélisation, Simulation, SRSMC, UMR CNRS 7565, Université de Lorraine, 57078 Metz, France.

V. Aquilanti
Dipartimento di Chimica, Università di Perugia, Via Elce di Sotto 8, I-06123 Perugia, Italy.

J.E. Avery
Niels Bohr Institute, University of Copenhagen, Copenhagen, Denmark.

A. Bouferguene
Faculté Saint Jean/University of Alberta, 8406-91 Street, Edmonton, Alberte, Canada T6C 4G9.

D. Calderini
Scuola Normale Superiore di Pisa, Via Consoli del Mare, 2, I-56126 Pisa, Italy.

J.C. Cesco
Instituto de Matemática Aplicada, Universidad Nacional de San Luis, CONICET; Avda. Ej. de los Andes 950, 1er piso, CP D5700HHW San Luis, Argentina.

F.D. Colavecchia
División Colisiones Atómicas, Centro Atómico Bariloche and CONICET, Río Negro, Argentina.

C. Coletti
Dipartimento di Farmacia, Università G. d'Annunzio, Via dei Vestini, I-66100 Chieti, Italy.

C.C. Denner
Departamento de Matemática, FCEFQyN, Universidad Nacional de Río Cuarto, Ruta Nac. 36 Km 601, CP X5804BYA Río Cuarto, Argentina.

M. Ertürk
Department of Physics, Faculty of Arts and Sciences, Canakkale Onsekiz Mart University, Canakkale, 17100, Turkey.

S. Firat
Department of Physics, Faculty of Arts and Sciences, Canakkale Onsekiz Mart University, Canakkale, 17100, Turkey.

A.L. Frapiccini
Departamento de Física, Universidad Nacional del Sur and IFISUR-CONICET, Bahía Blanca, Argentina.

A.M. Frolov
Department of Chemistry, University of Western Ontario, Ontario, London, Canada, N6H 5B7.

G. Gasaneo
Departamento de Física, Universidad Nacional del Sur and IFISUR-CONICET, Bahía Blanca, Argentina.

F.E. Harris
Department of Physics, University of Utah, Salt Lake City, UT 84112, USA.
Quantum Theory Project, P.O. Box 118435, University of Florida, Gainesville, FL 32611, USA.

P.E. Hoggan
Institut Pascal, UMR CNRS 6602, BP 80026, 24 Avenue des Landais, 63171 AUBIERE CEDEX, France.

S.P. Mc Carthy
Department of Chemistry, University of New Brunswick, Fredericton, New Brunswick, Canada E3B 5A3.

D.M. Mitnik
Instituto de Astronomía y Física del Espacio (IAFE), CONICET and Departamento de Física, FCEyN, Universidad de Buenos Aires, Argentina.

F.S. Ortiz
Departamento de Física, FCEFQyN, Universidad Nacional de Río Cuarto, Ruta Nac. 36 Km 601, CP X5804BYA Río Cuarto, Argentina.

S. Özay
Department of Physics, Faculty of Science and Arts, University of Ondokuz Mayis, Atakum, Samsun 55139, Turkey.

E. Öztekin
Department of Physics, Faculty of Science and Arts, University of Ondokuz Mayis, Atakum, Samsun 55139, Turkey.

J.E. Pérez
Departamento de Física, FCEFQyN, Universidad Nacional de Río Cuarto, Ruta Nac. 36 Km 601, CP X5804BYA Río Cuarto, Argentina.

J.M. Randazzo
División Colisiones Atómicas, Centro Atómico Bariloche and CONICET, Río Negro, Argentina.

A.E. Rosso
Departamento de Matemática, FCEFQyN, Universidad Nacional de Río Cuarto, Ruta Nac. 36 Km 601, CP X5804BYA Río Cuarto, Argentina.

M.B. Ruiz
Department of Theoretical Chemistry, Friedrich-Alexander-University Erlangen-Nürnberg, Egerlandstraße 3, Erlangen 91058, Germany.

O.E. Taurian
Departamento de Física, FCEFQyN, Universidad Nacional de Río Cuarto, Ruta Nac. 36 Km 601, (CP X5804BYA) Río Cuarto, Argentina.

A.J. Thakkar
Department of Chemistry, University of New Brunswick, Fredericton, New Brunswick, Canada E3B 5A3.

N. Yükçü
Department of Technical Programs, Vocational School, Sinop University, Sinop, Turkey.

PART ONE

Exponential Type Orbitals

CHAPTER ONE

Fully Correlated Wavefunctions for Three- and Four-Body Systems

F.E. Harris[a,b] and V.V. Albert[c]
[a]Department of Physics, University of Utah, Salt Lake City, UT 84112, USA
[b]Quantum Theory Project, P.O. Box 118435, University of Florida, Gainesville, FL 32611, USA
[c]Department of Physics, Yale University, P.O. Box 208120, New Haven, CT 06520, USA

Contents

Abstract

There is particular value in having wavefunctions for few-body systems that are compact but nevertheless have the features needed to capture the essence of the correlation effects. These desiderata are inherently in conflict and cannot be simultaneously reconciled when the wavefunction is described in terms of orbitals. However, explicit use of the interparticle distances leads to more efficient expansions; when all these distances are included as exponentials it is found that the expansions converge extremely rapidly, but only when the nonlinear parameters occurring therein are very carefully optimized. The present contribution reports the use of such expansions. For typical three-body systems (the He isoelectric series), with wavefunctions containing four terms (configurations), ground-state energies are obtained with errors between 32 and 38 microhartrees, and other properties, including those arising entirely from electron correlation, are given with very small errors. For a four-body system (the Li atom), a four-term wavefunction yields a ground-state energy within 350 microhartrees of the correct value.

1. INTRODUCTION

Ever since the advent of quantum mechanics, physicists and chemists have pursued the elusive goal of identifying descriptions of many-particle systems that are both reasonably accurate and sufficiently compact that

Advances in Quantum Chemistry, Volume 67
ISSN 0065-3276, http://dx.doi.org/10.1016/B978-0-12-411544-6.00001-7

the essential content of the description can be easily understood. It is well known that for atoms and molecules, an (antisymmetrized) orbital product, even if optimized by the Hartree–Fock procedure, lacks the accuracy needed to provide a good picture of effects that depend upon the interaction between the electrons. This failure suggests the possible utility of coordinate systems other than that in which each electron is described by its distance and angular position relative to a fixed center (an atomic nucleus).

The selection of optimum coordinates for few-body problems has a long history, extending at least as far back as the work of Jacobi and Radau in the nineteenth century. More recently suggestions have been made for the use of hyperspherical coordinates for this purpose,[1] and the use of more general internal (*shape*) coordinates has received attention. A general review of that topic and a survey of its current status can be found in the work of Littlejohn and Rensch[2] and the many citations provided therein. Some of these efforts have led to proposals that may be too complicated to be easily used in quantum-mechanical contexts, but they certainly point to the possible advantages of approaches in which the interparticle distances are used as coordinates.

In fact, as early as 1929, Hylleraas[3] studied the He atom using a wavefunction constructed by appending powers of the interelectron distance to an orbital product of $1s^2$ type. He obtained surprisingly good results for the He ground-state energy, with calculational methods that were relatively simple, in part because the volume element for the internal coordinates (r_1, r_2, r_{12}) has the form $r_1 r_2 r_{12}\, dr_1\, dr_2\, dr_{12}$. We note in passing that this simple form does not extend to the shape coordinates of a four-body system; there the volume element is extremely complicated.

The success of the use of the interelectron distance as a coordinate has led to a proliferation of calculations in which powers of various r_{ij} are appended to orbitals containing exponentials in the r_i (this approach is often referred to as a *Hylleraas method*). Since direct computation in the shape coordinates may be difficult, it has become customary to expand the r_{ij} in terms of the orbital coordinates, using formulas first reported by Sack.[4] The present state of this approach is nicely summarized in recent review articles by Drake[5] and King.[6] These papers examine both three- and four-body problems (e.g., the He isoelectronic series and the Li atom).

For three-body problems, the Hylleraas computation receiving the most attention was by Pekeris,[7] who reported a highly accurate study for He-like systems involving 1078 terms; a later study by Frankowski and Pekeris[8] that also included powers of $(r_1^2 + r_2^2)^{1/2}$ and $\ln(r_1 + r_2)$ produced an even more accurate wavefunction with 246 terms. It is fair to say that, while a computational tour de force, this effort did not really reach the goal of combining accuracy with easy visualization of the wavefunction. In fact, the situation corresponds well to an observation made by Mulliken[9] (and resuscitated in a recent paper by Scully et al.[10]):

> *...the more accurate the calculations become the more the concepts tend to vanish into thin air.*

An alternative to the original Hylleraas approach for He is to utilize exponentials in all three interparticle distances; the first work of this type was reported by Delves and Kalotas[11] and by Thakkar and Smith.[12] Thakkar and Smith found that with 66 terms of this type, a He wavefunction could give the ground-state energy within a microhartree while also yielding near-quantitative values of other properties. More recently, extremely accurate studies on He-like systems have been reported, including that of Frolov[13] on H^- and He, an even more precise study of He by Schwartz,[14] and computations on the He-like positive ions with nuclear charges $Z = 3$ through 10 by Frolov and Smith.[15] While these investigations quantitatively reproduced the energy and other properties of He-like systems, the wavefunctions which did so contained thousands of terms, making an understanding of their features less than apparent and underscoring Mulliken's concern.

The totally symmetric formulation provided by fully exponentially correlated wavefunctions has the interesting feature that it is appropriate for "nonadiabatic systems," i.e., those containing particles of arbitrary relative mass. Finite-mass nuclei and exotic "atoms" (e.g., those including muons) can be treated nonperturbatively and directly without the introduction of "mass-polarization" corrections.

For four-body problems, the system most studied has been the Li atom; an early and extensive Hylleraas-method study was reported by Larsson in 1968,[16] followed more recently by many highly accurate investigations using similar methods.[17–24] A similar study has also been reported for the five-body system Be.[25] Another approach for four-body systems that has received attention is the use of exponentially correlated Gaussians (i.e., the inclusion of quantities $e^{\alpha r_{ij}^2}$ in the wavefunction). Such functions, referred to as a *Gaussoid basis*, have been used for a number of four-body problems[26–28] and even for the five-body system Be.[29] The use of Gaussoids is practical for arbitrary numbers of particles (see, e.g., Ref. 30), but the convergence is slow.

Another possible approach for four-body problems is the use of fully correlated exponentials. Due to the difficulty of evaluating the necessary integrals, that approach was deemed impractical until Fromm and Hill[31] reported a closed-form analytic expression for a basic four-body integral from which all quantities needed for energy computations could be obtained. Additional analysis provided by one of the present authors[32] has now made computations of this type easier, but it remained a challenge to treat wavefunctions that contained powers of the distances as well as the exponential correlations. A preliminary calculation on the Li atom (using numerical integrations)[33] indicated the importance of including powers of distances (as expected from the fact that the zero-order description of the electronic structure of Li is $1s^2 2s$), and that observation prompted the development of recursive methods for reaching higher quantum numbers.[34] One of us has also previously reported preliminary computations on the ultimate nonadiabatic four-body system, the positronium "molecule," namely $e^+e^-e^+e^-$.[35]

None of the work thus far cited has really approached the goal of providing compact, yet reasonably accurate wavefunctions for few-body systems. A different approach is represented by recent work by Kleinekathöfer, Patil, Tang, and Toennies[36] (hereinafter KPTT). These authors elected to create compact few-body wavefunctions by choosing a functional form that enforced proper behavior at both the small- and large-r limits. The KPTT wavefunction is more complicated than the exponentially correlated functions, but it does have the virtue that it does not contain arbitrary parameters.

The present work is a study that examines the extent to which optimized compact (few-term) exponentially correlated wavefunctions can faithfully represent three- and four-body Coulomb systems. It differs from previous studies by others in that the nonlinear parameters in the wavefunctions have been carefully optimized, with the optimization revealing regularities and trends that were not previously demonstrated. We summarize some work on three-body systems that has, in part, been previously reported, and greatly extend our previous work[37] on the four-body system. Comparison with the work cited above indicates that a pragmatic approach using relatively simple correlated functions is capable of reaching higher accuracy than any other method yet explored, coming closer to (but not totally reaching) the goal of accuracy with easy visualization.

2. WAVEFUNCTIONS

A three-body system has a total of nine spatial coordinates, of which three are needed for center-of-mass translation, three for overall rotation about the center of mass, leaving three internal coordinates to provide a complete description of the relative positions of the particles (its shape coordinates). These shape coordinates can be the three interparticle distances.

We have chosen to use a fully exponentially correlated shape wavefunction for a three-body system of the general form

$$\varphi_n(r_{12}, r_{13}, r_{23}) = r_{12}^p r_{13}^q r_{23}^s e^{-\alpha r_{12} - \beta r_{13} - \gamma r_{23}}. \tag{1}$$

This equation could have been written in an alternate form with $r_{12} \to r_1$, $r_{13} \to r_2$, and $r_{23} \to r_{12}$; we did not do so because that choice of notation makes the symmetry of the form less obvious. The subscript n labels the parameter set associated with φ, i.e., the values of the integers p, q, s and the real (or possibly complex) nonlinear parameters α, β, γ. It is not necessary that all of α, β, γ have positive real parts; a bound-state wavefunction results if the real parts of $\alpha + \beta$, $\alpha + \gamma$, and $\beta + \gamma$ are all positive. These conditions cause the wavefunction to vanish asymptotically under all possible dissociations of the three-body system. The ranges of the three variables r_{ij} are coupled in that they must satisfy a "triangle condition": their lengths must be such that they can form a triangle.

To the shape wavefunction one must append the overall angular dependence associated with the complete set of orientations of the shape triangle; these orientations can be described by a set of three Euler angles, with each angular wavefunction normally chosen to be a quantum state of definite angular momentum. Because the angular orientations are more complicated than those needed to identify the position of a point particle (the orientation of a triangle requires three angular coordinates, while the angular position of a particle requires only two), the analysis and its implementation require more sophistication than is needed to treat the angular dependence of orbital products. This is an old problem, first discussed by Breit in 1930,[38] with more recent analyses by Schwartz[39] and Bhatia and Temkin.[40] A comprehensive discussion of the topic can be found in a book by Biedenharn and Louck.[41]

In the problems to be treated in the present contribution, we deal only with S states, for which the angular wavefunction is simply a constant. If the system contains identical particles, we must also consider their spin and permutational symmetry; since we are studying the ground states of two-electron atoms, we must append the anstisymmetric two-electron singlet spin state and require the overall wavefunction to be antisymmetric under interchange of the two electron coordinates. Our overall multiconfiguration wavefunction is therefore (writing $\mathcal{A}$ to denote the electron antisymmetrizer)

$$\Psi = \sum_n \mathcal{A}\left[\left(\frac{\uparrow\downarrow - \downarrow\uparrow}{\sqrt{2}}\right)\varphi_n\right]. \tag{2}$$

Continuing to four-body systems, and limiting consideration to S states, the shape wavefunction involves $12 - 3 - 3 = 6$ internal coordinates, which can be chosen to be the six interparticle distances. Note that the congruence between the numbers of shape coordinates and interparticle distances does not remain valid when there are five or more particles; that geometric fact limits applicability of the methods discussed in this paper to three- and four-body systems.

We consider fully exponentially correlated shape wavefunctions for four-body systems of the general form

$$\varphi_n(r_{12}, r_{13}, r_{14}, r_{23}, r_{24}, r_{34}) = r_{12}^p r_{13}^q r_{14}^s r_{23}^u r_{24}^v r_{34}^w e^{-\alpha r_{12}-\beta r_{13}-\gamma r_{14}-\lambda r_{23}-\mu r_{24}-\nu r_{34}}. \tag{3}$$

The values of the nonlinear parameters must be such that the wavefunction approaches zero asymptotically when the system dissociates in any way. Thus, individual exponential parameters can be negative, but the following sums of parameters must have positive real parts: $\alpha + \beta + \gamma$, $\alpha + \lambda + \mu$, $\beta + \lambda + \nu$, $\gamma + \mu + \nu$, $\beta + \gamma + \lambda + \mu$, $\alpha + \gamma + \lambda + \nu$, and $\alpha + \beta + \mu + \nu$. The ranges of the r_{ij} must be such that their six lengths can form a tetrahedron. Since a tetrahedron and its mirror image can both be constructed from the same set of r_{ij}, the complete shape space requires also a two-valued chirality indicator. Unless a problem involving chirality is under consideration, this doubling of the shape space can be ignored when evaluating matrix elements.

Our current application is to Li-like systems, which contain three electrons. The ground state of this set of problems is a spin doublet. But because the three-electron spin space consists of two independent doublets Θ_1 and Θ_2 with different energies (due to their influence on the permutational symmetry of the spatial wavefunction), we need to retain both spin states and optimize the contribution of each to the overall space-spin wavefunction. Our overall multiconfiguration wavefunction therefore has the form

$$\Psi = \sum_n \mathcal{A}\left[\left(C_{n1}\Theta_1 + C_{n2}\Theta_2\right)\varphi_n\right]. \tag{4}$$

3. OPERATORS AND MATRIX ELEMENTS

The optimum wavefunctions are determined by the variation method, which requires matrix elements of the Hamiltonian operator written in the shape coordinates. If the N particles of a Coulomb system have the respective masses m_i and charges q_i (in units of the electron mass and charge), the kinetic energy operator (in the relative coordinates r_{ij}) has for S states the form (for general N)[42,43]

$$\mathsf{T} = -\frac{1}{2}\sum_i \frac{1}{m_i}\sum_{j\neq i}\left[\frac{\partial^2}{\partial r_{ij}^2} + \frac{2}{r_{ij}}\frac{\partial}{\partial r_{ij}} + \sum_{k\neq i,j}\cos\Theta_{ijk}\frac{\partial^2}{\partial r_{ij}\partial r_{ik}}\right] \tag{5}$$

with

$$\cos\Theta_{ijk} = \frac{r_{ij}^2 + r_{ik}^2 - R_{jk}^2}{2r_{ij}r_{ik}}. \tag{6}$$

The Hamiltonian H is

$$\mathsf{H} = \mathsf{T} + \sum_{i<j}\frac{q_i q_j}{r_{ij}}, \tag{7}$$

and the units of H are hartree atomic units (1 hartree $\approx$ 27.20 eV). When any term of H is applied to a shape wavefunction φ_n, its effect will at most be to reduce the powers of some of the r_{ij} by no more than one unit and cause multiplication by some of the nonlinear parameters of φ_n. Matrix elements of H for shape wavefunctions $\mathcal{A}\varphi_m$ and $\mathcal{A}\varphi_n$ therefore reduce to linear combinations of integrals I_3 (three-body) or I_4 (four-body) of the general types

$$I_3(p,q,s) = \int r_{12}^p r_{13}^q r_{23}^s e^{-\alpha r_{12}-\beta r_{13}-\gamma r_{23}}\, \mathrm{d}\tau_s, \tag{8}$$

$$I_4(p,q,s,u,v,w) = \int r_{12}^p r_{13}^q r_{14}^s r_{23}^u r_{24}^v r_{34}^w e^{-\alpha r_{12}-\beta r_{13}-\gamma r_{14}-\lambda r_{23}-\mu r_{24}-\nu r_{34}}\, \mathrm{d}\tau_s. \tag{9}$$

The powers of the r_{ij} (i.e., p, q, s, u, v, w) are related to those in the wavefunctions φ_m, φ_n and all are integers greater than or equal to -1; the nonlinear parameters (not shown explicitly as arguments of I_N) are sums of corresponding parameters of φ_m and a term of $\mathcal{A}\varphi_n$. The quantities $d\tau_s$ are the shape-space volume element; for the three-body case it is simply $r_{12}r_{13}r_{23}\, dr_{12}\, dr_{13}\, dr_{23}$, causing the overall powers of all r_{ij} in the integration to be nonnegative. The resulting three-body integrals I_3 are therefore elementary; their values have been presented by a number of authors.[44,45]

For the four-body case the volume element is extremely complicated, but it contains a factor offsetting the possibility that powers of some r_{ij} may be -1. Evaluation of I_4 directly from Eq. (9) is impractical, but by a method involving Fourier representation and multiple contour integrations Fromm and Hill[31] obtained a closed form for the basic integral $I_4(-1,-1,-1,-1,-1,-1)$. Differentiation of this basic integral with respect to its nonlinear parameters enables the generation of any I_4. However, the repeated differentiations required to reach all needed I_4 are laborious and, even on a computer, time-consuming. It is therefore advisable to generate the needed integrals recursively[34] and write computer programs that hard-code (rather than compute) the index arithmetic. Details of the matrix-element computation, including application to properties other than the energy, can be found in earlier work.[46]

4. WAVEFUNCTION OPTIMIZATION

Once a form for Ψ (i.e., a set of powers of r_{ij} for each configuration) has been selected, it is then necessary to find the set of nonlinear parameters that optimizes the expectation value of the energy,

$$E = \frac{\langle \Psi | \mathrm{H} | \Psi \rangle}{\langle \Psi | \Psi \rangle}. \tag{10}$$

The optimizations were carried out by a conjugate gradient method. However, when the number of parameters in the optimization is large the procedure becomes numerically inefficient and repeated conjugate-gradient cycles become necessary. This difficult behavior is already apparent for the three-body problems at three and four configurations (9 and 12 nonlinear parameters) and for the four-body problem even at two configurations (12 parameters). The behavior is to be expected, as the energy should become less basis-dependent as the number of configurations is increased.

Our experience indicates that many of the supposedly optimized three-body wavefunctions in the literature are in fact far from optimized; the lack of optimization is evident not only because our few-term wavefunctions have better energies than those reported by others, but also from the fact that more complete optimization leads to previously unobserved regularities in trends across the members of an isoelectronic series.

At this point we assert that one should look with great skepticism at claims that hundreds or thousands of nonlinear parameters have been optimized in an inherently ill-conditioned electronic-structure problem.

5. RESULTS: THREE-BODY PROBLEMS

The helium atom and its isoelectronic series provide an excellent set of three-body problems for the application of exponentially correlated wavefunctions. In all the computations reported here, the nuclei were assumed to be of infinite mass. This assumption, though unnecessary, was adopted to facilitate comparison with earlier computations by others.

We begin by examining single-configuration ground-state wavefunctions for He, with the results shown in Table 1.1. The energy labeled as "Exact" is from the work of Schwartz.[14] It is well known that an energy improvement is obtained by allowing the two 1s orbitals to have different screening parameters; we note that splitting the $1s^2$ shell decreases the error in the ground-state energy by about a factor of 2. The most striking feature of the computations is the large gain obtained by use of an optimum exponentially correlated wavefunction. Not only is there a nearly tenfold improvement in the energy; qualitatively reasonable correlation-dependent properties are also obtained.

Table 1.1 Optimized single-configuration He ground-state wavefunctions based on $\varphi = e^{-\alpha r_1 - \beta r_2 - \gamma r_{12}}$. Left, closed-shell ($1s^2$); center, split-shell ($1s1s'$); right, exponentially correlated. All quantities in hartree atomic units; ν_1 is the nuclear-electron cusp, ν_2 is the electron–1electron cusp

	$\alpha = \beta,\ \gamma = 0$	$\alpha \neq \beta,\ \gamma = 0$	$\alpha \neq \beta,\ \gamma \neq 0$	Exact
E	−2.847656	−2.875661	−2.899534	−2.903724
$E - E_{\text{exact}}$	0.055618	0.028063	0.004190	
$\langle \mathbf{r}_1 \cdot \mathbf{r}_2 \rangle$	0.000000	0.000000	−0.085010	−0.064737
$\langle \mathbf{p}_1 \cdot \mathbf{p}_2 \rangle$	0.000000	0.000000	0.192818	0.159069
$\langle \delta(\mathbf{r}_1) \rangle$	1.529613	1.751235	1.777836	1.810429
$\langle \delta(\mathbf{r}_{12}) \rangle$	0.191202	0.164767	0.129494	0.106345
$\langle \delta(\mathbf{r}_1)\delta(\mathbf{r}_2) \rangle$	2.340	2.010	2.000	1.869
ν_1	−1.688	−1.910	−1.962	−2.000
ν_{12}	0.000	0.000	+0.207	+0.500

Table 1.2 Helium ground-state properties from fully converged exponentially correlated wavefunctions of 1–4 terms. All quantities in hartree atomic units; ν_1 is the nuclear-electron cusp, ν_2 is the electron–electron cusp

	1 cfg	2 cfg	3 cfg	4 cfg	Exact
E	−2.899534	−2.903270	−2.903629	−2.903688	−2.903724
$E - E_{\text{exact}}$	0.004190	0.000454	0.000095	0.000036	
$\langle \mathbf{r}_1 \cdot \mathbf{r}_2 \rangle$	−0.085010	−0.064522	−0.064373	−0.064662	−0.064737
$\langle \mathbf{p}_1 \cdot \mathbf{p}_2 \rangle$	0.192818	0.160746	0.158308	0.158902	0.159069
$\langle \delta(\mathbf{r}_1) \rangle$	1.777836	1.783896	1.810488	1.807626	1.810429
$\langle \delta(\mathbf{r}_{12}) \rangle$	0.129494	0.111481	0.108729	0.108367	0.106345
$\langle \delta(\mathbf{r}_1)\delta(\mathbf{r}_2) \rangle$	2.000	1.803	1.849	1.885	1.869
ν_1	−1.962	−1.967	−2.001	−1.993	−2.000
ν_{12}	+0.207	+0.368	+0.427	+0.436	+0.500

Continuing now to fully optimized wavefunctions with 2, 3, and 4 configurations, we note from Table 1.2 that the energy and other properties are given with surprising accuracy; the four-configuration wavefunction reproduces the exact energy (obtained from a massive fully-converged computation)[14] to within 36 microhartrees. As mentioned in the section on wavefunction optimization, this parameterization is far more effective than any previously published by others, and was obtained only after a lengthy and careful optimization.

The quality of the wavefunction optimization can also be assessed using the results of computation on the isoelectronic series $\text{H}^-, \text{He}, \text{Li}^+, \ldots$ at the four-configuration level. We studied all members through $Z = 10$ (Ne^{+8}). The wavefunctions of this isoelectronic series show great regularity in their parameter values; we provide in Figure 1.1 one example, the values of α, β, and γ [defined in Eq. (8)] of the leading configuration in the fully optimized four-term wavefunction. Similar plots for the parameters of the other three sets of nonlinear parameters have been published elsewhere.[47] The energies calculated for the isoelectronic series, shown in Table 1.3, also exhibit great regularity, with the energy error decreasing with increasing Z, consistent with the expectation that a larger nuclear charge should strengthen the electron–nucleus correlation and diminish the effect of the electron–electron interaction. This regularity is completely absent from the most accurate previous work on the He isoelectronic series, that of Frolov and Smith,[15] and is a signal that the optimization performed by those authors was not complete.

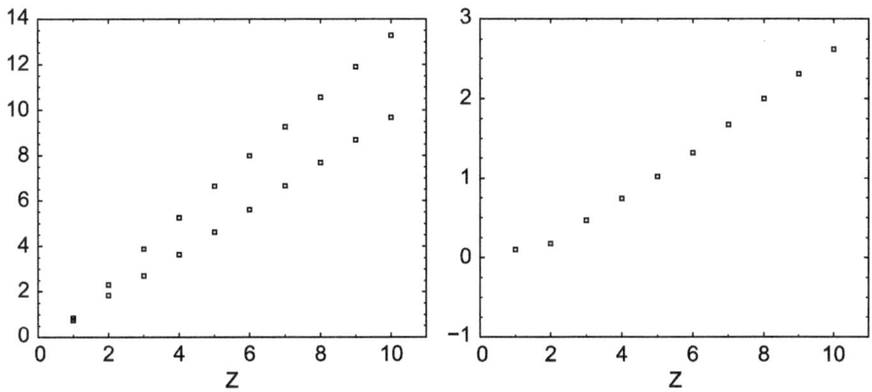

Figure 1.1 Nonlinear parameters α and β (left), and γ (right) for the leading configuration in fully optimized ground-state exponentially correlated wavefunctions for the He isoelectronic series for $Z = 1$ through 10.

Table 1.3 Calculated ground-state energies, He isoelectronic series. Energies in hartrees

	E **(4 cfg)**	***E*** **(Exact)**	***E*** **−** ***E*** **(Exact)**
H^-	−0.5277131	−0.5277510	0.0000379
He	−2.9036883	−2.9037244	0.0000361
Be^{+2}	−13.6555322	−13.6555662	0.0000340
C^{+4}	−32.4062132	−32.4062466	0.0000334
O^{+6}	−59.1565622	−59.1565951	0.0000329
Ne^{+8}	−93.9067737	−93.9068065	0.0000328

A rather different approach to compact wavefunctions is that taken by Kleinekathöfer et al. (KPTT).[36] KPTT's main premise is that the limiting small- and large-r behavior could form a starting point for constructing a good compact wavefunction. The four-term correlated wavefunctions of the present study are comparable in complexity to the KPTT wavefunction, and it is of interest to compare these two wavefunctions. Table 1.4 makes such a comparison for the energy and one additional observable, the momentum correlation $2\langle \mathbf{p}_1 \cdot \mathbf{p}_2 \rangle$. It is obvious that the pragmatically determined exponentially correlated function provides a better description of these three-body systems than does the more "first-principles" KPTT wavefunction.

Table 1.4 Comparison of four-term exponentially correlated wavefunction and the wavefunction of Kleinekathöfer et al. (KPTT). Data in hartree atomic units

		$E - E_{exact}$	$2\langle \mathbf{p}_1 \cdot \mathbf{p}_2 \rangle$	
		μhartree	Calculated	Exact
H^-	KPTT	2100	0.0628	0.0658
	This work	38	0.0659	
He	KPTT	3720	0.309	0.318
	This work	36	0.318	
Li^+	KPTT	5000	0.555	0.578
	This work	35	0.578	
Be^{+2}	KPTT	5700	0.852	0.841
	This work	34	0.841	

6. RESULTS: FOUR-BODY PROBLEMS

The four-body problem studied in this research was the Li atom (with infinite-mass nucleus). Prior to the present work, the most rapidly convergent computations on the Li ground state were by the Hylleraas method; an early study by Larsson[16] reported the convergence rate systematically as individual configurations were added. In quoting Larsson's results (and in discussion of those obtained here), errors were characterized relative to the highly precise recent computations by Puchalski and Pachucki.[20]

In the Hylleraas method, the leading configuration is built from a $1s^2 2s$ orbital product; it is not a very good approximation to the exact wavefunction and cannot describe electron correlation. Adding configurations in the order of their influence on the energy, the next three Hylleraas configurations permit a split-shell orbital description of the two-electron core but do not involve an interelectron coordinate. Configurations 5 and 6 are the first containing such a coordinate, but both address only the largest pair interaction, namely that between the two core orbitals. The result is that the first few terms in the Hylleraas expansion are not particularly effective for optimizing the energy. On the other hand, the flexibility inherent in the exponentially correlated function enables even an initial configuration to address all these energetic issues. A comparison of few-term expansions by the Hylleraas and exponentially correlated methods is presented in

Table 1.5. A further indication of the difference in the convergence rates of the two expansions is provided by the observation that it requires 30 terms of the Hylleraas expansion to reach the energy that can be attained by six terms of the exponentially correlated expansion, which brings the computed ground-state energy to within 125 microhartrees of the exact value.

An examination of the optimum choices for the powers of the r_{ij} in the exponentially correlated wavefunction shows that they have only a partial correspondence with those found optimum in the Hylleraas computation. These "prefactors" are listed in Table 1.6. As expected, the leading term is a fully correlated version of the 1s1s′2s atomic configuration. The second term can be viewed as an adjustment to permit the 2s orbital to have the hydrogenic, rather than the Slater-orbital form. Then the third term corresponds to a similar adjustment of the exponential 1s–2s correlation, which by numerical experimentation was found to be energetically slightly more important than an adjustment to the 1s–1s correlation. A more extended qualitative analysis is probably not particularly illuminating; the essence of our findings is that the initial convergence of the exponentially correlated expansion is faster than that of any other method thus far investigated.

Table 1.5 Ground-state energy of Li atom. Hylleraas: calculations by Larsson;[16] This research: exponentially correlated calculations. Energies in hartrees

Cfgs	$E_{\text{Hylleraas}}$	$E - E_{\text{exact}}$	$E_{\text{Thisresearch}}$	$E - E_{\text{exact}}$
1	−7.417907	0.060153	−7.471305	0.006755
2			−7.474682	0.003378
3	−7.430046	0.048014	−7.476987	0.002073
4	−7.444731	0.033329	−7.477723	0.000337
6	−7.474021	0.004039	−7.477936	0.000124
30	−7.477935	0.000125		

Table 1.6 Four-body wavefunctions. Prefactor multiplies exponentially correlated wavefunction. *E* in hartrees

Cfgs	Prefactor	*E*	Error (μH)
1	1s1s′2s	−7.471305	6755
2	⋯+1s1s′1s″	−7.474682	3378
3	⋯+1s1s′r_{13}	−7.476987	2073
4	⋯+1s1s′1s″	−7.477723	337

7. CONCLUSION

While the numerical effort needed to find an optimum exponentially correlated few-body wavefunction may be demanding, the resulting wavefunctions converge rapidly toward exact values and may provide a platform for the development of qualitative insights. It should be cautioned, however, that rapid initial convergence is not identical to rapid convergence at extreme accuracy, and the research direction exhibited in the current contribution may not represent an optimum approach for high-precision benchmark computations.

ACKNOWLEDGMENTS

Some of the work reported here was carried out with support from the US National Science Foundation, Grants PHY-0303412 and PHY-0601758, and was in addition supported by an NSF Graduate Research Fellowship to V.V.A. F.E.H. also thanks the organizers of the Workshop *Molecular Electronic Structure at Troy* for support of his participation.

REFERENCES

1. Öhrn, Y.; Linderberg, J. Hyperspherical Coordinates in Four Particle Systems. *Mol. Phys.* **1983,** *49*, 53–64.
2. Littlejohn, R. G.; Reinsch, M. Internal or Shape Coordinates in the N-body Problem. *Phys. Rev. A* **1995,** *52*, 2035–2051.
3. Hylleraas, E. Neue Berechnung der Energie des Heliums im Grundzustande, sowie des tiefsten Terms von Ortho-Helium. *Z. Phys.* **1929,** *54*, 347–366.
4. Sack, R. A. Generalization of Laplace's Expansion to Arbitrary Powers and Functions of the Distance between Two Points. *J. Math. Phys.* **1964,** *5*, 245–251.
5. Drake, G. W. F. High Precision Theory of Atomic Helium. *Phys. Scr.* **1999,** *T83*, 83–92.
6. King, F. W. Progress on High Precision Calculations for the Ground State of Atomic Lithium. *J. Mol. Struct. (Theochem)* **1997,** *400*, 7–56.
7. Pekeris, C. L. 1^1S, 2^1S, and 2^3S States of H^- and of He. *Phys. Rev.* **1962,** *126*, 1470–1476.
8. Frankowski, K.; Pekeris, C. L. Logarithmic Terms in the Wave Functions of the Ground State of Two-electron Atoms. *Phys. Rev.* **1966,** *146*, 46–49.
9. Mulliken, R. S. Molecular Scientists and Molecular Science: Some Reminiscences. *J. Chem. Phys.* **1965,** *43*, S2–S11.
10. Scully, M. O.; Allen, R. E.; Dou, Y.; Kapale, K. T.; Kim, M.; Chen, G.; Svidzinsky, A. Molecular Calculations with Two-center Correlated Orbitals. *Chem. Phys. Lett.* **2004,** *389*, 385–392.
11. Delves, L. M.; Kalotas, T. Variational Calculations for Three-body S-states with Coulomb Interactions. *Aust. J. Phys.* **1968,** *21*, 431–438.
12. Thakkar, A. J.; Smith, V. H., Jr. Compact and Accurate Integral-transform Wave Functions. I. The 1^1S State of the Helium-like Ions from H^- through Mg^{10+}. *Phys. Rev. A* **1977,** *15*, 1–15.

13. Frolov, A. M. Two-stage Strategy for High-precision Variational Calculations. *Phys. Rev. A* **1998,** *57,* 2436–2439.
14. Schwartz, C. Experiment and Theory in Computations of the He Atom Ground State. *Int. J. Mod. Phys. E* **2006,** *15,* 877–888.
15. Frolov, A. M.; Smith, V. H., Jr. Exponential Representation in the Coulomb Three-body Problem. *J. Phys. B* **2004,** *37,* 2917–2932.
16. Larsson, S. Calculations on the ^2S Ground State of the Lithium Atom Using Wave Functions of Hylleraas Type. *Phys. Rev.* **1968,** *169,* 49–54.
17. Puchalski, M.; Pachucki, K. Relativistic, QED, and Finite Nuclear Mass Corrections for Low-lying States of Li and Be^+. *Phys. Rev. A* **2008,** *78,* 052511.
18. Pachucki, K.; Puchalski, M. Singular Hylleraas Three-electron Integrals. *Phys. Rev. A* **2008,** 77, 032511.
19. Puchalski, M.; Moro, A. M.; Pachucki, K. Isotope Shift of the $3^2S_{1/2}$–$2^2S_{1/2}$ Transition in Lithium and the Nuclear Polarizability. *Phys. Rev. Lett.* **2006,** *97,* 133001.
20. Puchalski, M.; Pachucki, K. Ground-state Wave Function and Energy of the Lithium Atom. *Phys. Rev. A* **2006,** *73,* 022503.
21. Pachucki, K.; Puchalski, M. Extended Hylleraas Three-electron Integral. *Phys. Rev. A* **2005,** *71,* 032514.
22. Pachucki, K.; Puchalski, M.; Remiddi, E. Recursion Relations for the Generic Hylleraas Three-electron Integral. *Phys. Rev. A* **2004,** *70,* 032502.
23. Thakkar, J.; Koga, T.; Tanabe, T.; Teruya, H. Compact Hylleraas-type Wavefunctions for the Lithium Isoelectronic Sequence. *Chem. Phys. Lett.* **2002,** *366,* 95–99.
24. King, F. W.; Ballegeer, D. G.; Larson, D. J.; Pelzl, P. J.; Nelson, S.; Prosa, T. J.; Hinaus, B. M. Hylleraas-type Calculations of the Relativistic Corrections for the Ground State of the Lithium Atom. *Phys. Rev. A* **1998,** *58,* 3597–3603.
25. King, F. W. High-precision Calculations of the Hyperfine Constants for the Low-lying Excited ^2S States of Be^+. *J. Phys. Chem. A* **2009,** *113,* 4110–4116.
26. Kinghorn, D. B.; Poshusta, R. D. Nonadiabatic Variational Calculations on Dipositronium Using Explicitly Correlated Gaussian Basis Functions. *Phys. Rev. A* **1993,** *47,* 3671–3681.
27. Kozlowski, P. M.; Adamowicz, L. Nonadiabatic Variational Calculations for the Ground State of the Positronium Molecule. *Phys. Rev. A* **1993,** *48,* 1903–1908.
28. Frolov, A. M.; Wardlaw, D. M. Compact Variational Wave Functions for Bound States in Three-electron Atomic Systems. *J. Exp. Theor. Phys.* **2009,** *108,* 583–592.
29. Frolov, A. M.; Wardlaw, D. M. Compact Wave Functions for Four-electron Atomic Systems. *Phys. Rev. A* **2008,** *78,* 042506.
30. Harris, F. E.; Monkhorst, H. J. Recurrence Relations for Fully Correlated Gaussians with Odd Powers of Interparticle Coordinates. *Int. J. Quantum Chem.* **2006,** *106,* 3186–3189.
31. Fromm, D. M.; Hill, R. N. Analytic Evaluation of Three-electron Integrals. *Phys. Rev. A* **1987,** *36,* 1013–1044.
32. Harris, F. E. Analytic Evaluation of Three-electron Integrals with Slater Wave Functions. *Phys. Rev. A* **1997,** *55,* 1820–1831.
33. Guevara, N. L.; Harris, F. E.; Turbiner, A. V. An Accurate Few-parameter Ground State Wave Function for the Lithium Atom. *Int. J. Quantum Chem.* **2009,** *109,* 3036–3040.
34. Harris, F. E. Recurrence Formulas for Fully Exponentially Correlated Four-body Wavefunctions. *Phys. Rev. A* **2009,** *79,* 032517.
35. Harris, F. E.; Smith, V. H., Jr. Highly Compact Wavefunctions for Two-electron Systems. *Adv. Quantum Chem.* **2005,** *48,* 407–419.
36. Kleinekathöfer, U.; Patil, S. H.; Tang, K. T.; Toennies, J. P. Boundary-condition-determined Wave Function for the Ground State of Helium and Isoelectronic Ions. *Phys. Rev. A* **1996,** *54,* 2840–2849.

37. Albert, V. V.; Guevara, N. L.; Sabin, J. R.; Harris, F. E. Few-parameter Exponentially Correlated Wavefunctions for the Ground State of Lithium. *Int. J. Quantum Chem.* **2009,** *109*, 3791–3797.
38. Breit, G. Separation of Angles in the Two-electron Problem. *Phys. Rev.* **1930,** *35*, 569–578.
39. Schwartz, C. L. Lamb Shift in the Helium Atom. *Phys. Rev.* **1961,** *123*, 1700–1705.
40. Bhatia, A. K.; Temkin, A. Symmetric Euler-angle Decomposition of the Two-electron Fixed-nucleus Problem. *Rev. Mod. Phys.* **1964,** *36*, 1050–1064.
41. Biedenharn, L. C.; Louck, J. D. *Angular Momentum in Quantum Physics: Theory and Application;* Addison-Wesley: Reading, MA, 1981.
42. Rebane, T. K. To the Energy Calculation of Coulomb System. *Opt. Spektrosk.* **1993,** *75*, 945–953.
43. Harris, F. E. Current Methods for Coulomb Few-body Problems. *Adv. Quantum Chem.* **2004,** *47*, 129–155.
44. Drake, G. W. F. High Precision Calculations for Helium. In *Atomic, Molecular, and Optical Physics Handbook;* Drake, G. W. F., Ed.; AIP: Woodbury, NY, 1996; pp 154–171.
45. Harris, F. E.; Frolov, A. M.; Smith, V. H., Jr. Singular and Non-singular Three-body Atomic Integrals for Exponential Wavefunctions. *J. Chem. Phys.* **2004,** *121*, 6323–6333.
46. Harris, F. E.; Frolov, A. M.; Smith, V. H., Jr. Exponential Variational Expansion in Relative Coordinates for Highly Accurate Bound State Calculations in Four-body Systems. *J. Chem. Phys.* **2003,** *119*, 8833–8841.
47. Harris, F. E.; Smith, V. H., Jr. Systematics of the Ground States. In *Symmetry, Spectroscopy and SCHUR, Proceedings of the Prof. Brian G. Wybourne Commemorative Meeting;* King, R. C., Bylicki, M., Karwowski, J., Eds.; Nicolaus Copernicus University Press: Torun, Poland, 2006; pp 127–137.

CHAPTER TWO

Electron and Electron-Pair Number and Momentum Densities for Low-Lying States of He, H^-, and Li^+

A. J. Thakkar and S. P. Mc Carthy
Department of Chemistry, University of New Brunswick, Fredericton, New Brunswick, Canada E3B 5A3

Contents

Abstract

Electron and electron-pair number and momentum distributions play a fundamental role in chemistry; hence, benchmark calculations for small systems are important. A brief review of such computations for the low-lying states of the helium atom and isoelectronic ions is presented. Although very highly accurate calculations have been reported previously for the electron number density and the electron-pair intracule and extracule densities in position space, there is a lack of highly accurate computations of their momentum space counterparts. Fresh calculations using explicitly correlated wave functions expanded in well-optimized Gaussian geminals are reported to fill this lacuna. A consolidated set of tables of moments of these densities in both spaces is compiled for future reference.

1. INTRODUCTION

Reduced density kernels, more commonly known as reduced density matrices, provide a means by which dimensionality can be reduced without loss of the physically most relevant information in the N-electron

Advances in Quantum Chemistry, Volume 67
ISSN 0065-3276, http://dx.doi.org/10.1016/B978-0-12-411544-6.00002-9

wave function.[1–5] Since the usual electronic Hamiltonian for an atom or molecule contains only one- and two-electron terms, the second-order reduced density matrix (2-RDM) suffices for almost all purposes. If only probability distributions and multiplicative operators are of interest, then it suffices to work with the diagonal element of the 2-RDM which is just the electron-pair density. Quite often, the spin dependence is not of interest and the focus can then be restricted to the spin-traced electron-pair density Γ. The latter can be obtained from the N-electron wave function Ψ by integration (contraction) in the following manner:

$$\Gamma(\mathbf{r}_1, \mathbf{r}_2) = c_2 \int \left|\Psi(\mathbf{x}_1, \mathbf{x}_2, \ldots, \mathbf{x}_N)\right|^2 d\sigma_1\, d\sigma_2\, d\mathbf{x}_3 \cdots d\mathbf{x}_N \qquad (1)$$

in which the combined space-spin coordinate, $\mathbf{x}_j = (\mathbf{r}_j, \sigma_j)$, is composed of the position vector $\mathbf{r}_j$ and spin σ_j of electron j. The normalization of Γ is fixed by c_2. Common choices of c_2 include unit normalization $c_2 = 1$, McWeeny's normalization[4] $c_2 = N(N-1)$, and Löwdin's normalization[3] $c_2 = N(N-1)/2$. In this work, we use the Löwdin normalization. The above formulation of Γ as a contraction of the N-electron probability distribution, the square of the N-electron wave function, clearly highlights Γ as the spatial probability distribution for a pair of electrons. Hence, Γ is non-negative everywhere. Moreover, Γ is symmetric with respect to interchange of $\mathbf{r}_1$ and $\mathbf{r}_2$ because electrons are indistinguishable. The electron-pair number density Γ is a function of special interest to quantum chemists because the concept of electron pairs is fundamental to chemistry.

It is not easy to visualize $\Gamma(\mathbf{r}_1, \mathbf{r}_2)$ because it is a function of six variables. One way to proceed is by further reduction of dimensionality. The simplest reduction is to contract to the one-electron number density ρ by integrating Γ over one of the electrons as follows:

$$\rho(\mathbf{r}) = \frac{c_1}{c_2} \int \Gamma(\mathbf{r}, \mathbf{r}_2) d\mathbf{r}_2 \qquad (2)$$

in which c_1 determines the normalization of ρ. The usual choices are either $c_1 = 1$ or $c_1 = N$. In this work, we use the latter so that ρ is normalized to the number of electrons. X-ray scattering is the primary experimental technique for measuring the electron number density.[6] The electron density is, of course, the fundamental quantity in density functional theory[7,8] and in Bader's theory of molecular structure.[9] Both experimental and ab initio electron densities are used widely for the purposes of interpretation and understanding.[6,8–10]

However, electron-pair information is more clearly obtained from contractions of Γ that retain true pair character. Complementary three-dimensional contractions of the electron-pair density are the intracule density given by

$$I(\mathbf{u}) = \int \Gamma(\mathbf{r}_1, \mathbf{r}_2)\delta(\mathbf{u} - \mathbf{r}_1 + \mathbf{r}_2)\mathrm{d}\mathbf{r}_1\, \mathrm{d}\mathbf{r}_2, \tag{3}$$

and the extracule density defined by

$$E(\mathbf{R}) = \int \Gamma(\mathbf{r}_1, \mathbf{r}_2)\delta[\mathbf{R} - (\mathbf{r}_1 + \mathbf{r}_2)/2]\mathrm{d}\mathbf{r}_1\, \mathrm{d}\mathbf{r}_2, \tag{4}$$

in which $\delta(\mathbf{r})$ is the three-dimensional Dirac delta function. The intracule density is the probability density for the interelectronic vector $\mathbf{u} = \mathbf{r}_1 - \mathbf{r}_2$. It was first alluded to by Debye[11] and first defined using wave functions by Coulson and Neilson.[12] Coleman[13] first defined the extracule density which is the probability density for $\mathbf{R} = (\mathbf{r}_1 + \mathbf{r}_2)/2$, the position vector of the electron pair's center-of-mass. The first quantum mechanical calculations of the radial intracule density were reported by Coulson and Neilson[12] for He, and the first calculations of the extracule density were made by Thakkar and Moore[14] for He, Li, Be, H_2, and LiH. More recent work on the intracule and extracule densities can be traced through reviews given by Thakkar,[15] Boyd and Ugalde,[16] Koga,[17] Valderrama et al.,[18] Poater et al.,[19] and Gill et al.[20] Total X-ray and high-energy electron scattering experiments[21] can be used to probe the intracule density and, in the special case of two-electron systems, even the correlation contribution to the intracule density.[22] However, the overwhelming majority of the available information on the intracule density has come from ab initio computations.

An alternative perspective[23,24] of electronic structure is provided by the momentum-space (p-space) representation[25,26] instead of the usual position-space (r-space) representation. In p-space, wave functions are expressed in terms of the momenta ($\mathbf{p}_i$) and spins of the particles instead of their positions and spins. The combined space-spin coordinate $\mathbf{x}_i = (\mathbf{r}_i, \sigma_i)$ is replaced by a combined momentum-spin coordinate $\mathbf{y}_i = (\mathbf{p}_i, \sigma_i)$. Thus, a p-space wave function $\Phi(\mathbf{y}_1, \mathbf{y}_2, \ldots, \mathbf{y}_N)$ replaces its r-space counterpart $\Psi(\mathbf{x}_1, \mathbf{x}_2, \ldots, \mathbf{x}_N)$. Exactly the same information can be extracted from the wave function in either representation because the two wave functions are related by a $3N$-dimensional, norm-preserving, Fourier transform:

$$\Phi = (2\pi)^{-3N/2} \int \mathrm{e}^{-\mathrm{i}(\mathbf{p}_1\cdot\mathbf{r}_1+\cdots+\mathbf{p}_N\cdot\mathbf{r}_N)}\Psi \mathrm{d}\mathbf{r}_1 \mathrm{d}\mathbf{r}_2 \cdots \mathrm{d}\mathbf{r}_N. \tag{5}$$

Reduced density matrices can be defined[15,23,27] in p-space in exact analogy with their r-space counterparts. Thus, $\Pi_2(\mathbf{p}_1, \mathbf{p}_2)$, the p-space pair density,[15,23] can be obtained from the N-electron, p-space wave function Φ as follows:

$$\Pi_2(\mathbf{p}_1, \mathbf{p}_2) = c_2 \int \left|\Phi(\mathbf{y}_1, \mathbf{y}_2, \ldots, \mathbf{y}_N)\right|^2 \mathrm{d}\sigma_1\, \mathrm{d}\sigma_2\, \mathrm{d}\mathbf{y}_3 \cdots \mathrm{d}\mathbf{y}_N. \tag{6}$$

Integration over the momentum of one electron leads to the one-electron momentum density Π as follows:

$$\Pi(\mathbf{p}) = \frac{c_1}{c_2} \int \Pi_2(\mathbf{p}, \mathbf{p}_2) \mathrm{d}\mathbf{p}_2. \tag{7}$$

Positron annihilation, binary $(e, 2e)$ or electron momentum spectroscopy, and inelastic (Compton) scattering of X-rays, γ-rays, and high-energy electrons are all experimental avenues to information about the one-electron momentum density Π.[28,29] In principle, (γ, 2e) and (e, 3e) scattering experiments[30] could be used to probe the p-space pair density but, in practice, almost all of our knowledge of the electron-pair momentum density comes from ab initio calculations.

Momentum-space versions of the intracule and extracule densities are defined in a manner entirely analogous to that used in r-space. Thus, contractions of the electron-pair momentum density can be made[15] to find the p-space intracule density defined by

$$I^p(\mathbf{v}) = \int \Pi_2(\mathbf{p}_1, \mathbf{p}_2)\delta(\mathbf{v} - \mathbf{p}_1 + \mathbf{p}_2) \mathrm{d}\mathbf{p}_1\, \mathrm{d}\mathbf{p}_2, \tag{8}$$

and the p-space extracule density defined by

$$E^p(\mathbf{P}) = \int \Pi_2(\mathbf{p}_1, \mathbf{p}_2)\delta[\mathbf{P} - (\mathbf{p}_1 + \mathbf{p}_2)/2] \mathrm{d}\mathbf{p}_1\, \mathrm{d}\mathbf{p}_2, \tag{9}$$

in which $\mathbf{v} = \mathbf{p}_1 - \mathbf{p}_2$ is the interelectronic momentum vector, and $\mathbf{P} = (\mathbf{p}_1 + \mathbf{p}_2)/2$ is the central momentum vector of the electron pair. The first calculations of the p-space intracule density were made by Banyard and Reed,[31] and the first calculations of the p-space extracule density were made by Wang and Smith.[32]

The r- and p-space RDMs of each order are related by Fourier transforms[23,27] because the r- and p-space wave functions Ψ and Φ are Fourier transforms of each other. However, it is important to realize that the diagonal elements of the RDMs in the two spaces are *not* related by Fourier transformation.[23,27] In particular, the electron and electron-pair number densities are not Fourier transforms of their momentum counterparts. Similarly, the intracule and extracule densities in the two spaces are not Fourier transforms of each other.

Benchmark high-accuracy calculations of the r- and p-space one-electron and electron-pair densities for at least a few simple systems are highly desirable. The helium atom and its isoelectronic ions are perhaps the simplest two-electron systems. In this article, we review what is known about the electron and electron-pair number and momentum densities for the ground state and low-lying excited states of the helium isoelectronic series. Those densities that have not been reported previously are calculated to complete the picture. A consolidated set of tables of moments of these densities in both spaces is compiled for future reference. Atomic units are used throughout.

2. EXPLICITLY CORRELATED WAVE FUNCTIONS

The non-relativistic and infinite nuclear mass approximations are made throughout. This section contains a discussion of the wave functions for two-electron atoms under such approximations.[33] Since spin can be factored out from a two-electron wave function, it suffices to discuss spin-free r-space wave functions that depend only on the position vectors, $\mathbf{r}_1$ and $\mathbf{r}_2$, of the two electrons. Letting $\hat{P}_{12}$ be the operator that interchanges $\mathbf{r}_1$ and $\mathbf{r}_2$, note that the Pauli principle requires that the spatial wave function $\Psi(\mathbf{r}_1, \mathbf{r}_2)$ be symmetric with respect to $\hat{P}_{12}$ for singlet states and antisymmetric with respect to $\hat{P}_{12}$ for triplet states. Similarly, the spin-free wave function in p-space, Φ, depends only upon the electronic momentum vectors $\mathbf{p}_1$ and $\mathbf{p}_2$ and has analogous behavior with respect to permutation of $\mathbf{p}_1$ and $\mathbf{p}_2$.

The relative simplicity of the two-electron one-center problem is such that calculations can be carried out to much higher accuracy than is currently possible with correlated wave function theory that is based, implicitly or explicitly, upon expansions in Slater determinants. The most accurate wave functions for the helium isoelectronic sequence require an explicit dependence on the interelectronic distance. Such explicitly correlated wave functions were pioneered by Hylleraas.[34,35] English translations of four of Hylleraas' papers can be found in Ref. 36. The wave functions for the spherically symmetric S-states depend only upon the side lengths of the triangle formed by the nucleus and the two electrons. These lengths are the interelectronic distance r_{12}, and the two electron-nucleus distances r_1 and r_2. In particular, we can expand the S-state wave functions in the general r-space form:

$$\Psi = \left(1 \pm \hat{P}_{12}\right) \sum_{i=1}^{K} c_i \psi_i(r_1, r_2, r_{12}), \tag{10}$$

where c_i is a variational expansion coefficient for the geminal basis function ψ_i and K is the number of geminals used. The plus sign is required for singlet states and the minus sign for triplet states. Explicit dependence of the geminals ψ_i on r_{12} means that they cannot be factored into a product of orbitals. Angular factors needed for states of higher angular momentum can be determined by making use of the invariance of the Hamiltonian to rotation about any axis through the nucleus. For P-states, the $M_L = 0$ wave function can be written in the general form:

$$\Psi = \left(1 \pm \hat{P}_{12}\right) \sum_{i=1}^{K} c_i z_1 \psi_i(r_1, r_2, r_{12}), \tag{11}$$

where z_1 is the z-coordinate of electron 1.

The geminals appearing in the above expansions, Eqs. (10) and (11), are quite often written in so-called Hylleraas form as

$$\psi_i(r_1, r_2, r_{12}) = r_1^{j_i} r_2^{k_i} r_{12}^{\ell_i} e^{-\alpha r_1 - \beta r_2}, \tag{12}$$

where the powers j_i, k_i, and ℓ_i are non-negative integers. It is usual to include all geminals with powers satisfying a condition like $0 \leq j_i + k_i + \ell_i \leq \Omega$. The integer Ω provides a systematic means of choosing and ordering geminals with different powers and determines the expansion length. More efficient schemes for systematically selecting the powers have been suggested and used.[37–39] Many Hylleraas-type wave functions are expressed in these triangular coordinates but other coordinates are used too. In fact, Hylleraas[34,35] himself chose the elliptic coordinates given by $s = r_1 + r_2$, $t = r_1 - r_2$, and $u = r_{12}$. Coolidge and James[40] chose the perimetric coordinates defined by $x = r_1 + r_2 - r_{12}$, $y = -r_1 + r_2 + r_{12}$, and $z = r_1 - r_2 + r_{12}$. Later, Pekeris used perimetric coordinates and expansions containing as many as 1078 terms in his ground-breaking calculations;[41–46] they led to energies now known to be no more than a few nanohartrees above the exact ones.

Several variants of the Hylleraas scheme described above have been used in high-accuracy calculations. One approach is to use the elliptic coordinates and allow additional fractional[47–49] and negative[50] powers of s and u; this line of inquiry leads to more compact wave functions with 100-term expansions which yield ground state energies accurate to a few tenths of a picohartree.[51,52] A second approach,[53,54] suggested by the Fock expansion[55–59] of the true wave function, is to include negative powers of s, half-integer powers of $s^2 + t^2$, and powers of $\ln s$. A third approach[60–63] to obtaining relatively compact Hylleraas-type wave functions is to forgo all powers but allow many different scale parameters by using the Slater-Hylleraas geminals[35,64,65] defined by

$$\psi_i(r_1, r_2, r_{12}) = \mathrm{e}^{-\alpha_i r_1 - \beta_i r_2 - \gamma_i r_{12}}. \tag{13}$$

The exponential parameters $(\alpha_i, \beta_i, \gamma_i)$ were chosen in a pseudo-random manner from a variationally optimized rectangular parallelepiped or cuboid to avoid expensive full non-linear optimization.[60] A fourth approach[66–68] is to double the Hylleraas basis using different non-linear scale parameters, α and β, in the two sets of terms. This method has been used to obtain definitive energies with subpicohartree accuracy for many states of the helium isoelectronic sequence.[68]

Hylleraas wave functions and their variants as described above can be extremely accurate. However, Fourier transforming them to momentum space is quite difficult.[69,70] This problem can be sidestepped[71–73] by using explicitly correlated wave functions constructed from Gaussian geminal basis functions of the Boys-Singer form[74,75] defined by

$$\psi_i(r_1, r_2, r_{12}) = \mathrm{e}^{-\alpha_i r_1^2 - \beta_i r_2^2 - \gamma_i r_{12}^2}. \tag{14}$$

Wave functions using Gaussian geminals are easy to transform to p-space. If the geminals are well optimized, then such wave functions can predict most, but not all, properties of interest with sufficient accuracy. In recent years, full optimization of all the non-linear parameters in basis functions like the Gaussian geminals and Slater-Hylleraas geminals has become feasible because of the explosive growth in computer resources and the development of efficient algorithms. One such algorithm[76] is based on varying one geminal at a time and solving the generalized matrix eigenvalue problem at each step in this partial optimization problem using rank-one matrix updates.

Table 2.1 Comparison of the Gaussian geminal energies E_{GG} with the exact energies E_{ex} for the ^{1}S states. The error in the Gaussian geminal energies is $\Delta E = E_{GG} - E_{ex}$

State	K	E_{GG}^{a}	E_{ex}^{b}	$10^9 \Delta E$
H^- (1^1S)	160	−0.5277510142	−0.5277510165	2.3
He (1^1S)	160	−2.9037243753	−2.9037243770	1.7
He (1^1S)	335	−2.9037243754	−2.9037243770	1.6
He (2^1S)	335	−2.1459740358	−2.1459740460	10.2
Li^+ (1^1S)	160	−7.2799134103	−7.2799134127	2.4
Li^+ (1^1S)	335	−7.2799134104	−7.2799134127	2.3
Li^+ (2^1S)	335	−5.0408767327	−5.0408767452	13.5

[a] K Gaussian geminals, this work.
[b] Nearly exact energies from Ref.68 except for Li^+ (2^1S) which is from Ref. 79.

3. CALCULATIONS

The focus of this article is on the He atom, the H^- anion, and the Li^+ cation. For He and Li^+, the ground 1^1S state and the four lowest excited states (2^3S, 2^1S, 2^3P, and 2^1P) are examined. Only the ground 1^1 S state is considered for H^- because all its singly excited states are unbound.[77]

When a sufficiently accurate value for a property that we wanted to tabulate was not, to our knowledge, available in the literature, we computed it from wave functions expanded in Gaussian geminals. This choice of geminal basis was made because most unavailable properties were p -space properties. The wave functions for the 2^3S, 2^3P, and 2^1P states of He and Li^+ are those we described in our earlier work on r-space extracule densities.[78] These wave functions contain 110 geminals for the ^{3}S-states and 175 geminals for the P-states. Wave functions for the ground 1^1S states were obtained as in our earlier work.[76] One hundred and sixty terms were sufficient to obtain energies no more than 2.4×10^{-9} above the "exact" ones;[51,52,68] see Table 2.1.

The 2^1S state is the second-lowest one of ^{1}S symmetry and the requisite energy upper bound is given by the second-lowest eigenvalue of the secular equation.[80,81] However, use of a geminal basis set with the scale parameters optimized for the ground 1^1S state energy leads to a poor energy for the 2^1S state; see, for example, Ref. 61. Thus, it is usual[80,82] to optimize separately the scale parameters for the 2^1S state knowing that the resulting geminal basis set will lead to a poor approximation for the ground state.[82]

Such a procedure in which different geminal basis sets are used for two states of the same symmetry is used commonly to this day.[61,63,68,79] The resulting lack of strict orthogonality between the two approximate wave functions does not normally cause any problems.

It is straightforward to optimize a Gaussian geminal basis set for the 2^1S state. Nevertheless, it seems natural to ask whether the geminal basis sets already optimized for the 1^1S, 2^3S, 2^3P, and 2^1P states can be used in some fashion to construct an accurate basis set for the 2^1S state. After a bit of thought and experimentation, we obtained excellent results without much computation in the following manner. We took the 175 Gaussian geminals optimized[78] for the 2^1P state and uniformly scaled all the α_i and β_i by a factor ζ and all the γ_i by a factor ω. Next we added, without change, the 160 geminals optimized for the ground 1^1S state to obtain a basis set of 335 geminals. The 2^1S energy in the 335 geminal basis set was then minimized with respect to ζ and ω. The best energies obtained in this manner are above the true ones by no more than 13.5 nanohartrees; see Table 2.1. The optimized scale factors were $\zeta = 0.865$ and $\omega = 1.2$ for He, and $\zeta = 0.80$ and $\omega = 1.1$ for Li^+. The same 335 geminal basis set was used to compute the ground state energies and they turn out to be only 0.1 nanohartrees lower than the ones obtained with the original 160 optimized geminals. For consistency, we use the 335 geminal wave functions for both the 1^1S and 2^1S states of He and Li^+ in the rest of this work. An added benefit of this approach is that the approximate wave functions for the 1^1S and 2^1S states are strictly orthogonal. This may turn out to be useful, in later work, for properties that depend on the transition density. For example, we plan to check whether the generalized oscillator strength for transitions from the 2^1S state calculated previously[83,84] using different geminal basis sets for the two states changes significantly at small values of the momentum transfer if a single geminal basis set is used instead.

4. POSITION-SPACE DENSITIES

The electron number density and the electron-pair intracule and extracule densities are all spherically symmetric for S-state atoms. However, different non-spherical densities are obtained for the P states depending on the value of M_L (-1, 0, or 1). Since these states with different M_L are degenerate, the physically meaningful densities for *free* atoms are obtained by averaging the densities over the manifold of degenerate states. The ensemble averaged density is identical to the spherically averaged density for any fixed M_L. In this work, attention is restricted to these averaged densities.

In this work, $\rho(r), I(u)$, and $E(R)$, respectively, are used to denote the number, intracule, and extracule densities for the S states and their spherically averaged counterparts for the P states. The corresponding radial probability densities are then given by

$$\rho_r(r) = 4\pi r^2 \rho(r), \tag{15}$$

$$I_r(u) = 4\pi u^2 I(u), \tag{16}$$

and

$$E_r(R) = 4\pi R^2 E(R). \tag{17}$$

The literature on densities abounds with special-case notation for spherically averaged and radial densities that varies from author to author and sometimes even from article to article. In this work, we avoid non-systematic notation because using it could make this article quite confusing, even for a specialist. For example, other notation which we have encountered includes n, γ, P_1, and D_1 for the electron number density, h for the spherically averaged intracule density, d for the spherically averaged extracule density, D, D_0, and P for the radial number density, and f, P, and P_{12} for the radial intracule density.

Many studies of the electron number densities of He and isoelectronic ions, based on Hylleraas-type wave functions, have been published. Surprisingly, the first such calculation of the radial number density ρ_r for the ground state of He made in 1929 by Bethe[65] has been overlooked almost entirely; we could find only six citations to it! Bethe computed ρ_r from a Hylleraas wave function and compared it with one from a Hartree wave function. Remarkably, his tabulated values[65] for the number density computed from a three-parameter Slater-Hylleraas wave function do not differ from those calculated from our 335-term wave function by more than 0.02 atomic units. In a subsequent 1929 paper[85] on the electron affinity of the hydrogen atom, Bethe reported and compared ρ_r for the ground states of H^- and Li^+ with that of He. Bethe also reported algebraic expressions for the densities. Later, Benesch[86] published algebraic expressions for the ground state densities corresponding to 20-parameter Hylleraas-type wave functions[87] for several members of the helium isoelectronic sequence. Increasingly more accurate number densities were reported later.[88,89] Koga et al.[39] reported algebraic densities from 204-term Hylleraas-type wave functions for the ground states of H^-, He, Li^+, Be^{2+}, B^{3+}, and Ne^{8+}. Even more accurate wave functions were used by de Saavedra et al.[90–92] who focused on accuracy at vanishingly small values of the electron-nucleus distance r and satisfaction of the electron-nucleus cusp condition:[93–95]

$$\lim_{r \to 0} \frac{\mathrm{d} \ln \rho(r)}{\mathrm{d}r} = -2Z. \tag{18}$$

Electron number densities for excited states have also been computed from Hylleraas-type wave functions. Perhaps the first such calculation was that of Coolidge and James[82] who reported ρ_r for the 2^1S state of He and compared the Hylleraas ρ_r with the Hartree–Fock ρ_r and a ρ_r computed from a wave function that accounts only for the radial electron correlation. Their paper[82] contains, to our knowledge, the first quantitative discussion of radial versus angular electron correlation. More accurate work was reported later for the excited S states,[88,92,96,97] P states,[96,97] and D states.[97,98] The most extensive study[97] considers all n^3S, n^1S, n^3P, n^1P, n^3D, and n^1D states up to $n = 6$ for He and all the two-electron cations from Li^+ to Ne^{8+}.

The issues examined in the above studies of ρ include the small r behavior and cusp conditions for ρ, the large r asymptotic behavior of ρ, monotonicity and convexity of ρ in the ground state, pseudo nodes in the excited states, differences between the correlated ρ and its Hartree–Fock counterpart, and variations in the density with respect to changes in spin multiplicity, nuclear charge Z, angular momentum L, and degree of excitation n. Density differences between triplet and singlet states have attracted the most attention. Ever since the early work of Hartree and Hartree[99] on the excited ^{1}P and ^{3}P states of Be, it has been known that the triplet state has a more compact electron distribution than the singlet state. This runs counter to what a first-order perturbation theory argument predicts and is of crucial relevance to the understanding of Hund's first rule. The voluminous literature on the latter subject can be traced through the early work[100–106] and the bibliographies compiled in more recent work.[107–109]

The radial number density $\rho_r(r)$ for the five lowest states of He computed from our Gaussian geminal wave functions is shown in Figure. 2.1. When one examines a graph of a probability distribution, it is natural to look for the locations of the maxima and minima and the values of the

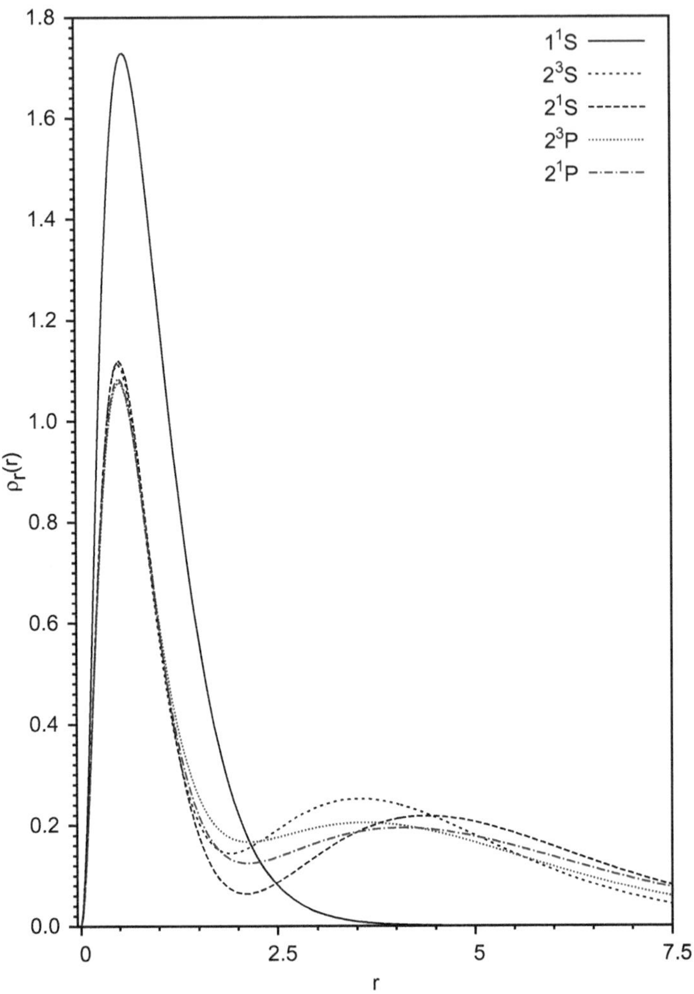

Figure 2.1 Radial one-electron number densities for the ground and four-lowest excited states of He. For color version of this figure, the reader is referred to the online version of this chapter.

function at those points. To facilitate this and to deliver the information more compactly and more precisely, Table 2.2 lists the locations of the maxima and minima of ρ_r and its values at the extrema for H^-, He, and Li^+.

The location of the outermost maximum of ρ_r is often taken as a measure of the radius of the atom or ion, and the locations of the minima of ρ_r are sometimes used to demarcate shell boundaries. Observe that this radius of the helium atom increases in the order:

Table 2.2 Extrema of radial electron and electron-pair distributions in position space

State	r_e	$\rho_r(r_e)$	u_e	$I_r(u_e)$	R_e	$E_r(R_e)$
H^- (1^1S)	1.177	0.639	3.168	0.206	1.338	0.425
He (1^1S)	0.566	1.729	1.097	0.625	0.502	1.298
He (2^3S)	0.495	1.112	3.680	0.243	1.823	0.481
	1.907	0.144				
	3.538	0.252				
He (2^1S)	0.502	1.120	0.854	0.034	0.397	0.071
	2.102	0.064	1.647	0.016	0.796	0.032
	4.382	0.218	4.498	0.213	2.241	0.422
He (2^3P)	0.505	1.077	3.678	0.206	1.865	0.395
	2.141	0.167				
	3.595	0.205				
He (2^1P)	0.501	1.085	4.187	0.191	2.063	0.382
	2.128	0.125				
	4.065	0.194				
Li^+ (1^1S)	0.363	2.812	0.664	1.024	0.313	2.105
Li^+ (2^3S)	0.327	1.697	2.160	0.428	1.071	0.852
	1.183	0.290				
	2.065	0.453				
Li^+ (2^1S)	0.333	1.711	0.514	0.082	0.246	0.170
	1.265	0.180	0.981	0.042	0.478	0.090
	2.388	0.408	2.494	0.394	1.239	0.779
Li^+ (2^3P)	0.340	1.622	1.927	0.412	0.959	0.765
	1.416	0.412				
	1.710	0.417				
Li^+ (2^1P)	0.336	1.634	2.140	0.370	1.047	0.763
	1.323	0.342				
	2.014	0.389				

$$1^1\mathrm{S} < 2^3\mathrm{S} < 2^3\mathrm{P} < 2^1\mathrm{P} < 2^1\mathrm{S} \tag{19}$$

confirming that the triplet states are smaller than the corresponding singlets. The number density is also characterized by its moments defined by:

$$\langle r^k \rangle = \int r^k \rho(\mathbf{r}) d\mathbf{r} = \int_0^\infty r^k \rho_r(r) dr, \tag{20}$$

for $k \geq -2$. Our choice of Löwdin normalization implies that

$$\langle r^0 \rangle = \int_0^\infty \rho_r(r)\mathrm{d}r = 2 \tag{21}$$

for two-electron species. The electron-nucleus attraction is given by $-Z\langle 1/r \rangle$, the diamagnetic susceptibility[33,110] is proportional to $\langle r^2 \rangle$, and the even moments $\langle r^{2k} \rangle$ with $k > 0$ determine the small momentum transfer expansion[111] of the X-ray form factor. The $k = 1$ and 2 moments are the ones most often reported in the literature. The best available values of the moments with $-2 \leq k \leq 4$ are collected in Tables 2.3–2.5, respectively, for H^-, He, and Li^+. The tables also list values of the number density at the nucleus, $\rho(0)$, which is associated via a sum rule to the oscillator strength distribution[33] and determines the leading term of the asymptotic expansion[112] of the X-ray form factor.

Many reports of the electron-pair intracule densities of He and its isoelectronic ions, based on Hylleraas-type wave functions, have been published. Coulson and Neilson[12] reported the first radial intracule density for the ground state of He. Curl and Coulson[113] followed up with calculations of I_r for the ground states of the two-electron ions. Benesch[86] published algebraic expressions for the ground state intracules predicted by 20-parameter Hylleraas-type wave functions.[87] The Kato electron-electron cusp condition[93] requires[112] the exact spherically averaged intracule density to satisfy a cusp condition as follows:

$$\lim_{u \to 0} \frac{\mathrm{d} \ln \, I(u)}{\mathrm{d}u} = 1. \tag{22}$$

This in turn implies[112] that the exact $I(u)$ must have a local minimum at $u = 0$. However, all wave functions of finite one-rank, including Gaussian

Table 2.3 Position-space properties for H^- in atomic units

$\rho(0)$	0.3291057	$I(0)$	0.00273799	$E(0)$	0.07300930
$\langle 1/r^2 \rangle$	2.23332565	$\langle 1/u^2 \rangle$	0.155104153	$\langle 1/R^2 \rangle$	0.9446237
$\langle 1/r \rangle$	1.36652354	$\langle 1/u \rangle$	0.311021502	$\langle 1/R \rangle$	0.7210836
$\langle r \rangle$	5.42035656	$\langle u \rangle$	4.41269450	$\langle R \rangle$	2.035945
$\langle r^2 \rangle$	23.8273994	$\langle u^2 \rangle$	25.2020253	$\langle R^2 \rangle$	5.613193
$\langle r^3 \rangle$	152.0	$\langle u^3 \rangle$	180.6	$\langle R^3 \rangle$	20.01550
$\langle r^4 \rangle$	1290	$\langle u^4 \rangle$	1590	$\langle R^4 \rangle$	89.23141

Extracular properties are from Koga and Matsuyama,[117] $\langle r^3 \rangle$ and $\langle u^3 \rangle$ from Koga et al.,[39] $\langle r^4 \rangle$ and $\langle u^4 \rangle$ from this work, and all remaining properties are from Drake.[68]

Table 2.4 Position-space properties for He in atomic units

	1^1S	2^3S	2^1S	2^3P	2^1P
$\rho(0)$	3.620859	2.640710	2.618921	2.517721	2.548786
$\langle 1/r^2 \rangle$	12.0348177	8.3408911	8.29388	8.029727	8.086067
$\langle 1/r \rangle$	3.37663360	2.30932831	2.2708154	2.26648484	2.24635502
$\langle r \rangle$	1.85894459	5.10092535	5.946122	5.34792322	5.82136846
$\langle r^2 \rangle$	2.38696599	22.9286432	32.17847	26.4234786	31.531302
$\langle r^3 \rangle$	3.93589	130.430183	216.121	169.75	218.52
$\langle r^4 \rangle$	7.9471	856.80	1651.5	1277.8	1757.2
$I(0)$	0.1063454	0	0.008648876	0	0.00073562
$\langle 1/u^2 \rangle$	1.46477092	0.088906	0.14372	0.09405686	0.08579791
$\langle 1/u \rangle$	0.945818449	0.268197855	0.24968265	0.266641309	0.245023869
$\langle u \rangle$	1.42207026	4.44753522	5.2696962	4.69995493	5.1383281
$\langle u^2 \rangle$	2.51643931	23.0461975	32.302379	26.642791	31.598508
$\langle u^3 \rangle$	5.30801	136.742509	224.0304	177.05	225.86
$\langle u^4 \rangle$	12.9813	916.39	1737.4	1349.6	1839.2
$E(0)$	1.462363	0	0.1130136	0.02841653	0
$\langle 1/R^2 \rangle$	7.309275	0.3694413	0.6684725	0.5403869	0.3107227
$\langle 1/R \rangle$	2.061251	0.5436366	0.5119874	0.5725175	0.4851530
$\langle R \rangle$	0.6684023	2.208255	2.619764	2.313104	2.564308
$\langle R^2 \rangle$	0.5643732	5.702771	8.013639	6.551041	7.866025
$\langle R^3 \rangle$	0.5742096	16.90003	27.77683	21.82479	28.07885
$\langle R^4 \rangle$	0.6830523	56.63445	107.7431	83.42354	114.3002

For the ground state, extracular properties are from Koga and Matsuyama,[117] $\langle r^k \rangle$ and $\langle u^k \rangle$ with $k > 2$ from Koga et al.,[39] and all remaining properties are from Drake.[68] For the 2^3S state, $\langle u^4 \rangle$ and the extracular properties are from McCarthy and Thakkar,[78] $\langle r^4 \rangle$ is from this work, and all the remaining properties are from de Saavedra et al.[92] For the 2^1S state, $\langle r^4 \rangle$, $\langle u^4 \rangle$, and the extracular properties are from this work, and all the remaining properties are from de Saavedra et al.[92] For the 2^3P and 2^1P states, $\rho(0)$, $I(0)$, and the $\langle r^k \rangle$ and $\langle u^k \rangle$ with $-1 \le k \le 2$ are from Schiff et al.,[45] the extracular and remaining intracular properties are from McCarthy and Thakkar,[78] and all other $\langle r^k \rangle$ are from this work.

geminal wave functions, are unable to reproduce this. Only wave functions that have an explicit dependence on odd powers of u can obtain this feature.[112] This observation led to more accurate intracule densities being computed using Slater-Hylleraas geminals to examine small u behavior in the ground states of the helium-like ions.[89] Koga et al.[39] reported algebraic densities from 204-term Hylleraas-type wave functions for the ground states of H^-, He, Li^+, Be^{2+}, B^{3+}, and Ne^{8+}. Even more accurate wave functions were used by de Saavedra et al.[90–92] who focused on accuracy at small u and satisfaction of the cusp condition given by Eq. (22).

Table 2.5 Position-space properties for Li^+ in atomic units

	1^1S	2^3S	2^1S	2^3P	2^1P
$\rho(0)$	13.70402	9.12758	9.0376	8.470126	8.613142
$\langle 1/r^2 \rangle$	29.8552474	19.20352	19.08870	18.18157	18.35251
$\langle 1/r \rangle$	5.37584879	3.561169	3.519292	3.52143804	3.49315908
$\langle r \rangle$	1.14554830	2.9877868	3.28842	2.8367334	3.0438640
$\langle r^2 \rangle$	0.892558022	7.547182	9.3902	6.930810	8.083254
$\langle r^3 \rangle$	0.882556	24.018	33.444	22.167	27.772
$\langle r^4 \rangle$	1.05920	87.994	135.77	83.707	111.61
$I(0)$	0.5337225	0	0.064270	0	0.00984
$\langle 1/u^2 \rangle$	4.08223279	0.2621487	0.5487871	0.3360452	0.3653111
$\langle 1/u \rangle$	1.56771956	0.4620526	0.4761232	0.50888277	0.49277510
$\langle u \rangle$	0.862315375	2.561469	2.84445	2.4282528	2.592469
$\langle u^2 \rangle$	0.927064803	7.583692	9.4371	7.046632	8.07421
$\langle u^3 \rangle$	1.188566	25.516	35.285	23.837	29.230
$\langle u^4 \rangle$	1.766039	96.309	146.83	92.198	120.41
$E(0)$	6.057510	0	0.70201	0.2198190	0
$\langle 1/R^2 \rangle$	18.92794	1.079763	2.445845	2.111560	1.172242
$\langle 1/R \rangle$	3.323612	0.9341443	0.9746768	1.127468	0.9483138
$\langle R \rangle$	0.4128926	1.272745	1.412141	1.179268	1.302810
$\langle R^2 \rangle$	0.2145128	1.877668	2.335826	1.703748	2.023076
$\langle R^3 \rangle$	0.1337669	3.153754	4.362673	2.892233	3.648845
$\langle R^4 \rangle$	0.09715118	5.948869	9.077969	5.623505	7.496418

For the ground state, extracular properties are from Koga and Matsuyama,[117] $\langle r^k \rangle$ and $\langle u^k \rangle$ with $k > 2$ from Koga et al.,[39] and all remaining properties are from Drake.[68] For the 2^3S state, $\rho(0)$ and the $\langle r^k \rangle$ and $\langle u^k \rangle$ with $k = 1$ and 2 are from Accad et al.,[46] the extracular and remaining intracular properties are from McCarthy and Thakkar,[78] and all other $\langle r^k \rangle$ are from this work. For the 2^1S state, $\rho(0)$ and the $\langle r^k \rangle$ and $\langle u^k \rangle$ with $k = 1$ and 2 are from Accad et al.,[46] $I(0)$ and $E(0)$ are from Thakkar,[79] and all remaining properties are from this work. For the 2^3P and 2^1P states, $\rho(0)$, $I(0)$, and the $\langle r^k \rangle$ and $\langle u^k \rangle$ with $-1 \leq k \leq 2$ are from Accad et al.,[46] the extracular and remaining intracular properties are from McCarthy and Thakkar,[78] and all other $\langle r^k \rangle$ are from this work.

Boyd and Coulson[114] reported the first accurate radial intracule densities for excited states, the 2^3S and 2^1S states of helium. This was followed by a study of the lowest S and P states of the helium-like ions,[96,115] and an understanding of the higher-order interelectronic cusp condition[116] that applies to intracules for states of maximum multiplicity. An extensive study[97] considered all n^3S, n^1S, n^3P, n^1P, n^3D, and n^1D states up to $n = 6$ for He and all the two-electron cations from Li^+ to Ne^{8+}. A further study focusing on the small u behavior of the excited S states was reported.[92] The issues examined in the above studies include the small u behavior of $I(u)$, the Coulomb hole defined as the difference between the correlated radial intracule $I_r(u)$ and its Hartree–Fock counterpart, and variations in $I(u)$ and $I_r(u)$ with respect to changes in spin multiplicity, nuclear charge Z, angular momentum L, and degree of excitation n. Not surprisingly, all the studies included an analysis of differences between $I_r(u)$ for triplet and singlet states in connection with the interpretation of Hund's rules.

The radial intracule density $I_r(u)$ for the five lowest states of He computed from our Gaussian geminal wave functions is shown in Figure 2.2. Table 2.2 lists the locations of the maxima and minima of $I_r(u)$ and its values at the extrema for H^-, He, and Li^+.

The intracule moments are defined by

$$\langle u^k \rangle = \int u^k I(\mathbf{u}) \mathrm{d}\mathbf{u} = \int_0^\infty u^k I_r(u) \mathrm{d}u, \tag{23}$$

in which $k \geq -2$. Our choice of normalization is such that

$$\langle u^0 \rangle = \int_0^\infty I_r(u) \mathrm{d}u = 1 \tag{24}$$

for two-electron species. The electron-electron repulsion is given by $\langle 1/u \rangle$ and the even moments $\langle u^{2k} \rangle$ with $k > 0$ determine the small momentum transfer expansion[111] of the total scattered X-ray intensity. The best available values of the moments with $-2 \leq k \leq 4$ are collected in Tables 2.3–2.5, respectively, for H^-, He, and Li^+. The tables also list values of the electronic coalescence density, $I(0)$, which determines the leading term of the asymptotic expansion[112] of the total scattered X-ray intensity.

Figure 2.2 and Tables 2.3–2.5 show that the interelectronic separation is smaller in the triplets than in the related singlets for He although a reversal[96] to the hydrogenic limit can already be seen for the 2^1S and 2^3S states of Li^+. All the studies referred to above discuss the interpretation of Hund's rule in connection with the related observation that the interelectronic repulsion can be greater in the triplet than the corresponding singlet.

By contrast with the number and intracule densities, very little attention has been paid to the extracule density. We are aware of only five studies using Hylleraas-type wave functions. Three of these were devoted to $E(R)$ for the ground state of He and its isoelectronic ions.[76,117,118] Very accurate values of the counter-balance density $E(0)$ were reported for the n^1S-states

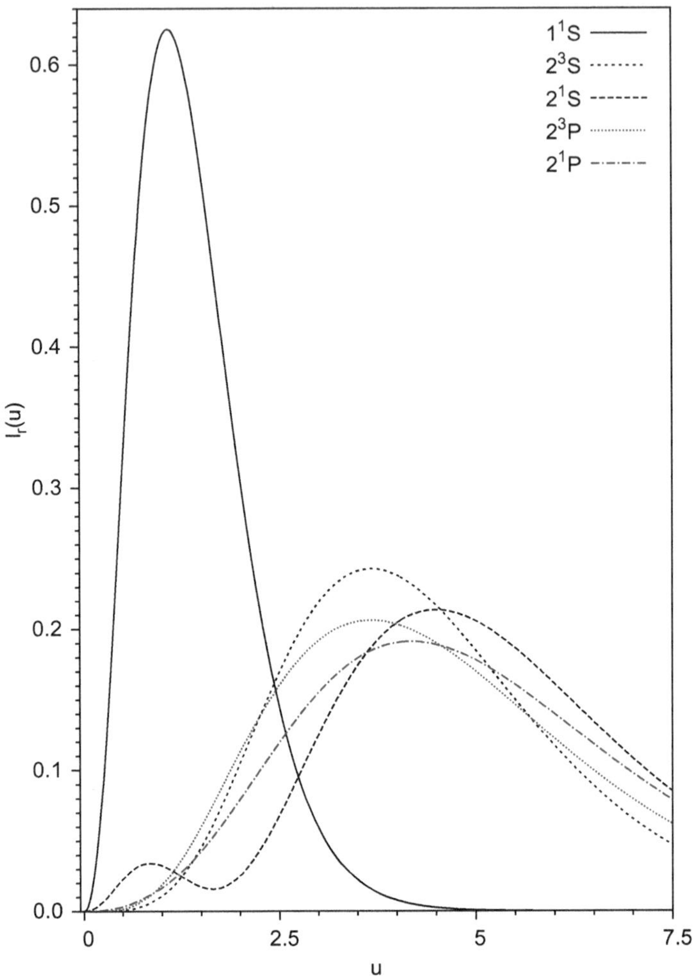

Figure 2.2 Radial electron-pair intracule-distance densities for the ground and four-lowest excited states of He. For color version of this figure, the reader is referred to the online version of this chapter.

($n < 7$) of He and the cations from Li^{+} to Ne^{8+}.[79] The extracule densities for the 2^3S, 2^3P, and 2^1P states of He and Li^{+} were reported.[78]

The radial extracule density $E_r(R)$ for the five lowest states of He computed from our Gaussian geminal wave functions is shown in Figure 2.3. Table 2.2 lists the locations of the maxima and minima of $E_r(R)$ and its values at the extrema for H^{-}, He, and Li^{+}. The extracular moments are defined by

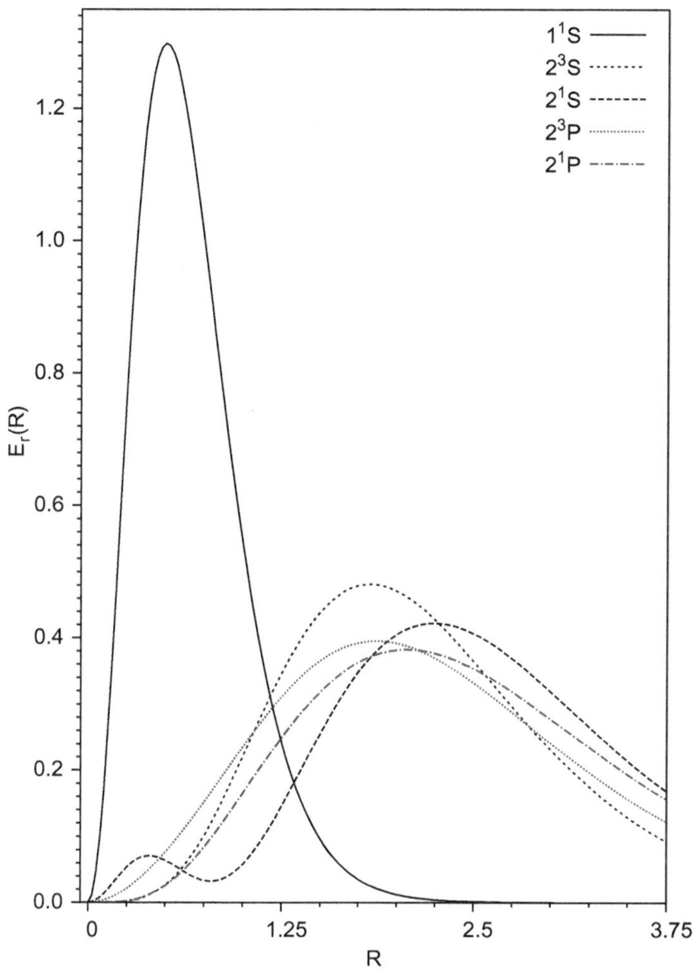

Figure 2.3 Radial electron-pair extracule-distance densities for the ground and four-lowest excited states of He. For color version of this figure, the reader is referred to the online version of this chapter.

$$\langle R^k \rangle = \int R^k E(\mathbf{R}) \mathrm{d}\mathbf{R} = \int_0^\infty R^k E_r(R) \mathrm{d}R \tag{25}$$

in which $k \geq -2$. Moreover,

$$\langle R^0 \rangle = \int_0^\infty E_r(R) \mathrm{d}R = 1 \tag{26}$$

for the two-electron species considered in this work. The best available values of the moments with $-2 \leq k \leq 4$ are collected in Tables 2.3–2.5, respectively, for H^-, He, and Li^+. The tables also list values of the counter-balance density, $E(0)$.

Some of the most interesting points discussed in the papers mentioned above[78,79,117] are the approximate isomorphism[119]

$$E(R) \approx 8I(2R) \tag{27}$$

and its corollary[120]

$$\langle u^k \rangle / \langle R^k \rangle \approx 2^k. \tag{28}$$

The discussions all focus on how well the above relations are satisfied quantitatively.[79,78,117] They conclude that Eq. (27) is weakest at small values of R and consequently Eq. (28) has the largest errors when $0 < k$. Small values of R should be less important for $E_r(R)$ because of the R^2 factor in Eq. (17). Hence, we examine the radial counterpart of Eq. (27) which is simply

$$E_r(R/2) \approx 2I_r(R). \tag{29}$$

Equation (29) tells us that $E_r(R)$ and $I_r(u)$ should have essentially the same shape. Moreover, it should be possible to nearly superpose the two curves if the vertical scale for E_r is twice that for I_r and the horizontal (distance) scale for E_r is half that for I_r. A comparison of Figure 2.3 with Figure 2.2 shows that this behavior is fulfilled in a qualitative sense. Moreover, Eq. (29) is satisfied semiquantitatively for the five lowest-energy states of the two-electron species He and Li^+. This is demonstrated by noting that ratios of corresponding pairs of extremizing distances, u_e/R_e, from Table 2.2 range between 1.97 and 2.19 and the ratios of corresponding extreme values, $E_r(R_e)/I_r(u_e)$, range between 1.86 and 2.14. The isomorphism of Eq. (29) is weaker for H^- where $u_e/R_e = 2.37$.

5. MOMENTUM-SPACE DENSITIES

The electron momentum density $\Pi(p)$ and the electron-pair momentum intracule $I^p(v)$ and extracule $E^p(P)$ densities are all spherically symmetric for S-state atoms. As with their r-space counterparts, the spherically averaged (equivalently, ensemble averaged) densities are considered for the P-states. The radial probability densities are then given by

$$\Pi_r(p) = 4\pi p^2 \Pi(p), \tag{30}$$

$$I_r^p(v) = 4\pi v^2 I^p(v), \tag{31}$$

and

$$E_r^p(P) = 4\pi P^2 E^p(P). \tag{32}$$

Other commonly used notation which we have encountered includes γ_p for the spherically averaged electron momentum density and I for the radial electron momentum density.

The literature on p-space densities obtained from wave functions depending explicitly on r_{12} is rather sparse. To our knowledge, the first such calculation is that of Benesch[69] who used quadrature to calculate $\Pi_r(p)$ for He from the few-parameter Slater-Hylleraas wave functions of Bonham and Kohl.[121] Both one- and two-electron p-space properties were computed from Gaussian geminal wave functions[73] for He and H^-. Monte Carlo methods were used to obtain p-space properties from explicitly correlated wave functions for He.[118,122,123] More accurate Gaussian geminal wave functions were used to compute moments of I_r^p and E_r^p for He.[76] There does not seem to be any published work on p-space properties for Li^+. All these calculations pertain to the ground state. Since we are unaware of any such calculations for excited states using explicitly correlated wave functions, we mention configuration interaction (CI) computations of p-space densities[124] for the 2^3S, 2^3P, and 2^1P states of He. Since there are so few previous studies, some comparisons will be made with the best values most of which were calculated for this work from our Gaussian geminal wave functions.

The radial momentum density $\Pi_r(p)$ for the five lowest states of He computed from our Gaussian geminal wave functions is shown in Figure 2.4. Table 2.6 lists the locations of the maxima and minima of $\Pi_r(p)$ and its values at the extrema for H^-, He, and Li^+.

Table 2.6 and Figure 2.4 show that the momentum at which the global maximum of Π_r occurs for He increases in the order

$$2^1\text{S} < 2^3\text{S} < 2^1\text{P} < 2^3\text{P} < 1^1\text{S} \tag{33}$$

since the electrons move more quickly in the compact triplets than in the corresponding singlets. Comparison of the above order in p-space with the r-space size ordering in Eq. (19) provides another illustration of the reciprocity between the two spaces.[23,125] For the ground state of helium, Benesch's tabulated values[69] of Π_r obtained from a compact Slater-Hylleraas

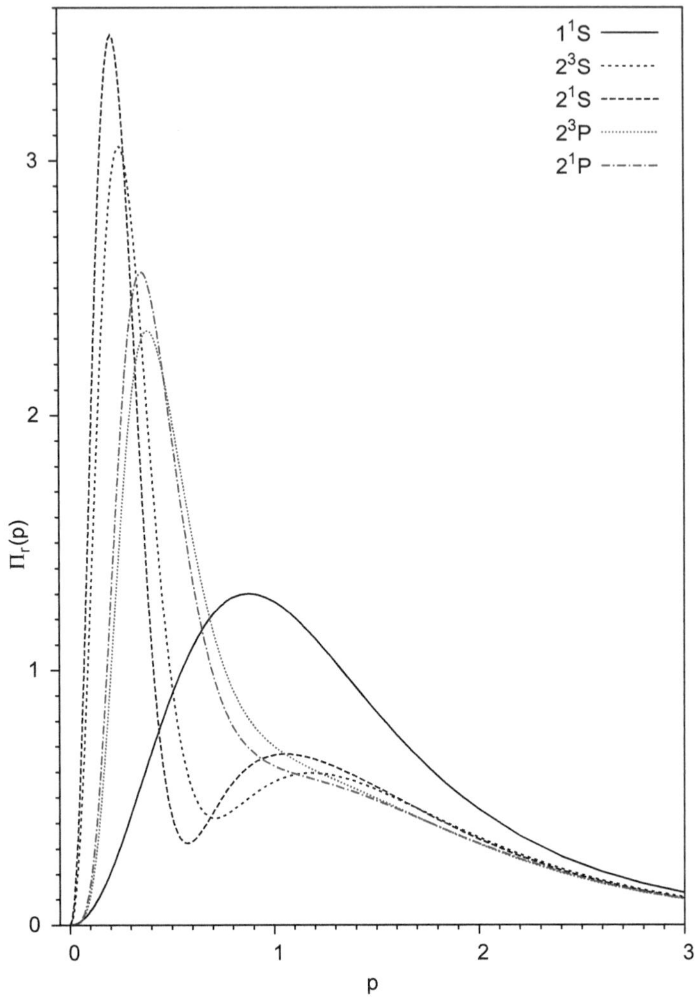

Figure 2.4 Radial one-electron momentum densities for the ground and four-lowest excited states of He. For color version of this figure, the reader is referred to the online version of this chapter.

wave function[121] differ from ours by an average of 0.8% whereas the results of a previous 40-term Gaussian geminal calculation[73] differ from our more accurate results by an average of only 0.02%. We note in passing that $\Pi(p)$ (not shown) is a monotonically decreasing function for the 1^1S, 2^3S, and 2^1S states but non-monotonic for the 2^3P and 2^1P states. At the orbital level, this can be understood[126–128] in terms of the relative contributions of *s*- and *p*-momentals to $\Pi(p)$ near $p = 0$.

Table 2.6 Extrema of radial electron and electron-pair densities in momentum space

State	p_e	$\Pi_r(p_e)$	v_e	$I_r^p(v_e)$	P_e	$E_r^p(P_e)$
H^- (1^1S)	0.232	3.326	0.569	1.085	0.310	1.974
He (1^1S)	0.877	1.300	1.481	0.452	0.781	0.847
He (2^3S)	0.248	3.055	1.252	0.573	0.633	1.137
	0.712	0.420				
	1.183	0.596				
He (2^1S)	0.210	3.492	1.277	0.465	0.631	0.935
	0.576	0.321				
	1.055	0.670				
He (2^3P)	0.386	2.332	1.273	0.550	0.606	0.978
He (2^1P)	0.357	2.562	1.242	0.490	0.621	1.096
Li^+ (1^1S)	1.461	0.811	2.397	0.281	1.244	0.538
Li^+ (2^3S)	0.443	1.736	1.996	0.381	1.006	0.756
	1.203	0.310				
	1.864	0.406				
Li^+ (2^1S)	0.395	1.870	1.467	0.292	0.758	0.589
	1.037	0.271				
	1.738	0.456				
Li^+ (2^3P)	0.781	1.280	2.092	0.362	0.851	0.644
Li^+ (2^1P)	0.724	1.367	1.742	0.316	1.018	0.726

The moments of the electron momentum density are defined by

$$\langle p^k \rangle = \int p^k \Pi(\mathbf{p}) d\mathbf{p} = \int_0^\infty p^k \Pi_r(p) dp, \tag{34}$$

for $4 \geq k \geq -2$ where the upper bound for k holds for the exact $\Pi(p)$ because of its p^{-8} asymptotic behavior.[126,129] Our normalization requires that $\langle p^0 \rangle = 2$ for two-electron species. The $\langle p^k \rangle$ moments are of considerable physical interest.[28,130] The peak of the Compton profile[28] is, in the impulse approximation, simply one-half of $\langle p^{-1} \rangle$. The $\langle p^k \rangle$ moments with $4 \geq k \geq 1$ are proportional to moments of the Compton profile,[28] and $\langle p^{-2} \rangle$ can be related to the Compton profile by an integral identity.[130] The electronic kinetic energy is half of $\langle p^2 \rangle$ and the relativistic mass-velocity correction to the energy[33] is proportional to $\langle p^4 \rangle$. Some other relationships between the $\langle p^k \rangle$ and physical quantities are discussed elsewhere.[23,131]

The moments of $\Pi_r(p)$ with positive and even values of k, that is $\langle p^2 \rangle$ and $\langle p^4 \rangle$, can be obtained directly from r-space wave functions without the need to construct p-space wave functions or densities. The virial theorem, $T = -E$, relates the electronic kinetic energy $T = \langle p^2 \rangle/2$ to the total energy, and hence $\langle p^2 \rangle = -2E$ is available for all Hylleraas-type wave functions. A value of $\langle p^4 \rangle$ is often reported for Hylleraas-type wave functions. This is fortunate because $\langle p^4 \rangle$ is predicted less accurately[73] by Gaussian geminal wave functions for which $\Pi(p)$ decays as e^{-cp^2} instead of the correct p^{-8}.

The best available values of the moments with $-2 \leq k \leq 4$ are collected in Tables 2.7–2.9, respectively, for H^-, He, and Li^+. The tables also list values of $\Pi(0)$. Comparison with the Hartree–Fock values[132] for ground-state He shows that electron correlation effects on $\langle p^k \rangle$ average to 1% with a maximum of 2.4% for $\langle p^4 \rangle$. The effects of correlation on the moments are roughly twice as small for ground-state Li^+ with the Hartree–Fock values[133] in error by an average of 0.4% and a maximum of 1% for $\langle p^4 \rangle$. By contrast, correlation effects are significantly larger for the H^- anion; the Hartree–Fock values[133] of $\langle p^k \rangle$ for H^- have errors of 21%, 7.4%, 1.5%, 7.6%, 12%, and 12% for $k = -2, -1, 1, 2, 3,$ and 4, respectively.

The radial intracular and extracular momentum densities computed from our Gaussian geminal wave functions are shown for the five lowest states of He in Figures 2.5 and 2.6, respectively. Table 2.6 lists the locations of the maxima and minima of $I_r^p(v)$ and $E_r^p(P)$ and their values at the extrema for H^-, He, and Li^+. The p-space versions of Eqs. (27) and (29) are the isomorphism[119]

$$E^p(P) \approx 8I^p(2P) \tag{35}$$

and its radial counterpart

$$E_r^p(P/2) \approx 2I_r^p(P). \tag{36}$$

Comparison of Figures 2.5 and 2.6 shows that $E_r^p(P)$ and $I_r^p(v)$ have essentially the same shape as expected from the approximate isomorphism of Eq. (36). However, ratios of corresponding pairs of extremizing momenta, v_e/P_e, from Table 2.6 range between 1.71 and 2.46 and the ratios of corresponding extreme values, $E_r^p(P_e)/I_r^p(v_e)$, range between 1.78 and 2.30. Evidently, the isomorphism of Eq. (36) is not as satisfactory an approximation as Eq. (29), its r-space counterpart. Comparison of Figure 2.5 and 2.2 suggests that reciprocity between p- and r-space is more subtle when pair densities are considered.

Table 2.7 Momentum-space properties for H^- ($1\,^1S$) in atomic units

$\Pi(0)$	19.5286	$I^p(0)$	0.872138	$E^p(0)$	5.63990
$\langle 1/p^2 \rangle$	44.85370	$\langle 1/v^2 \rangle$	5.204689	$\langle 1/P^2 \rangle$	17.86233
$\langle 1/p \rangle$	6.513704	$\langle 1/v \rangle$	1.726642	$\langle 1/P \rangle$	3.189362
$\langle p \rangle$	1.113893	$\langle v \rangle$	0.8398755	$\langle P \rangle$	0.4520355
$\langle p^2 \rangle$	1.055502	$\langle v^2 \rangle$	0.9897425	$\langle P^2 \rangle$	0.2803154
$\langle p^3 \rangle$	1.659805	$\langle v^3 \rangle$	1.712835	$\langle P^3 \rangle$	0.2441266
$\langle p^4 \rangle$	4.925117	$\langle v^4 \rangle$	5.260	$\langle P^4 \rangle$	0.3527

$\langle p^4 \rangle$ value from Drake[68] and all other values from this work.

Table 2.8 Momentum-space properties for He in atomic units

	$1\,^1S$	$2\,^3S$	$2\,^1S$	$2\,^3P$	$2\,^1P$
$\Pi(0)$	0.443127	11.4017	17.8553	0.112256	0.0968813
$\langle 1/p^2 \rangle$	4.099605	31.61036	41.98740	9.373848	10.78572
$\langle 1/p \rangle$	2.138688	5.100616	5.715535	3.373297	3.596273
$\langle p \rangle$	2.814585	2.084204	2.057421	2.182743	2.145606
$\langle p^2 \rangle$	5.807449	4.350459	4.291948	4.266328	4.247686
$\langle p^3 \rangle$	18.40954	14.36954	14.10449	13.66297	13.76131
$\langle p^4 \rangle$	108.1761	83.67108	82.23739	79.29675	80.23402
$I^p(0)$	0.0495701	0	0.182209	0	0.166390
$\langle 1/v^2 \rangle$	0.7955089	0.6988708	1.540148	0.7039086	1.475357
$\langle 1/v \rangle$	0.6866642	0.7250734	0.8771514	0.7227121	0.8712453
$\langle v \rangle$	2.029114	1.805835	1.747312	1.821286	1.726659
$\langle v^2 \rangle$	5.489310	4.335574	4.272938	4.395473	4.155597
$\langle v^3 \rangle$	20.41748	14.68325	14.79603	14.80983	14.23001
$\langle v^4 \rangle$	124.8	85.14	86.34	84.18	82.85
$E^p(0)$	0.338000	0	1.43354	1.26754	0
$\langle 1/P^2 \rangle$	2.847758	2.757794	6.111287	5.851031	2.955566
$\langle 1/P \rangle$	1.296806	1.440618	1.750006	1.746429	1.473408
$\langle P \rangle$	1.075149	0.9072921	0.8763464	0.8605648	0.9014989
$\langle P^2 \rangle$	1.531397	1.091336	1.077739	1.034296	1.084944
$\langle P^3 \rangle$	2.932852	1.846185	1.878014	1.769941	1.833710
$\langle P^4 \rangle$	8.708	5.337	5.476	5.141	5.255

$\langle p^4 \rangle$ values from Drake[68] for the $1\,^1S$ state, from Pekeris[42] for the $2\,^3S$ and $2\,^1S$ states, and from Schiff et al.[45] for the $2\,^3P$ and $2\,^1P$ states. All other values are from this work.

Table 2.9 Momentum-space properties for Li^+ in atomic units

	1^1S	2^3S	2^1S	2^3P	2^1P
$\Pi(0)$	0.0977909	2.02178	2.70044	0.0329823	0.0288323
$\langle 1/p^2 \rangle$	1.515496	10.06830	12.02583	2.648885	2.952387
$\langle 1/p \rangle$	1.309976	2.929225	3.132795	1.848871	1.943410
$\langle p \rangle$	4.507865	3.274665	3.245825	3.499128	3.440492
$\langle p^2 \rangle$	14.55983	10.22145	10.08175	10.05543	9.986702
$\langle p^3 \rangle$	70.54748	50.30814	49.23878	46.91354	47.25144
$\langle p^4 \rangle$	621.0943	436.6057	427.4855	402.1	408.2919
$I^p(0)$	0.0117487	0	0.0510834	0	0.0433562
$\langle 1/v^2 \rangle$	0.3053114	0.2775087	0.6896066	0.2599841	0.6369284
$\langle 1/v \rangle$	0.4260016	0.4611384	0.5892133	0.4457231	0.5795662
$\langle v \rangle$	3.254093	2.789677	2.658487	2.873419	2.602079
$\langle v^2 \rangle$	13.98188	10.18632	10.03834	10.66568	9.484338
$\langle v^3 \rangle$	81.04314	51.82009	53.42943	53.92142	49.09535
$\langle v^4 \rangle$	743.4	447.4	467.7	448.3	428.0
$E^p(0)$	0.0840051	0	0.400348	0.325491	0
$\langle 1/P^2 \rangle$	1.129906	1.096795	2.729362	2.489107	1.090814
$\langle 1/P \rangle$	0.8184877	0.9168596	1.174648	1.154334	0.9095035
$\langle P \rangle$	1.695393	1.401280	1.332817	1.298784	1.419290
$\langle P^2 \rangle$	3.784445	2.564148	2.531291	2.361295	2.622267
$\langle P^3 \rangle$	11.25310	6.518258	6.792732	6.098322	6.648352
$\langle P^4 \rangle$	50.98	28.06	29.79	26.49	27.93

$\langle p^4 \rangle$ values from Drake[68] for the 1^1S state, from Pekeris[43] for the 2^3S and 2^1S states, and from Schiff et al.[45] for the 2^1P state. All other values are from this work.

The intracular momentum moments are defined by

$$\langle v^k \rangle = \int v^k I^p(\mathbf{v}) d\mathbf{v} = \int_0^\infty v^k I_r^p(v) dv, \tag{37}$$

and the extracular momentum moments are defined by

$$\langle P^k \rangle = \int P^k E^p(\mathbf{P}) d\mathbf{P} = \int_0^\infty P^k E_r^p(P) dP \tag{38}$$

for $k \geq -2$. Our normalization implies that $\langle v^0 \rangle = \langle P^0 \rangle = 1$ for two-electron species. Using simple identities[73] that follow from the definitions of $\mathbf{v}$ and $\mathbf{P}$, the $k = 2$ moments can be found from $\langle p^2 \rangle$ and $\langle \mathbf{p}_1 \cdot \mathbf{p}_2 \rangle$. The latter is often reported for use in the mass-polarization correction to the energy and in a sum rule of the oscillator strength distribution.[33] For

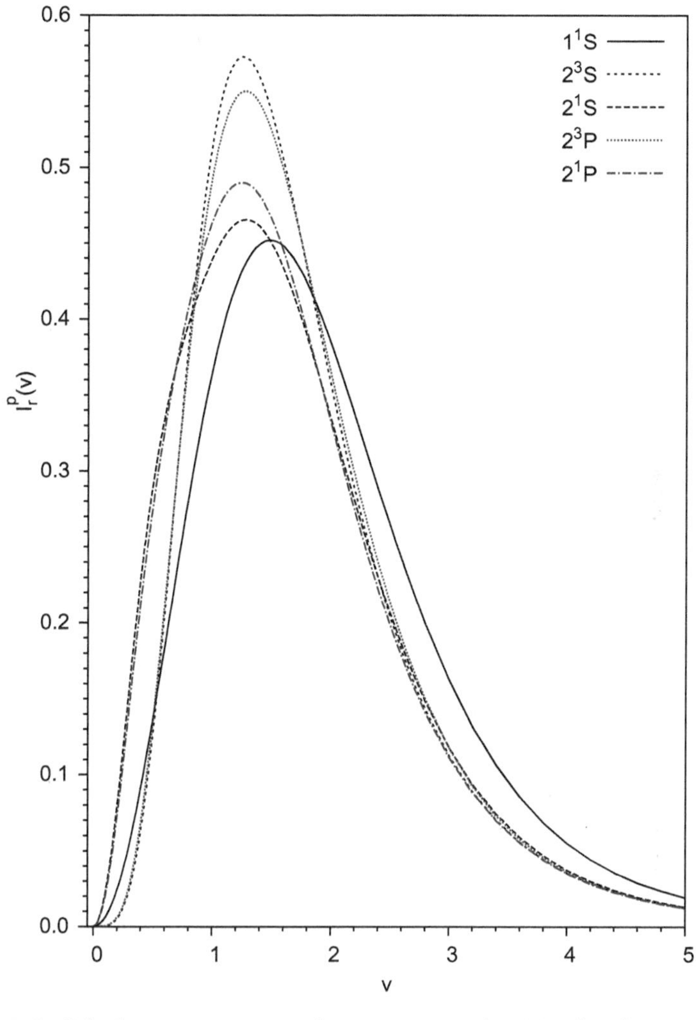

Figure 2.5 Radial electron-pair intracule-momentum densities for the ground and four-lowest excited states of He. For color version of this figure, the reader is referred to the online version of this chapter.

two-electron systems and Löwdin normalization, the required identities are $\langle v^2 \rangle = \langle p^2 \rangle - \langle \mathbf{p}_1 \cdot \mathbf{p}_2 \rangle$ and $4\langle P^2 \rangle = \langle p^2 \rangle + \langle \mathbf{p}_1 \cdot \mathbf{p}_2 \rangle$. The best available values of the moments with $-2 \leq k \leq 4$ are collected in Tables 2.7, 2.8, and 2.9, respectively, for H^-, He, and Li^+. The tables also list values of $I^p(0)$ and $E^p(0)$. The values of $\langle p^2 \rangle$, $\langle P^2 \rangle$, and $\langle v^2 \rangle$ predicted by our Gaussian geminal wave functions agree to a few parts in 10^8 with those extracted from Hylleraas-type values[42,43,45,68] of E and $\langle \mathbf{p}_1 \cdot \mathbf{p}_2 \rangle$ using the virial theorem and the identities mentioned above.

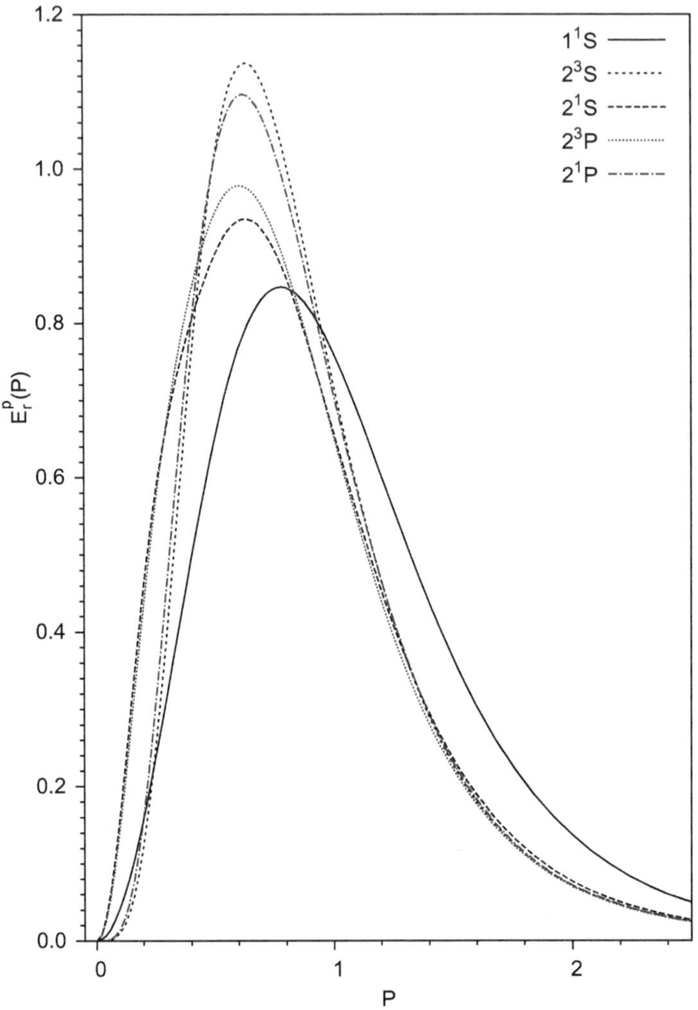

Figure 2.6 Radial electron-pair extracule-momentum densities for the ground and four-lowest excited states of He. For color version of this figure, the reader is referred to the online version of this chapter.

Consider the effects of electron correlation on $\langle v^k \rangle$ and $\langle P^k \rangle$ for He by comparing available Hartree–Fock values[119,124,134] with the best values listed in Table 2.8. In the ground state, the errors of the Hartree–Fock values[134] of $\langle v^k \rangle$ range from 0.2% for $\langle v^{-1} \rangle$ to 7.3% for $\langle v^3 \rangle$. Larger correlation effects are found for $\langle P^k \rangle$ with errors ranging from 4.3% for $\langle P \rangle$ to 14% for $\langle P^{-2} \rangle$ in the ground 1^1S state. Correlation effects are noticeably smaller in the excited states. In the 2^3P and 2^1P states, the errors in the Hartree–Fock

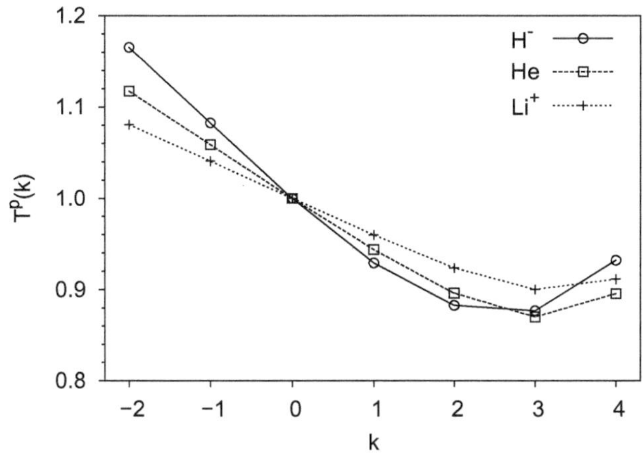

Figure 2.7 The ratio $T^p(k)$ for the ground states of He, H^-, and Li^+.

values[119] of $\langle \nu^k \rangle$ and $\langle P^k \rangle$ do not exceed 1% except for $k = -2$ where the errors are still less than 2.5%. The errors in the Hartree–Fock values[124] of $\langle \nu^k \rangle$ for $-2 \leq k \leq 2$ are less than 0.3% in the 2^3S state.

The corollary[120] to Eq. (35)

$$\langle \nu^k \rangle / \langle P^k \rangle \approx 2^k \tag{39}$$

is perhaps best examined with the help of the ratios

$$T^p(k) = 2^{-k} \langle \nu^k \rangle / \langle P^k \rangle. \tag{40}$$

Consider the ratios for the ground state shown in Figure 2.7. Observe that Eq. (39) can be in error by as much as 17%. Moreover, Eq. (39) is worse for $k < 0$ than for $k > 0$ and deviations from it are generally larger for H^- than for He. The ratios for the excited states are shown in Figure 2.8. The ratio is within 2% of unity for the 2^3S and 2^1S states of both He and Li^+ but very large deviations are seen in the 2^3P and 2^1P states for $k < 0$.

6. CONCLUDING REMARKS

The several forms of correlated wave functions, with explicit dependence on the interelectronic distance, available for the low-lying states of the helium atom and its isoelectronic ions were reviewed. Computations of one- and two-electron densities in *r*- and *p*-space from

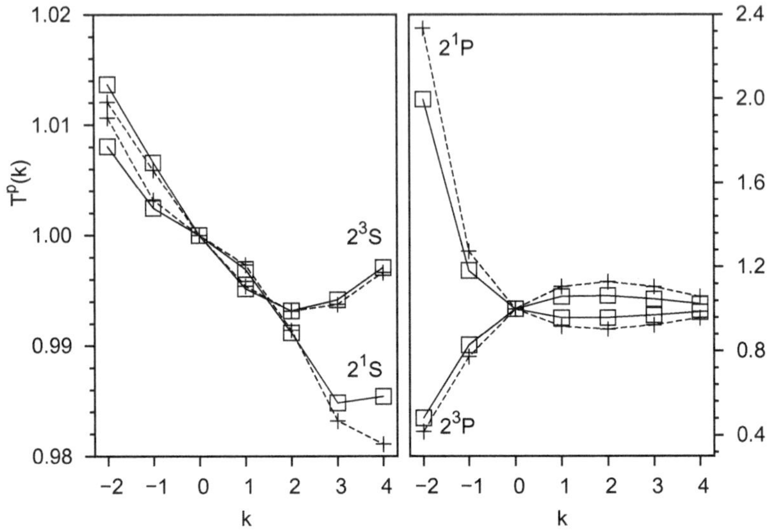

Figure 2.8 The ratio $T^p(k)$ for excited states of He (squares) and Li^+ (+). Note that the vertical scale is different in the two panels.

these wave functions were summarized. Finally, a consolidated set of tables of moments and critical points of those densities was assembled. A few *r*-space items and most of the *p*-space values in those tables were not available in the literature and were computed specifically for this article using wave functions expanded in Gaussian geminals. We think that the tables will serve as a valuable reference for a substantial amount of time. We hope that the gestalt of the complete set of tables will enable some readers to derive new insights into the electronic structure of the low-lying states of H^-, He, and Li^+.

ACKNOWLEDGMENT

Writing this article reminded one of us (AJT) of his lengthy collaboration with the late Vedene Smith. This work is dedicated to his memory. The Natural Sciences and Engineering Research Council of Canada supported this work.

REFERENCES

1. Husimi, K. Some Formal Properties of the Density Matrix. *Proc. Phys. Math. Soc. Japan* **1940,** *22,* 264–314.
2. McWeeny, R. On the Basis of Orbital Theories. *Proc. R. Soc. London, Ser. A* **1955,** *232,* 114–135.

3. Löwdin, P.-O. Quantum Theory of Many-Particle Systems. I. Physical Interpretation by Means of Density Matrices, Natural Spin-Orbitals, and Convergence Problems in the Method of Configuration Interaction. *Phys. Rev.* **1955,** *97*, 1474–1489.
4. McWeeny, R. Some Recent Advances in Density Matrix Theory. *Rev. Mod. Phys.* **1960,** *32*, 335–369.
5. Davidson, E. R. *Reduced Density Matrices in Quantum Chemistry;* Academic: New York, 1976.
6. Coppens, P. *X-ray Charge Densities and Chemical Bonding;* Oxford University Press: New York, 1997.
7. Parr, R. G.; Yang, W. *Density-Functional Theory of Atoms and Molecules;* Oxford University Press: New York, 1989.
8. March, N. H. *Electron Density Theory of Atoms and Molecules;* Academic: New York, 1992.
9. Bader, R. F. W. *Atoms in Molecules: A Quantum Theory;* Clarendon: Oxford, 1990.
10. Sukumar, N. *A Matter of Density: Exploring the Electron Density Concept in the Chemical, Biological, and Materials Sciences;* Wiley: New York, 2012.
11. Debye, P. The Scattering of X-Rays. *J. Math. Phys.* **1925,** *4*, 133–147.
12. Coulson, C. A.; Neilson, A. H. Electron Correlation in the Ground State of Helium. *Proc. Phys. Soc., London* **1961,** *78*, 831–837.
13. Coleman, A. J. Density Matrices in the Quantum Theory of Matter: Energy, Intracules and Extracules. *Int. J. Quantum Chem. Symp.* **1967,** *1*, 457–464.
14. Thakkar, A. J.; Moore, N. J. Electronic Extracule Densities. *Int. J. Quantum Chem. Symp.* **1981,** *15*, 393–400.
15. Thakkar, A. J. Extracules, Intracules, Correlation Holes, Potentials, Coefficients and all that. In *Density Matrices and Density*; Erdahl, R. M., Smith, V. H., Jr., Eds.; Reidel: London, 1987., pp 553–581.
16. Boyd, R. J.; Ugalde, J. M. Analysis of Wave Functions in Terms of One-electron and 2-electron Density Functions. In *Computational Chemistry, Part A*; Fraga, S., Ed.; Elsevier: Amsterdam, 1992; pp 273–299.
17. Koga, T. Electron-Pair Densities of Atoms. In *Many-Electron Densities and Reduced Density Matrices*; Cioslowski, J., Ed.; Kluwer Academic/Plenum: New York, 2000; pp 267–298.
18. Valderrama, E.; Ugalde, J. M.; Boyd, R. J. Intracule and Extracule Densities: Historical Perspectives and Future Prospects. In *Many-Electron Densities and Reduced Density Matrices;* Cioslowski, J., Ed.; Kluwer Academic/Plenum: New York, 2000; pp 231–248.
19. Poater, J.; Solà, M.; Duran, M.; Robles, J.; Fradera, X. Bielectronic Densities: Analysis and Applications in Molecular Structure and Reactivity. In Review Reviews of Modern Quantum Chemistry. A Celebration of the Contributions of Robert G. Parr; Sen, K. D., Ed.; World Scientific: Singapore, 2002; Vol. 1, pp 831–870.
20. Gill, P. M. W.; O'Neill, D. P.; Besley, N. A. Two-Electron Distribution Functions and Intracules. *Theor. Chem. Acc.* **2003,** *109*, 241–250.
21. Bonham, R. A.; Fink, M. *High Energy Electron Scattering*; Van Nostrand Reinhold: New York, 1974.
22. Thakkar, A. J.; Tripathi, A. N.; Smith, V. H., Jr; Molecular X-Ray- and Electron-Scattering Intensities. *Phys. Rev. A* **1984,** *29*, 1108–1113.
23. Thakkar, A. J. The Momentum Density Perspective of the Electronic Structure of Atoms and Molecules. *Adv. Chem. Phys.* **2004,** *128*, 303–352.
24. Thakkar, A. J. Electronic Structure: The Momentum Perspective. In *Theory and Applications of Computational Chemistry: The First 40years;* Dykstra, C. E., Frenking, G., Kim, K. S., Scuseria, G. E., Eds.; Elsevier: Amsterdam, Holland, 2005; pp 483–505, Chapter 19.
25. Dirac, P. A. M. The Physical Interpretation of the Quantum Dynamics. *Proc. R. Soc. London, Ser. A* **1927,** *113*, 621–641.

26. Dirac, P. A. M. *The Principles of Quantum Mechanics*, 4th ed.; Oxford University Press: London, 1958.
27. Thakkar, A. J.; Simas, A. M.; Smith, V. H., Jr. Internally Folded Densities. *Chem. Phys.* **1981,** *63*, 175–183.
28. Williams, B. G., Ed.; *Compton Scattering: The Investigation of Electron Momentum Distributions*; McGraw-Hill: New York, 1977.
29. Weigold, E.; McCarthy, I. E. *Electron Momentum Spectroscopy*; Kluwer Academic: New York, 1999.
30. Levin, V. G.; Neudatchin, V. G.; Pavlitchenkov, A. V.; Smirnov, Y. F. A Study of the Electron Correlations in the H_2 Molecule Using the Double Photoionisation Process $(\gamma, 2e)$. *J. Phys. B: At. Mol. Phys.* **1984,** *17*, 1525–1536.
31. Banyard, K. E.; Reed, C. E. Momentum Distributions for Two-Electron Systems: Electron Correlation and the Coulomb Hole. *J. Phys. B: At. Mol. Phys.* **1978,** *11*, 2957–2967.
32. Wang, J. H.; Smith, V. H., Jr. One-Electron and Electron Pair Densities of First-Row Hydrides in Momentum-Space. *J. Chem. Phys.* **1993,** *99*, 9745–9755.
33. Bethe, H. A.; Salpeter, E. E. *Quantum Mechanics of One- and Two-electron Atoms*; Springer-Verlag: Berlin, 1957.
34. Hylleraas, E. A. Über den Grundzustand des Heliumatoms. *Z. Phys.* **1928,** *48*, 469–494.
35. Hylleraas, E. A. Neue Berechnung der Energie des Heliums im Grundzustande, Sowie des Tiefsten Terms von Ortho-Helium. *Z. Phys.* **1929,** *54*, 347–366.
36. Hettema, H., Ed.; *Quantum Chemistry, Classic Scientific Papers;* World Scientific: Singapore, 2000.
37. Midtdal, J. Some Theoretical Investigations for Closed States of the Negative Hydrogen Ion. Part II. *Phys. Math. Univ. Osloensis* **1966,** *32*, 3–28.
38. Thakkar, A. J. Comments on Explicitly Correlated Wave Functions for the Ground State of the Helium Atom. *Phys. Rev. A* **1977,** *16*, 1740–1742.
39. Koga, T.; Kasai, Y.; Thakkar, A. J. Accurate Algebraic Densities and Intracules for Heliumlike Ions. *Int. J. Quantum Chem.* **1993,** *46*, 689–699.
40. Coolidge, A. S.; James, H. M. On the Convergence of the Hylleraas Variational Method. *Phys. Rev.* **1937,** *51*, 855–859.
41. Pekeris, C. L. Ground State of Two-Electron Atoms. *Phys. Rev.* **1958,** *112*, 1649–1658.
42. Pekeris, C. L. 1^1S, 2^1S, and 2^3S States of H^- and of He. *Phys. Rev.* **1962,** *126*, 1470–1476.
43. Pekeris, C. L. 1^1S, 2^1S, and 2^3S States of Li^+. *Phys. Rev.* **1962,** *126*, 143–145.
44. Pekeris, C. L. Excited S States of Helium. *Phys. Rev.* **1962,** *127*, 509–511.
45. Schiff, B.; Lifson, H.; Pekeris, C. L.; Rabinowitz, P. $2^{1,3}P$, $3^{1,3}P$, and $4^{1,3}P$ States of He and the 2^1P State of Li^+. *Phys. Rev.* **1965,** *140*, A1104–A1121.
46. Accad, Y.; Pekeris, C. L.; Schiff, B. S and P States of the Helium Isoelectronic Sequence Up to $Z = 10$. *Phys. Rev. A* **1971,** *4*, 516–536.
47. Schwartz, H. M. Ground-State Solution of the Nonrelativistic Equation for Helium. *Phys. Rev.* **1956,** *103*, 110–111.
48. Schwartz, H. M. Ritz-Hylleraas Solutions of the Ground State of Two-Electron Atoms Involving Fractional Powers. *Phys. Rev.* **1960,** *120*, 483–487.
49. Schwartz, C. Ground State of the Helium Atom. *Phys. Rev.* **1962,** *128*, 1146–1148.
50. Kinoshita, T. Ground State of the Helium Atom. *Phys. Rev.* **1957,** *105*, 1490–1502.
51. Thakkar, A. J.; Koga, T. Ground-State Energies for the Helium Isoelectronic Series. *Phys. Rev. A* **1994,** *50*, 854–856.
52. Thakkar, A. J.; Koga, T. Variational Calculations for Helium-Like Ions Using Generalized Kinoshita-Type Expansions. *Theor. Chem. Acc.* **2003,** *109*, 36–39.
53. Frankowski, K.; Pekeris, C. L. Logarithmic Terms in the Wave Functions of the Ground State of Two-Electron Atoms. *Phys. Rev.* **1966,** *146*, 46–49.

54. Freund, D. E.; Huxtable, B. D.; Morgan, J. D., III; Variational Calculations on the Helium Isoelectronic Sequence. *Phys. Rev. A* **1984,** *29*, 980–982.
55. Gronwall, T. H. The Helium Wave Equation. *Phys. Rev.* **1937,** *51*, 655–660.
56. Bartlett, J. H., Jr; The Helium Wave Equation. *Phys. Rev.* **1937,** *51*, 661–669.
57. Fock, V. A. On the Schrödinger Equation of the Helium Atom. *Izv. Akad. Nauk SSSR, Ser. Fiz.* **1954,** *18*, 161–174.
58. Faddeev, L. D., Khalfin, L. A., Komarov, I. V., Eds.; *V. A. Fock — Selected Works*; Chapman & Hall/CRC: Boca Raton, 2004.
59. Myers, C. R.; Umrigar, C. J.; Sethna, J. P.; Morgan, J. D. Fock's Expansion, Kato's Cusp Conditions, and the Exponential Ansatz. *Phys. Rev. A* **1991,** *44*, 5537–5546.
60. Thakkar, A. J.; Smith, V. H., Jr. Compact and Accurate Integral-Transform Wave Functions. I. The $1\,^1S$ State of the Helium-Like Ions from H^- Through Mg^{10+} . *Phys. Rev. A* **1977,** *15*, 1–15.
61. Thakkar, A. J.; Smith, V. H., Jr. Compact and Accurate Integral-Transform Wave Functions. II. The $2\,^1S$, $2\,^3S$, $2\,^1P$, and $2\,^3P$ States of the Helium-Like Ions from He Through Mg^{10+}. *Phys. Rev. A* **1977,** *15*, 16–22.
62. Thakkar, A. J. The Generator Coordinate Method Applied to Variational Perturbation Theory. Multipole Polarizabilities, Spectral Sums, and Dispersion Coefficients for Helium. *J. Chem. Phys.* **1981,** *75*, 4496–4501.
63. Cann, N. M.; Thakkar, A. J. Oscillator Strengths for S–P and P–D Transitions in the Helium-Like Ions. *Phys. Rev. A* **1992,** *46*, 5397–5405.
64. Slater, J. C. The Normal State of Helium. *Phys. Rev.* **1928,** *32*, 349–360.
65. Bethe, H. Vergleich der Elektronenverteilung im Heliumgrundzustand nach verschiedenen Methoden. *Z. Phys.* **1929,** *55*, 431–436.
66. Drake, G. W. F. High Precision Variational Calculations for the $1s^2\,^1S$ State of H^- and the $1s^2\,^1S$ and $1s2s\,^3S$ States of Helium. *Nucl. Instr. Meth. Phys. Res. B* **1988,** *31*, 7–13.
67. Drake, G. W. F.; Yan, Z. C. Energies and Relativistic Corrections for the Rydberg States of Helium: Variational Results and Asymptotic Analysis. *Phys. Rev. A* **1992,** *46*, 2378–2409.
68. Drake, G. W. F. High Precision Calculations for Helium. In *Atomic, Molecular and Optical Physics Handbook*; Drake, G. W. F., Ed.; American Institute of Physics: New York, 1996; pp 154–171.
69. Benesch, R. Momentum Distributions and Compton Profiles from Two-Electron Atomic Wavefunctions Containing Exponential Correlation Terms. *J. Phys. B* **1976,** *9*, 2587–2594.
70. Sapirstein, J. Momentum Space P-State Helium Wavefunctions and Helium Fine Structure. *J. Phys. B: At. Mol. Opt. Phys.* **2010,** *43*, 074015.
71. Jeziorski, B.; Szalewicz, K. High-accuracy Compton Profile of Molecular Hydrogen from Explicitly Correlated Gaussian Wave Function. *Phys. Rev. A* **1979,** *19*, 2360–2365.
72. Regier, P. E.; Thakkar, A. J. Approximate Solutions of the Momentum-Space Integral Schrödinger Equation for Two-Electron Atoms. *Phys. Rev. A* **1984,** *30*, 30–34.
73. Regier, P. E.; Thakkar, A. J. Momentum Space Properties of Two-Electron Atoms. *J. Phys. B: At. Mol. Phys.* **1985,** *18*, 3061–3071.
74. Boys, S. F. The Integral Formulae for the Variational Solution of the Molecular Many-Electron Wave Equation in Terms of Gaussian Functions with Direct Electronic correlation. *Proc. R. Soc. London, Ser. A* **1960,** *258*, 402–411.
75. Singer, K. The Use of Gaussian (Exponential Quadratic) Wave Functions in Molecular Problems. *Proc. R. Soc. London, Ser. A* **1960,** *258*, 412–420.
76. Thakkar, A. J. Accurate Electron-Pair, Momentum-Space Properties for the Helium Atom. *Chem. Phys. Lett.* **2003,** *381*, 80–85.

77. Power, J. D.; Somorjai, R. L. Upper and Lower Bounds to the Eigenvalues of Two-Electron Atoms: Proof that the $2^{1,3}$S States of H^- are not Bound. *Phys. Rev. A.* **1972,** *5,* 2401–2403.
78. McCarthy, S. P.; Thakkar, A. J. Electron-Pair Extracule Densities for Low-Lying Excited States of He and Li^+. *Int. J. Quantum Chem.* **2011,** *111,* 753–759.
79. Thakkar, A. J. Interelectronic Counter-Balance and Coalescence Densities for the $n\,^1$S ($n < 7$) States of the Helium Isoelectronic Sequence. *J. Mol. Struct. (Theochem)* **2003,** *633,* 257–262.
80. Hylleraas, E. A.; Undheim, B. Numerische Berechnung der 2S-Terme von Ortho- und Par-Helium. *Z. Phys.* **1930,** *65,* 759–772.
81. MacDonald, J. K. L. Successive Approximations by the Rayleigh-Ritz Variation Method. *Phys. Rev.* **1933,** *43,* 830–833.
82. Coolidge, A. S.; James, H. M. Wave Functions for $1s2s\;^1$S Helium. *Phys. Rev.* **1936,** *49,* 676–687.
83. Cann, N. M.; Thakkar, A. J. First Born Differential Cross-Sections for Electronic Excitation in the Helium Atom. *J. Elect. Spect. Relat. Phenom.* **2002,** *123,* 143–159.
84. Cann, N. M.; Thakkar, A. J. Generalized Oscillator Strengths for Electronic Excitation from the $2\,^1$S and $2\,^3$S Metastable States of the Helium Atom. *J. Elect. Spect. Relat. Phenom.* **2003**, *129,* 9–26.
85. Bethe, H. Berechnung der Elektronenaffinität des Wasserstoffs. *Z. Phys.* **1929,** *57,* 815–821.
86. Benesch, R. Algebraic Determination of Electron-Nuclear and Electron-Electron Distribution Functions from Hylleraas Type Wavefunctions. *J. Phys. B: At. Mol. Phys.* **1971,** *4,* 1403–1414.
87. Hart, J. F.; Herzberg, G. Twenty-Parameter Eigenfunctions and Energy Values of the Ground States of He and He-Like Ions. *Phys. Rev.* **1957,** *106,* 79–82.
88. Winkler, P.; Porter, R. N. Correlated Wavefunctions, Energies, and One-Electron Radial Densities for S States of the He Atom. *J. Chem. Phys.* **1974,** *61,* 2038–2046.
89. Thakkar, A. J.; Smith, V. H., Jr. Accurate Charge Densities and Two-Electron Intracule Functions for the Helium Like Ions. *J. Chem. Phys.* **1977,** *67,* 1191–1196.
90. de Saavedra, F. A.; Buendía, E.; Gálvez, F. J. Calculations of the One- and Two-Body Densities in Two-Electron Atomic Systems. *J. Phys. B: At. Mol. Opt. Phys.* **1994,** *27,* 4433–4441.
91. de Saavedra, F. A.; Buendía, E.; Gálvez, F. J. Single-Particle and Electron-Pair Densities at the Origin in the Ground State of Helium-Like Ions. *J. Phys. B: At. Mol. Opt. Phys.* **1994,** *27,* 5131–5137.
92. de Saavedra, F. A.; Porras, I.; Buendía, E.; Gálvez, F. J. Correlated One- and Two-Electron Densities of Low-Lying S-States in Helium-Like Atoms. *J. Phys. B: At. Mol. Opt. Phys.* **1995,** *28,* 3123–3133.
93. Kato, T. On the Eigenfunctions of Many-Particle Systems in Quantum Mechanics. *Commun. Pure Appl. Math.* **1957,** *10,* 151–177.
94. Steiner, E. Charge Densities in Atoms. *J. Chem. Phys.* **1963,** *39,* 2365–2366.
95. Hoffmann-Ostenhof, M.; Hoffmann-Ostenhof, T.; Stremnitzer, H. Electronic Wave Functions Near Coalescence Points. *Phys. Rev. Lett.* **1992,** *68,* 3857–3860.
96. Regier, P. E.; Thakkar, A. J. Charge Densities and Two-Electron Intracules for the Low-Lying Excited States of the Helium-Like Ions. *J. Phys. B: At. Mol. Phys.* **1984,** *17,* 3391–3403.
97. Cann, N. M.; Boyd, R. J.; Thakkar, A. J. Charge and Intracule Densities in Singly Excited Heliumlike Ions. *J. Chem. Phys.* **1993,** *98,* 7132–7139.
98. Cann, N. M.; Boyd, R. J.; Thakkar, A. J. Electron Correlation Effects in the Rydberg-Like $3\,^3$D and $3\,^1$D States of Helium-Like Ions. *Int. J. Quantum Chem.* **1993,** *48,* 1–14.

99. Hartree, D. R.; Hartree, W. Self-Consistent Field, with Exchange, for Beryllium. II. The (2s) (2p) ^3P and ^1P Excited States. *Proc. R. Soc. London, Ser. A* **1936,** *154*, 588–607.
100. Davidson, E. R. Single Configuration Calculations on Excited States of Helium II. *J. Chem. Phys.* **1965,** *42*, 4199–4200.
101. Kohl, D. A. On the Difference in Energy Between Singlet and Triplet States. *J. Chem. Phys.* **1972,** *56*, 4236–4238.
102. Katriel, J. Locality of the Effect of the Pauli Principle in Atoms. *Phys. Rev. A* **1972,** *5*, 1990–1992.
103. Boyd, R. J. Relative Sizes of High and Low Spin States of Atoms. *Nature* **1974,** *250*, 566–567.
104. Colpa, J. P.; Thakkar, A. J.; Smith, V. H., Jr.; Randle, P. An Analysis of Energy Differences in Atomic Multiplets in Connection with the Inequality Formulation of Hund's Rules. I. Frozen Versus Relaxed SCF Orbitals and Energies. *Mol. Phys.* **1975,** *29*, 1861–1875.
105. Katriel, J.; Pauncz, R. Theoretical Interpretation of Hund's Rule. *Adv. Quantum Chem.* **1977,** *10*, 143–185.
106. Shim, I.; Dahl, J. P. A New Interpretation of Hund's First Rule. *Theor. Chim. Acta* **1978,** *48*, 165–174.
107. Sajeev, Y.; Sindelka, M.; Moiseyev, N. Hund's Multiplicity Rule: From Atoms to Quantum Dots. *J. Chem. Phys.* **2008,** *128*, 061101.
108. Sako, T.; Paldus, J.; Ichimura, A.; Diercksen, G. H. F. Origin of Hund's Multiplicity Rule in Singly Excited Helium: Existence of a Conjugate Fermi Hole in the Lower Spin State. *Phys. Rev. A* **2011,** *83*, 032511.
109. Koga, T.; Matsuyama, H.; Thakkar, A. J. Interelectronic Angles: Rounding out a Geometric Picture of the Helium Atom. *Chem. Phys. Lett.* **2011,** *512*, 287–289.
110. van Vleck, J. H. *The Theory of Electric and Magnetic Susceptibilities;* Oxford University Press: London, 1932.
111. Thakkar, A. J.; Smith, V. H., Jr. Form Factors and Total Scattering Intensities for the Helium-Like Ions from Explicitly Correlated Wave Functions. *J. Phys. B: At. Mol. Phys.* **1978,** *11*, 3803–3820.
112. Thakkar, A. J.; Smith, V. H., Jr. The Electron-Electron Cusp Condition for the Spherical Average of the Intracule Matrix. *Chem. Phys. Lett.* **1976,** *42*, 476–481.
113. Curl, R. F., Jr.; Coulson, C. A. Coulomb Hole in the Ground State of Two-Electron Atoms. *Proc. Phys. Soc., London* **1965,** *85*, 647–652.
114. Boyd, R. J.; Coulson, C. A. Coulomb Hole in Some Excited States of Helium. *J. Phys. B: At. Mol. Phys.* **1973,** *6*, 782–793.
115. Thakkar, A. J. Coulomb Holes in the $2\,^3P$ and $2\,^1P$ States of the Helium Like Ions. *J. Phys. B: At. Mol. Phys.* **1987,** *20*, 3939–3945.
116. Thakkar, A. J. The Higher Order Electron-Electron Coalescence Condition for the Intracule Function for States of Maximum Spin Multiplicity. *J. Chem. Phys.* **1986,** *84*, 6830–6832.
117. Koga, T.; Matsuyama, H. Explicitly Correlated Extracule Densities for Two-Electron Atoms. *Int. J. Quantum Chem.* **1999,** *74*, 455–465.
118. Gálvez, F. J.; Buendía, E.; Sarsa, A. Correlated Electron Extracule Densities in Position and Momentum Spaces. *J. Chem. Phys.* **1999,** *111*, 3319–3326.
119. Koga, T.; Matsuyama, H. Electron-Pair Densities in Position and Momentum Spaces for Multi-Determinant Wavefunctions. *J. Phys. B: At. Mol. Opt. Phys.* **1998,** *31*, 3765–3777.
120. Koga, T.; Matsuyama, H. Electronic Extracule Moments of Atoms in Position and Momentum Spaces. *J. Chem. Phys.* **1998,** *108*, 3424–3430.
121. Bonham, R. A.; Kohl, D. A. Simple Correlated Wavefunctions for the Ground State of Heliumlike Atoms. *J. Chem. Phys.* **1966,** *45*, 2471–2473.

122. Sarsa, A.; Gálvez, F. J.; Buendía, E. Correlated Two-Electron Momentum Properties for Helium to Neon Atoms. *J. Chem. Phys.* **1999,** *110*, 5721–5727.
123. Sarsa, A.; Gálvez, F. J.; Buendía, E. Correlated One-Body Momentum Density for Helium to Neon Atoms. *J. Phys. B: At. Mol. Opt. Phys.* **1999,** *32*, 2245–2255.
124. Youngman, P. K.; Banyard, K. E. Momentum Space Correlation Effects in the 2^3S, 2^1P and 2^3P States of He. *J. Phys. B: At. Mol. Phys.* **1987,** *20*, 3313–3323.
125. Duncanson, W. E.; Coulson, C. A. Electron Momenta in Atoms. *Proc. Phys. Soc., London* **1948**, *60*, 175–183.
126. Thakkar, A. J. Asymptotic Behavior of Atomic Momentals. *J. Chem. Phys.* **1987,** *86*, 5060–5062.
127. Thakkar, A. J.; Wonfor, A. L.; Pedersen, W. A. Asymptotic Expansions of the Electron Momentum Densities of the Atoms from Hydrogen Through Lawrencium. *J. Chem. Phys.* **1987,** *87*, 1212–1215.
128. Koga, T.; Matsuyama, H.; Inomata, H.; Romera, E.; Dehesa, J. S.; Thakkar, A. J. Electron Momentum Densities of Atoms. *J. Chem. Phys.* **1998,** *109*, 1601–1606.
129. Kimball, J. C. Short-Range Correlations and the Structure Factor and Momentum Distribution of Electrons. *J. Phys. A: Math. Gen. Phys.* **1975,** *8*, 1513–1517.
130. Thakkar, A. J.; Simas, A. M.; Smith, V. H., Jr; Extraction of Momentum Expectation Values from Compton Profiles. *Mol. Phys.* **1980,** *41*, 1153–1162.
131. Pathak, R. K.; Sharma, B. S.; Thakkar, A. J. Approximate Relationships Between Density Power Integrals, Moments of the Momentum Density, and Interelectronic Repulsions in Diatomic Molecules. *J. Chem. Phys.* **1986,** *85*, 958–962.
132. Koga, T.; Thakkar, A. J. Moments and Expansion Coefficients of Atomic Electron Momentum Densities: Numerical Hartree-Fock Calculations for Hydrogen to Lawrencium. *J. Phys. B: At. Mol. Opt. Phys.* **1996,** *29*, 2973–2983.
133. Koga, T.; Romera, E.; Dehesa, J. S.; Matsuyama, H.; Thakkar, A. J. Expansion Coefficients and Moments of Electron Momentum Densities for Singly Charged Ions. *Theor. Chem. Acc.* **1999,** *103*, 70–76.
134. Koga, T. Correlated Electron-Pair Properties of the He Atom in Position and Momentum Spaces. *Chem. Phys. Lett.* **2001,** *350*, 135–140.

CHAPTER THREE

A Basis Set Composed of Only 1s Slater Orbitals and 1s Gaussian Orbitals to Perform Molecular Calculations, SCF-LCAO Approach

J.E. Pérez[a], O.E. Taurian[a,*], J.C. Cesco[b], A.E. Rosso[c], C.C. Denner[c], C.J. Alturria Lanzardo[c], F.S. Ortiz[a] and A. Bouferguene[d]

[a]Departamento de Física, FCEFQyN, Universidad Nacional de Río Cuarto, Ruta Nac. 36 Km 601, CP X5804BYA Río Cuarto, Argentina

[b]Instituto de Matemática Aplicada, Universidad Nacional de San Luis, CONICET; Avda. Ej. de los Andes 950, 1er piso, CP D5700HHW San Luis, Argentina

[c]Departamento de Matemática, FCEFQyN, Universidad Nacional de Río Cuarto, Ruta Nac. 36 Km 601, CP X5804BYA Río Cuarto, Argentina

[d]Faculté Saint Jean/University of Alberta, 8406-91 Street, Edmonton, Alberte, Canada T6C 4G9

Contents

Abstract

This work introduces molecular basis sets considering only 1s Slater orbitals and 1s Gaussian orbitals. The main purpose is the construction of a stable atomic model, to be useful for carrying out molecular calculations; among many possibilities, a tetrahedral structure is proposed. With these tetrahedral structures, together with a few floating 1s Gaussian orbitals, a molecular geometry optmization is performed. Test calculations, at the level of SCF-LCAO, for the molecules of *OH2* and *SH2*, are presented; they show the proper trend, i.e., very different optimized geometrical parameters.

* It is with deep regret that we communicate the death of Dr. Oscar E. Taurian [7/17/1947 Laboulaye (Arg.)-9/16/2012 Río Cuarto (Arg.)]. The present work is dedicated to his memory.

Advances in Quantum Chemistry, Volume 67
ISSN 0065-3276, http://dx.doi.org/10.1016/B978-0-12-411544-6.00003-0

1. INTRODUCTION

This work presents the design of a basis sets composed of 1s Slater Orbitals (1SO) and 1s Gaussian Orbitals (1GO) for approximating the electron molecular wave function. This study is motivated by the interesting results in the bibliography[1–6] and our results obtained with such a basis, applied to LiH system.[7]

In Ref. 7, the geometry and energy for the LiH molecule were obtained using basis sets composed of only 1SO and 1GO. Within the framework of the variational theorem, all the parameters were varied, obtaining good results in the SCF-LCAO approximation. However, if these particular basis sets are considered for studying, in general, larger molecular systems, the implementation must be changed because a full optimization would be impractical. Therefore, we present here the model of an atom, which, together with the incorporation of a very few floating 1GO into the molecular environments, will permit us to carry out molecular calculations in a systematic way.

In the new basis sets, the 1SO are located at the positions of the nuclei, ensuring the correct behavior of the wave function there, i.e., the asymptotic cusp condition.[8] Their orbital exponents are determined by using the variational theorem, considering the free atom.

The 1GO considered in the new basis sets are grouped according to the specific function they play in the molecule. Some of them are assembled to represent the atoms involved and others are floating orbitals to reinforce the bond description. The class of floating 1GO is located on the line connecting two bonded nuclei, their positions and orbital exponents can be obtained variationally. On the other hand, the class of 1GO representing the atoms, is determined following the procedure of most basis sets, i.e., modeling free atoms by using the variational theorem; once their orbital exponents and their positions on tetrahedral arrangements are obtained, they are introduced into the molecular environment.

The sequence of calculations that we perform within these mixed basis sets is:

(I) Minimizing the energy of the free atoms for obtaining the relative positions of 1GO in tetrahedral arrangements and their orbital exponents, together with the 1SO orbital exponents. See Sections 2 and 3.

(II) Minimizing the total energy of the molecule for obtaining its geometry. In this process, the structures of the tetrahedra remain rigidly fixed to the corresponding nuclear centers, their 1GO orbital exponents also remain fixed. In Section 4, as a first test of this model, we have determined the optimized geometry of the molecules *OH2* and *SH2*.

2. THE MODEL OF A FREE ATOM IN THIS CONTEXT

The model of free atoms we propose here takes elements of Refs. 9,10.

From Ref. 9, we take the tetrahedral arrangement of 1GO around a given nucleus. These four 1GO placed in this particular way provide through particular normalized linear combinations, which are truly "contractions," a representation of "1s, 2px, 2py, and 2pz" Gaussian orbitals, centered at the corresponding nucleus.

The normalized 1GO to be used in this work are given by

$$G_0\left(\vec{r},\vec{R}\right) = (2\alpha/\pi)^{3/4}\exp\left(-\alpha\left|\vec{r}-\vec{R}\right|^2\right), \tag{1}$$

where α is the orbital exponent and $\vec{R}$ is the point where the orbital is located. Equation (1) can be rearranged, see,[9] in the following form:

$$G_0\left(\vec{r},\vec{R}\right) = \exp\left(-\alpha\left|\vec{R}\right|^2\right) G_0\left(\vec{r},\vec{0}\right)\exp\left(-\alpha^{1/2}\sum_{i=1}^{3} r_i\varepsilon_i\right), \tag{2}$$

where $\vec{R} = \left(R_1, R_2, R_3\right)$, $\vec{r} = \left(r_1, r_2, r_3\right)$, and $\varepsilon_i = 2\alpha^{1/2}R_i$. In Eq. (2), expanding the function $\exp\left(-\alpha^{1/2}\sum_{i=1}^{3} r_i\varepsilon_i\right)$ in powers of the dimensionless parameters ε_i, it is obtained as:

$$G_0\left(\vec{r},\vec{R}\right) = \exp\left(-\sum_{i=1}^{3}\varepsilon_i^2/4\right)\times\left[G_0\left(\vec{r},\vec{0}\right) + \frac{1}{2}\sum_{i=1}^{3}\varepsilon_i G_i\left(\vec{r},\vec{0}\right) + o\left(\varepsilon_i^2\right)\right], \tag{3}$$

where the functions $G_i\left(\vec{r},\vec{0}\right)$ are the normalized "2p" Cartesian Gaussian functions located at $\vec{R} = \vec{0}$, which are given by

$$G_i\left(\vec{r},\vec{0}\right) = 2\alpha^{1/2} r_i G_0\left(\vec{r},\vec{0}\right). \tag{4}$$

In Ref. 9, using Eq. (3), it is discussed how some normalized linear combinations of 1GO centered around the nucleus can be related to the Gaussian "s", "p", "d", ... functions. As a result, they give the conditions under which several 1GOs arranged at the vertices of a tetrahedron, octahedron, etc., can approximate "s", "p", "d"... Gaussian functions located in the middle point of these regular solids, i.e., at the nucleus. Moreover, they also prove that the tetrahedral arrangement is the most economical way of representing a set of "1s" and "2p" Gaussian orbitals, i.e., these functions can be obtained with an octahedral disposition too, but introducing two more 1GO. Among these interesting possibilities, the tetrahedral arrangement is adopted in this work. Each atom cannot be described accurately by using

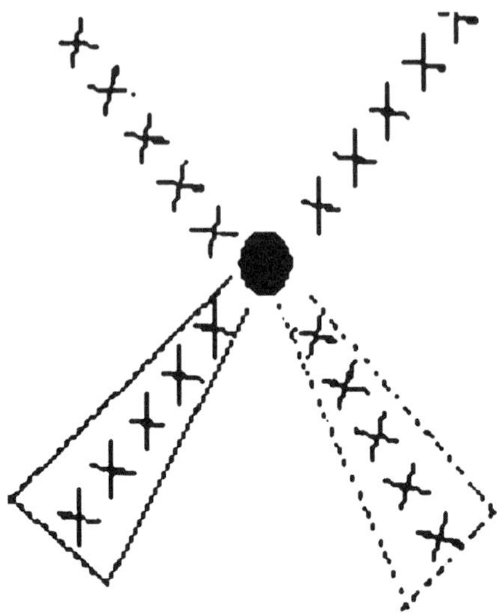

Figure 3.1 According to the notation, • represents an 1SO and × represents an 1GO; this figure is a sketch of Five aligned and concentric tetrahedra.

only one tetrahedral set of 1GO, i.e., a unique set of "1s" and "2p" Gaussian orbitals. So that, we need to consider a few concentric and aligned tetrahedra; see Figure 3.1 for the case of Five tetrahedra as we use them in this work. Their positions and sizes must be determined by using the variational theorem, considering the free atom.

From Ref. 10, we adopt the even tempered ansatz for the sequence of orbital exponents involved in the set of concentric and aligned tetrahedra around a given nucleus. This simplifying hypothesis is an important way of reducing the number of variational parameters. When we consider three or more aligned tetrahedra, only two parameters must be taken into consideration for optimization. Denoting these parameters a and b, respectively, then the orbital exponent for the ith tetrahedron, denoted by α_i, is given by $\alpha_i = ab^i$, $i = 1, 2, \ldots, nt$ where nt is the total number of aligned tetrahedra.

The details of the implementation of this model are given in Section 3.

3. EXAMPLES OF ATOMIC MODELS

In Figure 3.1, a scheme of the basis sets corresponding to an atom, for the case of 5 tetrahedra, is shown. When applying the variational theorem, the relation $\varepsilon = 2(\alpha/3)^{1/2}R$ must be considered, ε is the Taylor expansion parameter, see Eq. (3), R is the distance between the 1GO and the nucleus, and α is the orbital exponent of the corresponding 1GO, see Eq. (2). If the value of ε is fixed and it is the same for every tetrahedron, then the relation $\varepsilon = 2(\alpha/3)^{1/2}R$ acts like a constraint during the optimization process. Considering this criterion and having taken into account that the orbital exponents satisfy the even tempered ansatz, we show the total atomic energy at the level of HF-SCF as a function of ε, for oxygen atom in Figure 3.2 and sulfur atom in Figure 3.3, respectively. In these figures, for each value of ε, the values of 1SO orbital exponent, a and b, are varied until a minimum of the energy is reached.

When ε decreases, the tetrahedra move toward the nucleus, this tendency of collapsing is unavoidable, i.e., $\varepsilon = 2(\alpha/3)^{1/2}R$. However, as can be seen from Figures 3.2 and 3.3, there is a wide interval for which in spite of the phenomenon of collapsing, the curves are very smooth. There exists a threshold, from which the calculation gets unstable, i.e., $\varepsilon \approx 10^{-3}$. This behavior is observed because in our implementation, the 1GOs have not been "contracted" according to Ref. 9 as mentioned in Section 2, i.e., HF-SCF procedure assigns their coefficients through the usual LCAO

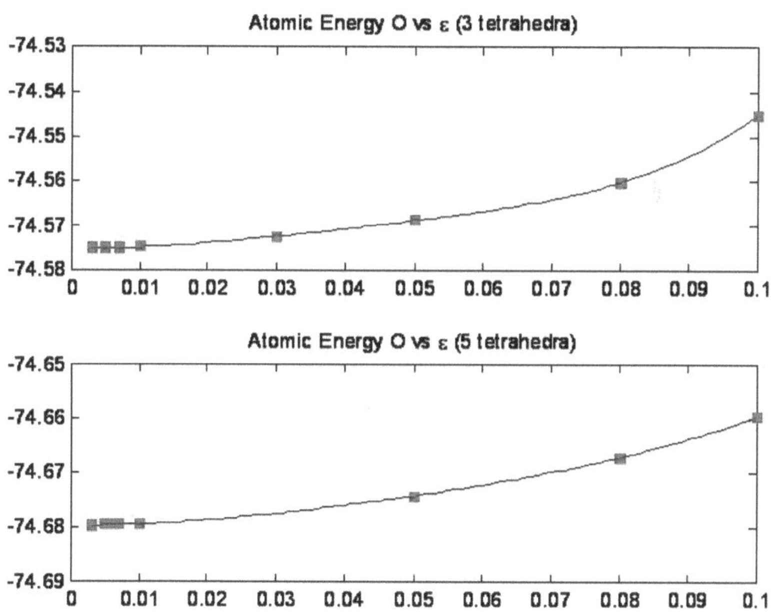

Figure 3.2 Oxygen free atom energy versus the parameter ε, for the cases of 3 and 5 tetrahedra, respectively. For each ε the corresponding a, b and the 1SO orbital exponent values are obtained variationally. For color version of this figure, the reader is referred to the online version of this chapter.

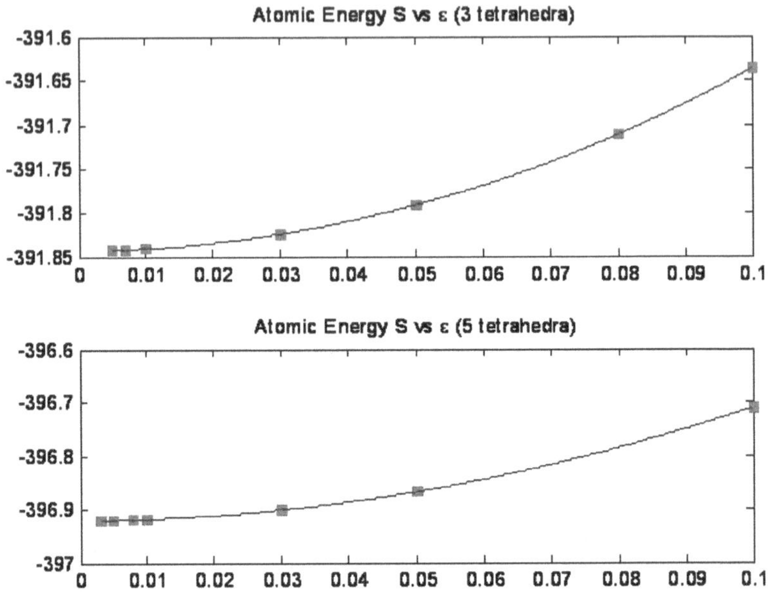

Figure 3.3 Sulfur free atom energy versus the parameter ε, for the cases of 3 and 5 tetrahedra, respectively. For each ε the corresponding *a*, *b* and the 1SO orbital exponent values are obtained variationally. For color version of this figure, the reader is referred to the online version of this chapter.

according to the molecular environment. We have chosen this "uncontracted?" option because it will give us much more flexibility when the atom be introduced into the molecule.

It is worth mentioning that, although we have not "contracted" the 1GOs and consequently we obtain an instability, from Figures 3.2 and 3.3, we can extrapolate the proper atomic model energy limit.

All the calculations presented in this work, at the level of SCF-LCAO, were performed with a code in the MatLab language and the optimization algorithm is based on the bfgswopt.m Keller´ routine (1997).

4. EXAMPLES OF MOLECULAR CALCULATIONS

In Figures 3.5 and 3.6 we show the total energy for *OH2* and *SH2* molecules, respectively, as a function of the bond angle, denoted by θ, i.e., the angle between the corresponding X–H bonds with X=O, S, see Figure 3.4. For each bond angle we have optimized bond distances (X–H) denoted by *d*, position and size of the floating 1GO and the 1SO orbital exponents for the Hydrogen atoms, see Figure 3.4.

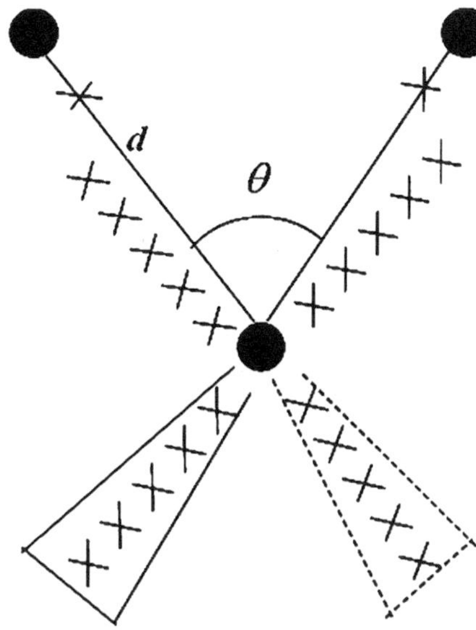

Figure 3.4 Using the same notation of Figure 3.1, this is a sketch of the molecules XH2, X=O, S. Here, the variables *d* and *θ* are shown explicitly and the location of the floating 1GO is also shown.

For all the calculations shown in Figures 3.5 and 3.6, the parameter ε remains constant with value 5×10^{-3}, which ensures that the molecular calculations be stable. The little error introduced with this choice can be estimated graphically from Figures 3.2 and 3.3, i.e., with respect to the extrapolated model energy value.

The results given in Table 3.1 clearly show the correct geometrical trend. The very different values of the d_{min} and θ_{min}, i.e., the corresponding values for the minimum energies, are a firm evidence of the great flexibility of the fixed tetrahedral structures. However, for the *SH2* system, the energy is not so good because we have not introduced more functions into the basis set.

5. CONCLUDING REMARKS

The introduction of a set of concentric and aligned tetrahedra for describing atoms, presented in this work, makes it possible that the systematic calculations of approximated electron wave functions by using only 1s orbitals be carried out.

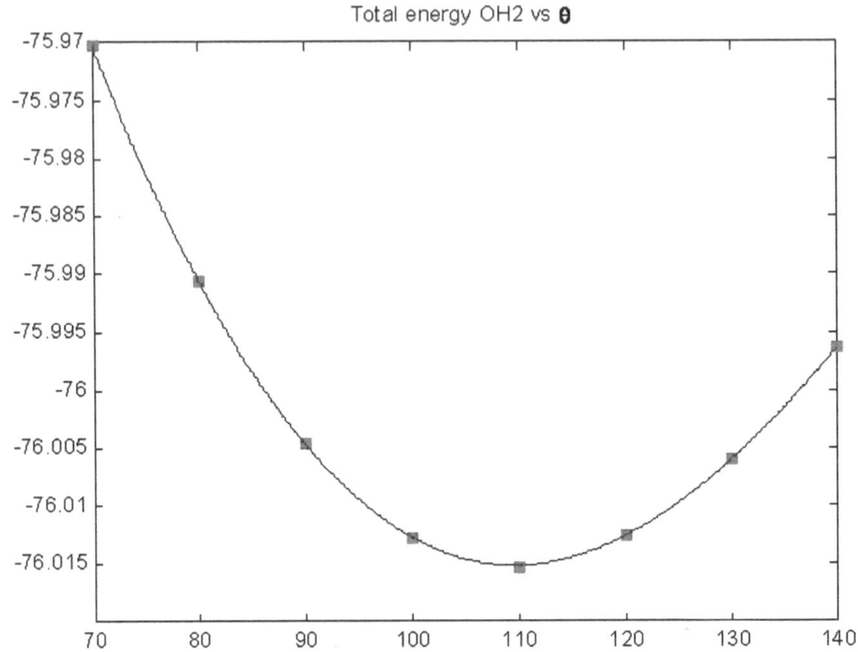

Figure 3.5 Total molecular *OH2* energy versus the bond angle θ is shown. For each value of θ, the parameter *d* is obtained variationally, together with the values for the location and the shape, on the line jointing the bonded nuclei, corresponding to the floating 1GO. All the calculations were done with $\varepsilon = 5 \times 10^{-3}$. For color version of this figure, the reader is referred to the online version of this chapter.

The model introduces a great simplification in the class of integrals to be dealt with, both the theoretical expressions to be used and the computing details.

The main achievement of this work is that it is possible to design a stable tetrahedral structure of an atom to be useful for molecular calculations.

In this context, it is possible to envisage future improvements in the obtained geometries, e.g., around the heavy nuclei introducing octahedral structures which would give "d" orbitals, also, introducing some 1GO more around the Hydrogen nuclei to reinforce the polarization given by the floating 1GO.

The results presented show that three aims can be achieved: (1) numerical stability in all the processes; (2) systematic way of introducing the proper atomic information into the molecule; and (3) a basis sets with adequate response to different molecular environment.

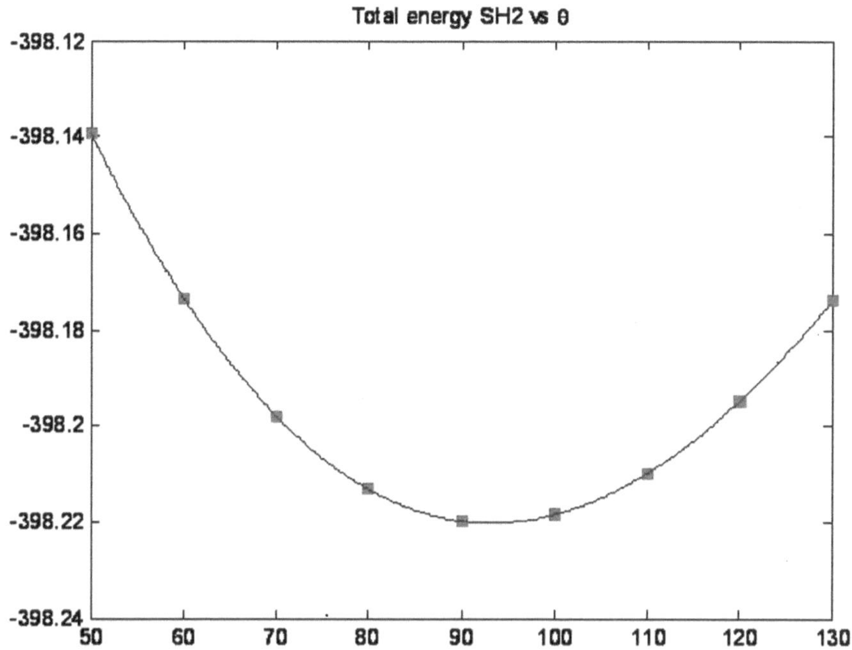

Figure 3.6 Total molecular *SH2* energy versus the bond angle θ is shown. For each value of θ, the parameter d is obtained variationally, together with the values for the location and the shape, on the line jointing the bonded nuclei, corresponding to the floating 1GO. All the calculations were done with $\varepsilon = 5 \times 10^{-3}$. For color version of this figure, the reader is referred to the online version of this chapter.

Table 3.1 The values corresponding to the minima of Figures 3. 5 and 3.6 are shown. Atomic units are used

Molecule		Experimental values[11,12]	Basis sets including polarization[13]	This work
OH2	θ (°)	104.5	106.4	109.7
	d	1.808	1.782	1.827
	Total energy		−76.068	−76.015
SH2	θ (°)	92.11	94.1	93.25
	d	2.5	2.518	2.56
	Total energy		−398.72	−398.22

ACKNOWLEDGMENTS

Grants from SeCyT Universidad Nacional de Río Cuarto and IMASL Universidad Nacional de San Luis-CONICET are appreciated.

REFERENCES

1. Bacskay, G. B.; Linnet, J. W. Mixed Basis Functions in Molecular Mechanics. *Theor. Chim. Acta (Berl.)* **1972,** *26,* 15.
2. Frost, A. A. A Floating Spherical Gaussian Orbitals Models for Molecular Structure, III First Row Hydrides. *J. Am. Chem. Soc.* **1968,** *72* (4), 1289.
3. Frost, A. A.; Rouse, R. A. A Floating Spherical Gaussian Orbitals Models for Molecular Structure, IV Hydrocarbon. *J. Am. Chem. Soc.* **1980,** 90.
4. Pakiari, A. H.; Linnet, J. W. Single Determinant Open-and-Closed Shell FSGO Calculation for Some Diatomic Molecules. *Int. J. Quantum Chem.* **1980,** *18* (3), 661.
5. Wilson, S. Distributed Basis Sets of s-Type Gaussian Functions for Molecular Electronic Structure Calculations. *Int. J. Quantum Chem.* **1996,** *60* (1), 47.
6. Brailsford, D. F.; Hall, G. G.; Hemming, N.; Martin, D. Floating and s-and-p Type Gaussian Orbitals. *Chem. Phys. Lett.* **1975,** *35* (4), 437.
7. Cesco, J. C.; Denner, C. C.; Giubergia, G. O.; Rosso, A. E.; Pérez, J. E.; Ortiz, F. S.; Taurian, O. E.; Contreras, R. H. Implementation of Atomic Basis Sets Composed of 1s Gaussian and 1s Slater-Type Orbitals to Carry Out Quantum Mechanics Molecular Calculations. *J. Comput. Chem.* **1999,** *20* (6), 604.
8. Kato, T. On the Eigenfunctions of Many-particle System in Quantum Mechanics. *Commun. Pure Appl. Math.* **1957,** *10* (2), 151.
9. Maggiora, G. M.; Christoffersen, R. E. Ab 10 (2) initio Calculations on Large Molecules Using Molecular Fragments, Generalization and Characteristics of Floating Spherical Gaussian Basis Sets. *J. Am. Chem. Soc.* **1976,** *98* (26), 8325.
10. Schmidt, M. W.; Ruedenberg, K. Effective Convergence to Complete Orbital Bases and to the Atomic H-F Limit through Systematic Sequences of Gaussian Primitives. *J. Chem. Phys.* **1979,** *71,* 3951.
11. Levine, I. N. *Quantum Chemistry,* 5th ed.; Prentice Hall: NJ, 2000; p 504.
12. Cook, R. L.; De Lucía, F. C.; Helminger, P. Molecular Force Field and Structure of Hydrogen Sulfide: Recent Microwave Results. *J. Mol. Struct.* **1975,** *28,* 237–246.
13. Hehre, W. J. *A Guide to Molecular Mechanics and Quantum Chemical Calculations;* Wavefunction Inc., 2003; p 491.

CHAPTER FOUR

On a Transformation for the Electrostatic Potential, Generated by the Product of Two 1s Slater Type Orbitals, Giving an Efficient Expression

J.E. Pérez[a], O.E. Taurian[a, 1], A. Bouferguene[b] and P.E. Hoggan[c]
[a]Departamento de Física, FCEFQyN, Universidad Nacional de Río Cuarto, Ruta Nac. 36 Km 601, (CP X5804BYA) Río Cuarto, Argentina
[b]Faculté Saint Jean/University of Alberta, 8406-91 street, Edmonton, Alberta, T6C 4G9, Canada
[c]Institut Pascal, UMR CNRS 6602, BP 80026, 24 Avenue des Landais, 63171 AUBIERE CEDEX, France

Contents

Abstract

This work presents new compact analytical formulae aimed at accurate evaluation of two-electron molecular integrals involving 1s functions. They revisit work by Shavitt and Karplus showing this can be generalized and use an integral representation of the electrostatic potential that introduces half-integer Bessel functions of the second kind and splits the oscillatory integral into two positive definite contributions for stable numerical evaluation.

[1] It is with deep regret that we communicate the death of Dr. Oscar E. Taurian [7/17/1947 Laboulaye (Arg.)-9/16/2012 Río Cuarto (Arg.)]. The present work is dedicated to his memory.

Advances in Quantum Chemistry, Volume 67

ISSN 0065-3276, http://dx.doi.org/10.1016/B978-0-12-411544-6.00004-2

1. INTRODUCTION

1.1 Context

Efficient evaluation of multi-center two-electron integrals is a bottleneck for molecular orbital calculations. Such integrals may arise from Hartree–Fock LCAO formalisms, in which case the most numerous integrals are four-center exchange terms which number about N^4 for N-atomic orbitals corresponding to the electron system of at least four atoms. The three-center Coulomb and exchange terms are also numerous. It has been shown by Karplus that all the products of atomic orbitals appearing in these integrals may be derived from that of 1s orbitals. This implies that the related electrostatic potential plays a fundamental role in the structure of such integrals and it will therefore focus our attention in this paper.

Robust formulae are required to evaluate the molecular integrals accurately and rapidly. The majority of electronic structure work uses Gaussian basis sets to represent the atomic orbitals, because the functions have a simple "addition theorem" that expresses their product as another Gaussian on a displaced center, even if the two initial atoms are distinct.

This work focuses on the electrostatic potential expressed in terms of analytical 1s atomic orbital expressions, i.e., exponential functions with negative arguments. These functions obey the Kato conditions (cusp at the nuclei and exponential decay) unlike the Gaussians. There are, therefore, certain applications for which exponential 1s orbitals must be used. Unfortunately, the product of two exponential functions has no simple closed form when the centers are different. We get an integral representation of the electrostatic potential from two 1s orbitals that may be evaluated accurately after partitioning the integral into two positive definite contributions (as can be seen below in Eq. (3) and those following). This new formula is obtained from one of the two expressions presented in Ref. 1, i.e., Eq (83) which involves improper integrals with oscillating integrands. It is transformed using an integral representation for Bessel functions of the second kind.[2] This provides a new expression which *avoids* improper integrals with oscillating functions. The new formula contrasts with the other expression in Ref. 1 i.e., Eq. (52) which explicitly depends on $k_2(z) = z^2 K_2(z)$ where $K_2(z)$ is the Bessel function of the second kind,[2] because it depends on the Bessel function of the second kind with *half-integer* index, i.e., it can be expressed in terms of elementary functions. In Section 2 this approach is presented. This work saves considerable computational effort and the numerical results are given in Table 4.1, where they correspond, to high accuracy with those of an independent analytical development published earlier by one of us.[4]

Table 4.1 The values are expressed in atomic units

X	Y	$\Phi(\vec{R}) \times 10^2$ Eq. (1) Reference	$\Phi_1(\vec{R}) \times 10^2$ Eq. (7)	$\Phi_2(\vec{R}) \times 10^2$ Eq. (8)	Relative error
0.5	0	4.030717025872722	4.454473002967213	0.423755977094476	4 (−15)
	0.5	2.869525298569841	2.919980325954056	0.050455027348216	1 (−16)
	1	1.782039311119250	1.783257715833022	1.218404713772(−3)	2 (−16)
	2	0.952573233847964	0.952573534718068	3.00870105 (−7)	−4 (−16)
0	0	7.659096006862764	17.48364420738961	9.824548200526198	8 (−14)
−0.5		3.425031452025301	3.522544268533491	9.751281650819(−2)	−6 (−16)
−1		1.849352568071357	1.850526102823503	1.173534752098(−3)	3 (−14)
−2		0.949135932043343	0.949136116971351	1.849280085403(−7)	−5 (−16)
1	0	2.082967230864852	2.092133216288379	9.165985423299(−3)	1 (−13)
1.5		1.358661689798895	1.358771740908086	1.100511091904(−4)	5 (−16)
2		1.007253852796482	1.007255232824035	1.380027553886(−6)	−7 (−16)
3		0.664152539572263	0.664152539795474	2.23211714177(−10)	−8 (−16)

2. METHOD

2.1 General considerations

The electrostatic potential Φ at $\overrightarrow{R}$ corresponding to a charge distribution given by the product of two 1s Slater Orbitals, which are located at $\overrightarrow{R_1}$ and $\overrightarrow{R_2}$ with orbital exponents α and β, respectively, see Ref. 1, is

$$\Phi(\overrightarrow{R}) = cte \int_0^{\infty} \mathrm{d}r \int_0^1 \mathrm{d}u\, u(1-u)\, K(Dz(r,u)) \frac{\sin\left(|\overrightarrow{P(u)} - \overrightarrow{R}|\, r\right)}{|\overrightarrow{P(u)} - \overrightarrow{R}|\, r}. \quad (1)$$

One of the purposes of this work is to perform the integration over the variable "r" in Eq. (1) analytically. To this end, the following integral representation, see page 376,[2] is used

$$K(D\,z) = \frac{K_{5/2}(D\,z)}{(D\,z)^{5/2}} = \frac{\Gamma(3)\,2^{5/2}}{\pi^{1/2}\,D^5} \int_0^{\infty} \mathrm{d}t \frac{\cos(D\,t)}{(t^2 + z^2)^3}, \quad (2)$$

where: $D = |\overrightarrow{R_1} - \overrightarrow{R_2}|$, z is $z(r,u) = [r^2\, u(1-u) + \alpha^2 u + \beta^2(1-u)]^{1/2}$, $cte = \left(\frac{32}{\pi}\right)^{1/2} (\alpha\beta)\, D^5, \overrightarrow{P(u)} = u\,\overrightarrow{R_2} + (1-u)\,\vec{R_1}$ and $K_{5/2}(x)$ is the Bessel function of the second kind.[2]

As far as the authors are aware, this application of Eq. (2) to the radial integral giving the electrostatic potential for two 1s orbitals is new and the analysis is presented in the next section for the first time.

2.2 Analytical treatment

Equations (1) and (2) give

$$\Phi(\overrightarrow{R}) = cte \frac{2^{5/2}\,\pi^{1/2}}{D^5} \int_0^1 \mathrm{d}u\, u(1-u) \int_0^{\infty} \mathrm{d}t \frac{\cos(D\,t)\,[u(1-u)]^{-3}}{p\,\gamma^6} \times \left[1 - \exp(-p\gamma)\left\{1 + \frac{5p\gamma}{8} + \frac{(p\gamma)^2}{8}\right\}\right], \quad (3)$$

where $p = \left|\overrightarrow{P(u)} - \overrightarrow{R}\right|$ and $\gamma = \frac{[t^2 + u\alpha^2 + (1-u)\beta^2]^{1/2}}{[u(1-u)]^{1/2}}$.

The Eq. (3) can be partitioned into two positive terms, $\Phi_1(\overrightarrow{R})$ and $\Phi_2(\overrightarrow{R})$, respectively. This is motivated by the need of ensuring the correct behavior in the limit when $p \to 0$, see Appendix A. Explicitly, the partition is

$$\Phi(\vec{R}) \equiv \Phi_1(\vec{R}) - \Phi_2(\vec{R}) =$$

$$= \left\{ cte \frac{2^{5/2}\pi^{1/2}}{D^5} \int_0^1 du\, u(1-u) \int_0^\infty dt \frac{\cos(D\,t)}{[u(1-u)]^3 \gamma^5} \frac{1-\exp(-\gamma p)}{\gamma p} \right\}$$

$$- \left\{ cte \frac{2^{5/2}\pi^{1/2}}{D^5} \int_0^1 du\, u(1-u) \int_0^\infty dt \frac{\cos(D\,t)\exp(-\gamma p)}{[u(1-u)]^3 \gamma^5} \left[\frac{5}{8} + \frac{\gamma p}{8}\right] \right\}. \tag{4}$$

By using the following Eqs. (5) and (6), taken from,[2,3] respectively, it is possible to obtain a final expression for $\Phi_1(\vec{R})$ and $\Phi_2(\vec{R})$, which are given in Eqs. (7) and (8), respectively.

$$\frac{\exp(-\varepsilon[t^2 + \alpha^2 u + \beta^2(1-u)]^{1/2})}{[t^2 + \alpha^2 u + \beta^2(1-u)]^{1/2}}$$

$$= \frac{2}{\pi^{1/2}} \int_0^\infty d\gamma \exp\left\{ -\gamma^2[t^2 + \alpha^2 u + \beta^2(1-u)] - \frac{\varepsilon^2}{4\gamma^2} \right\}, \tag{5}$$

$$\frac{1}{[t^2 + \alpha^2 u + \beta^2(1-u)]^{5/2}} = \frac{8}{3\pi^{1/2}}$$

$$\int_0^\infty dx\, x^4 \exp\{-x^2[t^2 + \alpha^2 u + \beta^2(1-u)]\}. \tag{6}$$

With $\varepsilon = \dfrac{p}{[u(1-u)]^{1/2}}$.

$$\Phi_1(\vec{R}) = cte \int_0^1 du \frac{u(1-u)}{p} \left\{ \frac{\pi\, K(D\,\delta)}{2} - \frac{2^{5/2}\pi^{1/2}\, Int(u)}{D^5} \right\}, \tag{7}$$

$$\Phi_2(\vec{R}) = cte \frac{2^{5/2}\pi^{1/2}}{D^5} \int_0^1 du \frac{u(1-u)}{p}$$

$$\times \left\{ \frac{2^{1/2}}{3\pi^{1/2}\delta^{3/2}} \int_0^{\pi/2} d\theta \frac{\cos^4(\theta)}{\sin^2(\theta)} \left\{ k_{3/2}\left(\delta \left[\frac{\varepsilon^2}{\sin^2(\theta)} + D^2 \right]^{1/2} \right) \frac{\varepsilon^2}{2} \right.\right.$$

$$\left.\left. + \frac{\varepsilon^2}{8} \frac{\varepsilon^2}{\sin^2\theta} \delta\, k_{1/2}\left(\delta \left[\frac{\varepsilon^2}{\sin^2(\theta)} + D^2 \right]^{1/2} \right) \right\} \right\}, \tag{8}$$

where

$$Int(u) = \frac{2^{1/2}}{3\pi^{1/2}} \int_0^{\pi/2} d\theta \ \cos(\theta)^4 \frac{k_{5/2}\left(\delta \left[\frac{\varepsilon^2}{\sin(\theta)^2} + D^2\right]^{1/2}\right)}{\delta^{5/2}}. \tag{9}$$

With $\delta = [\alpha^2 u + \beta^2(1-u)]^{1/2}$, $k_\nu(z) = z^\nu \ K_\nu(z)$ and $K_\nu(z)$ is the Bessel function of the second kind.[2]

3. RESULTS

Software, i.e., adaptive integration from MATHCAD or similar programs, setting the accuracy threshold to $10^{(-15)}$ can be used to compare analytical expressions (7) and (8) above with the reference cited, see Table 4.1 for a particular example.

$$\vec{R}_1 = (0,0,0), \quad \alpha = 7.7, \quad \vec{R}_2 = (1,0,0), \quad \beta = 1, \quad \vec{R} = (X,Y,0).$$

The reference value is evaluated following the procedure in,[4] i.e., in Eq. (1) the variable change: $w = p\,r$, is made and the resulting series accelerated.

The relative error is $[\Phi - (\Phi_1 - \Phi_2)]\,/\,\Phi$.

All the numerical integrals are evaluated with an adaptive procedure.

The negative numbers in parentheses should be interpreted as: $(-j) = \times 10^{-j}$.

4. CONCLUDING REMARKS

The results show that relative errors of $10^{-16} - 10^{-13}$ can be easily obtained. This trend, together with the clear behavior of the two contributions when $|\vec{R}|$ grows, i.e., Φ_1 begins to dominate asymptotically, makes the new expression suitable for applications.

Considering the asymptotic behavior and Eqs. (4), (7), and (9), it is important to remark that $\Phi(\vec{R})$ is given by a one-dimensional integral, i.e., Eq. (7) without the term $int(u)$.

We are working to generalize the present approach to four-center electron repulsion integrals.

APPENDIX A

Case with $\vec{p}(u) = \vec{0}$

Considering the following vectors,

$$\vec{R},\ \vec{R}_1,\ \vec{R}_2,\ \vec{D} = \vec{R}_2 - \vec{R}_1,\ \vec{p}(u) = \vec{R} - [\vec{R}_1 + u\vec{D}]. \tag{A.1}$$

Then, cases, which have a null p, are such that

$$\vec{p}(u_0) = \vec{R}_0 - [\vec{R}_1 + u_0\,\vec{D}] = \vec{0}, \tag{A.2}$$

where $\vec{R}_0$ is the particular $\vec{R}$ corresponding to u_0.

From Eq. (A.2), we obtain

$$u_0 = \frac{(\vec{R}_0 - \vec{R}_1)\cdot\vec{D}}{D^2}. \tag{A.3}$$

So that, the cases for which $\vec{R}$ is on the line connecting the centers 1 and 2, i.e., which are denoted by $\vec{R}_0$ in Eq. (A.2), give the following expression for the vector $\vec{p}$

$$\vec{p}(u) = (u_0 - u)\,\vec{D}. \tag{A.4}$$

The particular values 0 and 1 for u_0 are easier to treat because there are explicit cancellations, i.e., because the presence of the factor: $u(1-u)$; however, they require separate programs.

The numerical results are generally obtained (apart from the two previous special cases) by partitioning the integral over u in the Φ_1 contribution (for evaluating the reference value too), i.e., into two intervals $(0, u_0)$ and $(u_0, 1)$. This works well because the integrand is continuous at $u = u_0$, having a discontinuous derivative there.

ACKNOWLEDGMENTS

Grants from SeCyT Universidad Nacional de Río Cuarto are appreciated.

REFERENCES

1. Shavitt, I.; Karplus, M. Gaussian-Transform Method for Molecular Integrals. I. Formulation for Energy Integrals. *J. Chem. Phys.* **1965,** *43* (2), 398.
2. Abramowitz, M.; Stegun, I. *Handbook of Mathematical Functions*, 9th ed.; Gordon and Beach: New York, 1970.
3. Matsuoka, O.; Aoyama, T. Molecular Integral of Diamagnetic Contribution to Nuclear Spin–Spin Coupling Constant. *J. Chem. Phys.* **1980,** *73* (11), 5718.
4. Bouferguene, A.; Fares, M.; Rinaldi, D. Calculation of Multicenter Electron Integrals Over B-Functions Basis-Sets. *J. Chem. Phys.* **1994,** *100*, 8156.

CHAPTER FIVE

d-Dimensional Kepler–Coulomb Sturmians and Hyperspherical Harmonics as Complete Orthonormal Atomic and Molecular Orbitals

C. Coletti[a], D. Calderini[b] and V. Aquilanti[c]
[a]Dipartimento di Farmacia, Università G. d'Annunzio, Via dei Vestini, I-66100 Chieti, Italy
[b]Scuola Normale Superiore di Pisa, Via Consoli del Mare, 2, I-56126 Pisa, Italy
[c]Dipartimento di Chimica, Università di Perugia, Via Elce di Sotto 8, I-06123 Perugia, Italy

Contents

Abstract

Sturmian basis sets are increasingly finding applications to the description of atomic and molecular structure, because of their mathematical properties and their flexibility regarding the ability to describe features of specific physical problems. However, their nature

Advances in Quantum Chemistry, Volume 67
ISSN 0065-3276, http://dx.doi.org/10.1016/B978-0-12-411544-6.00005-4

and properties have not been fully exploited in quantum chemistry. In this work we present a classification of Kepler-Coulomb Sturmian sets, where notations, symmetry properties, useful formulae, and relationships are described in detail, so to provide support for their applications to physical problems. The mathematical solution of Schrödinger equation is given for these sets both in configuration and momentum space, where Sturmian eigenfunctions coincide with hyperspherical harmonics and connections between different sets manifestly appear as elements of angular momentum algebra. Applications of the considered sets and some of their generalizations are also briefly accounted for.

1. INTRODUCTION

When dealing with the theory of electronic structure of atoms and molecules, quantum chemistry employs basis sets for the solution of Schrödinger equation. Because the Schrödinger equation for the non-relativistic hydrogen atom is analytically solvable in a closed form, the use of hydrogen-like orbitals as expansion basis sets to calculate the electronic wavefunctions of atoms and molecules has been extensively considered. In order to tackle the obstacles encountered for the efficient computation of matrix elements (two-electron integrals in particular) involved in the calculations, preference has been given over the years to Gaussian orbitals, more convenient for such purposes.

However, recent progress concerning the use of Slater orbitals[1–6] has revived the interest in exponential type orbitals, more *natural*[7] from both physical and mathematical viewpoints. Indeed, the use of a procedure based on Gaussian orbitals suffers several disadvantages, including poor representation of long-range interactions and of the properties of the excited states, and the slow convergence in reproducing the cusps in the electronic densities at nuclei.

Among exponential type orbitals, Sturmian basis sets have been considered by several authors[8–11] both in their generalized form[8,11–13] and as Coulomb Sturmians[14–16,9,17,18] (attainable by appropriately setting the α factor equal to 1 in the expression of the generalized central potential $V(r) = b/r^{\alpha}$,[19,20] see the following) for applications to atomic and molecular structure.

There is a close relationship between Sturmians and Slater type orbitals (see for instance[21]) and the former can be expressed as a simple linear combination of the latter; therefore progress in the computation of integrals over Slater type orbitals can also be exploited for Sturmian basis sets.

Nevertheless, the full strength of the approach based on Sturmians (of Kepler–Coulomb type in particular) comes from the fact that their counterparts in momentum space are hyperspherical harmonics and their overlaps or superposition integrals can be easily reduced to sums over discrete analogs of classical orthogonal polynomials,[22] which, in turn, can be related to generalizations of vector coupling and recoupling coefficients ($3j$, $6j$, or $9j$ coefficients).[23]

As depicted in Ref. 23, the radial part of Coulomb Sturmians[24] appears as a step in the Askey scheme[25,26] of orthogonal polynomials, see Figure 5.1, and we will show that all the other steps in the "ladder" come into play when these basis sets are employed.

This paper aims at presenting properties, useful formulae, and exemplary illustrations of Coulomb-type Sturmian orbitals in configuration and momentum space. In the momentum (or reciprocal) space, as recent

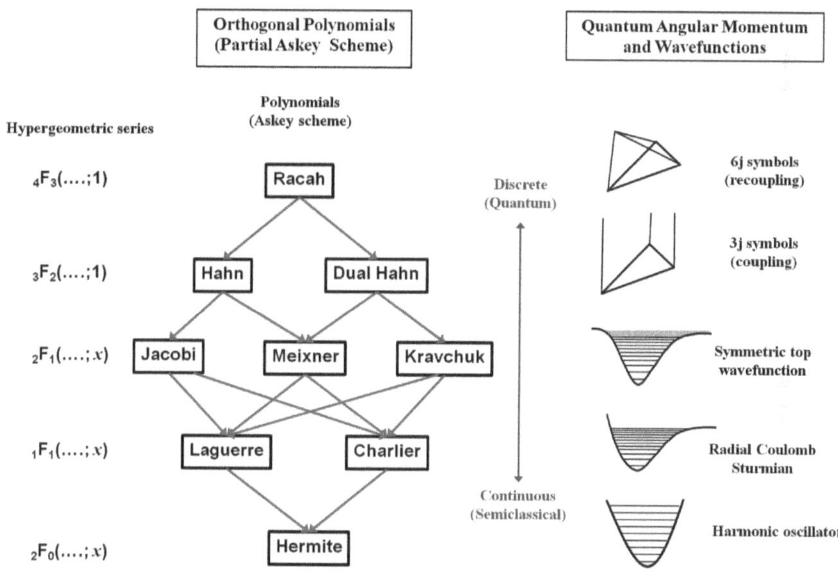

Figure 5.1 Comparison between a classification of orthogonal hypergeometric polynomials (a partial view of schemes in Refs. 22 and 25) and their current occurrence in quantum mechanics (see also Ref. 24). The first two downward connections are limiting relationships which physically correspond to semiclassical limits, while the remaining ones are progressive confinements of their range due to the form of the potential. Polynomials at the same level are identical, but the role of the three-term recurrence relation and of the finite difference equation varies - Hahn and dual Hahn differ in the role played by the "degree" and the "variable" parameters, which correspond to the dual role that angular momentum and projection quantum numbers play in vector coupling coefficients. Similarly for the Jacobi polynomials, for which the Meixner and Kravchuk polynomials are the "discrete" counterparts, and the Charlier polynomials are the "discrete" duals of Laguerre polynomials. For color version of this figure, the reader is referred to the online version of this chapter.

results show,[27–29] the mathematical solutions to fundamental expressions in quantum theory problems (mostly overlap integrals) benefit from several simplifications when exponential type orbitals are employed.

The work will thus develop by introducing basic notation, classifications, and relationships among alternative sets of Sturmian hydrogenic orbitals (Section 2), starting from the general d-dimensional case and introducing sets corresponding to different coordinate systems with a special emphasis on polar and parabolic Sturmians. The two-dimensional and, in particular, the physical three-dimensional cases will be derived from the general d-dimensional one and will be explicitly given. For the three-dimensional case, other "non-canonical" forms of hydrogenic Kepler–Coulomb Sturmians, corresponding to spheroidal and sphero-elliptic coordinates, will also be discussed.

The momentum space perspective (Section 3) will be used to shed light on the nature and symmetry properties of Kepler–Coulomb Sturmians and to obtain important mathematical connections (in terms of angular momentum algebra, i.e., of elements of Figure 5.1) between these sets and their momentum space counterparts, hyperspherical harmonics. All connections evidenced in direct space will thus be examined in terms of generators of d-dimensional spaces, where $d=2$ and $d=3$.

Section 4 summarizes applications of these sets and some of their generalizations to atomic and molecular structure problems, highlighting important relationships and integrals, of general use in quantum mechanical problems.

Additional remarks and an Appendix conclude the paper.

2. STURMIAN BASIS FUNCTIONS IN CONFIGURATION SPACE

Historically, the term "Sturmian" basis set was coined by Rotenberg[30] in 1962 to underline that these functions are solutions of a particular Sturm-Liouville problem, i.e., a differential equation closely related to Schrödinger equation. In fact, Rotenberg introduced Sturmian basis functions starting from Schrödinger's equation in the radial variable r for a central field problem in three-dimensional space. Sturmian wavefunctions were thus defined as solutions of the equation (in atomic units):

$$\left[\frac{1}{2r^2}\frac{\mathrm{d}}{\mathrm{d}r}r^2\frac{\mathrm{d}}{\mathrm{d}r} - \frac{l(l+1)}{r^2} + E_0 + a_{nl}V(r)\right]u_{nl}(r) = 0 \tag{1}$$

for $n = 1, 2, 3, \ldots$ and $l = 0, 1, \ldots, n-1$; E_0 is a fixed negative value and a_{nl} is the eigenvalue of the charge varied to meet the following boundary conditions: $u_{nl}(0) = 0$ and $u_{nl}(\infty) \to 0$.

In the important case of the Kepler–Coulomb potential,[31] to which we will refer throughout this work, Eq. (1) becomes:

$$\left[\frac{1}{2r^2}\frac{\mathrm{d}}{\mathrm{d}r}r^2\frac{\mathrm{d}}{\mathrm{d}r} - \frac{l(l+1)}{2r^2} + E_0 - \frac{Z_n}{r}\right]R_{nl}(r) = 0. \tag{2}$$

Equation (2) is similar to the radial part of Schrödinger equation for the hydrogen atom separated in polar coordinates, the main difference being that in the latter the nuclear charge is fixed and only a discrete spectrum of energies turns out to be allowed, while in the Sturm-Liouville equation the energy E_0 is fixed and the coupling force, i.e., the nuclear charge, is varied until boundary conditions are satisfied, generating a complete set of Kepler–Coulomb Sturmians, which are therefore the radial functions $R_{nl}(r)$ multiplied by the usual spherical harmonics, i.e., $u_{nlm}(\mathbf{r}) = R_{nl}(r)Y_{lm}(\vartheta, \varphi)$.

Indeed, the fact that Z_n can be varied generates an infinite number of $u_{nlm}(\mathbf{r})$, making this set complete. However, n remains numerable, that is Sturmian basis functions constitute a complete set, without the need of including continuum states, greatly simplifying all the mathematical machinery and as required for variational calculations using the constant energy orbitals defined by the same Eq. (2) involved when dealing with continuum basis sets.

Some simple manipulations of Eq. (2) (see, for instance, Ref. 21) lead to the following orthonormality relationship for Coulomb Sturmians:

$$c_{nlm}\int u_{nlm}(\mathbf{r})\frac{1}{r}u_{n'l'm'}(\mathbf{r})\mathrm{d}\mathbf{r} = \delta_{nn'}\delta_{ll'}\delta_{mm'}. \tag{3}$$

Such an orthonormality relation is different from that of usual hydrogenic orbitals, because it explicitly includes the Coulomb potential. For a detailed mathematical treatment of orthogonality relationships for these sets, the reader is referred to Ref. 32.

The explicit expression of the three-dimensional $u_{nlm}(\mathbf{r})$ will be given in the following, special case of general *d*-dimensional Coulomb Sturmians.

2.1 *d*-Dimensional polar Sturmians

d-Dimensional hydrogenic systems are increasingly being studied[33–38] both in configuration and momentum space because of their emerging importance in d-dimensional physics and chemistry,[39,40,9] quantum field theory,[41,42] and quantum cosmology,[43] just to mention a few examples.

The explicit form and properties of d-dimensional Sturmians in polar and parabolic were investigated in Ref. 33.

Let us consider the motion of the electron around the proton in d-dimensional space ($d \geq 2$). Letting $x_1, \ldots, x_d$ be the cartesian coordinates of the electron, we introduce polar coordinates defined by the following equations:

$$\begin{aligned}
x_1 &= \rho \sin \eta_{d-1} \cdots \sin \eta_3 \sin \eta_2 \sin \eta_1 \\
x_2 &= \rho \sin \eta_{d-1} \cdots \sin \eta_3 \sin \eta_2 \cos \eta_1 \\
x_3 &= \rho \sin \eta_{d-1} \cdots \sin \eta_3 \cos \eta_2 \\
&\vdots \\
x_{d-1} &= \rho \sin \eta_{d-1} \cos \eta_{d-2} \\
x_d &= \rho \cos \eta_{d-1}
\end{aligned} \tag{4}$$

where the set of Ω_{d-1} angles $\eta_{d-1}, \ldots, \eta_2, \eta_1$ defines a d-dimensional sphere whose hyperradius is ρ.[44]

The problem of a hydrogen atom in a space of dimension greater than three in hyperspherical coordinates can be found for example in Ref. 45,33. The Sturmian eigenfunctions at fixed energy are:

$$u_{n\lambda\kappa\cdots\ell m}(\rho, \Omega_{d-1}) = (2p_0)^{\frac{d}{2}} \frac{[(n-\lambda-1)!]^{\frac{1}{2}}}{[(2n+d-3)(n+\lambda+d-3)!^3]^{\frac{1}{2}}} (2p_0\rho)^{\lambda} \exp(-p_0\rho) L_{n-\lambda-1}^{2\lambda+d-2}(2p_0\rho) Y_{\lambda\kappa\ldots\ell m}(\Omega_{d-1}), \tag{5}$$

where

$$p_0 = \frac{2Z}{(2n+d-3)} \tag{6}$$

n is the principal quantum number, λ is the grand orbital angular momentum quantum number, κ, and the set of $d-3$ quantum numbers denoted by $\cdots \ell m$ are its projections. $L_{n-\lambda-1}^{2\lambda+d-2}$ is a Laguerre polynomial. Throughout the paper we will use Morse and Feschbach notation[46] for Laguerre polynomials; other notations and normalization factors can be found in different textbooks: connections among some of these definitions are given in Ref. 16.

$Y_{\lambda\kappa\ldots\ell m}$ is a hyperspherical harmonic of a d-dimensional sphere[44] which is conveniently expressed as:

$$Y_{\lambda\kappa\ldots\ell m}(\Omega_{d-1}) = \mathcal{N}_{\lambda\kappa} \sin^{\kappa} \eta_{d-3} C_{\lambda-\kappa}^{\kappa+\frac{d}{2}-1}(\cos\eta_{d-3}) Y_{\kappa\ldots\ell m}(\Omega_{d-2}), \quad (7)$$

where $Y_{\kappa\ldots\ell m}(\Omega_{d-2})$ is a hyperspherical harmonic of a (d-1)-dimensional sphere and $C_{\lambda-\kappa}^{\kappa+\frac{d}{2}-1}$ is a Gegenbauer polynomial whose normalization factor is:

$$\mathcal{N}_{\lambda\kappa} = 2^{\kappa+\frac{d}{2}-2}\left(\kappa + \frac{d}{2} - 2\right)! \left[\frac{(2\lambda + \mathrm{d} - 2)(\lambda - \kappa)!}{\pi(\lambda + \kappa + \mathrm{d} - 3)!}\right]^{\frac{1}{2}}$$

When $\kappa + \frac{d}{2} - 1$ is half integer (d odd) it is usual to identify the Gegenbauer polynomial with associated Legendre polynomials, as often done for the physical $d=3$ case.

The hyperspherical harmonics in Eq. (7) are eigenfunctions of the Beltrami–Laplace operator on the d dimensional sphere, $\Delta(\Omega_{d-1})$, with eigenvalues $\lambda(\lambda + d - 2)$.

2.2 *d*-Dimensional parabolic Sturmians

Parabolic coordinates, $\lambda_1,\ \lambda_2,\ \eta_{d-2},\ldots,\eta_1$, are connected to the Cartesian coordinates of d-dimensional space by the following relationships:

$$\begin{aligned}
x_1 &= \sqrt{\lambda_1\lambda_2}\sin\eta_{d-2}\cdots\sin\eta_3\sin\eta_2\sin\eta_1,\\
x_2 &= \sqrt{\lambda_1\lambda_2}\sin\eta_{d-2}\cdots\sin\eta_3\sin\eta_2\cos\eta_1,\\
x_3 &= \sqrt{\lambda_1\lambda_2}\sin\eta_{d-2}\cdots\sin\eta_2\cos\eta_1,\\
&\vdots\\
x_{d-1} &= \sqrt{\lambda_1\lambda_2}\cos\eta_{d-2},\\
x_d &= \frac{1}{2}(\lambda_1 - \lambda_2).
\end{aligned} \quad (8)$$

Polar and parabolic sets of coordinates are related through the following equations:

$$\begin{aligned}
\lambda_1 &= \rho(1 + \cos\eta_{d-1}),\\
\lambda_2 &= \rho(1 - \cos\eta_{d-1}),
\end{aligned} \quad (9)$$

where λ_1 and λ_2 range from 0 to $+\infty$.

The Schrödinger equation for a hydrogen atom in d-dimensional space in parabolic coordinates takes the form [33]:

$$\left\{\frac{4}{\lambda_1+\lambda_2}\left[\Delta(\lambda_1)+\Delta(\lambda_2)-\frac{\lambda_1+\lambda_2}{4\lambda_1\lambda_2}\Delta(\Omega_{d-2})\right]\right. \\ \left. +2E+\frac{4Z}{\lambda_1+\lambda_2}\right\}U_{n_1n_2\kappa\ldots\ell m}=0, \tag{10}$$

where $\Delta(\lambda_i)$ with $i = 1, 2$ is:

$$\Delta(\lambda_i)=\lambda_i^{-\frac{d-3}{2}}\frac{\partial}{\partial\lambda_i}\lambda_i^{\frac{d-1}{2}}\frac{\partial}{\partial\lambda_i} \tag{11}$$

and $\Delta(\Omega_{d-2})$ is the Beltrami–Laplace operator of (d-1)-dimensional sphere whose eigenfunctions are the hyperspherical harmonics $Y_{\kappa\ldots\ell m}$ and whose eigenvalues are $\kappa(\kappa+d-3)$.[44]

The solutions of the Schrödinger equation (10), i.e., the d-dimensional parabolic Sturmians, have the following expression:

$$U_{n_1n_2\kappa\ldots\ell m}(\lambda_1,\lambda_2,\Omega_{d-2}) \\ =2p_0^{\frac{d}{2}}\left[\frac{n_1!n_2!}{(2n_1+2n_2+2\kappa+d-1)[(n_1+\kappa+\frac{d-3}{2})!(n_2+\kappa+\frac{d-3}{2})!]^3}\right]^{\frac{1}{2}} \\ (p_0\lambda_1)^{\frac{\kappa}{2}}(p_0\lambda_2)^{\frac{\kappa}{2}}\times\exp\left[-p_0\left(\frac{\lambda_1+\lambda_2}{2}\right)\right]L_{n_1}^{\kappa+\frac{d-3}{2}} \\ (p_0\lambda_1)L_{n_2}^{\kappa+\frac{d-3}{2}}(p_0\lambda_2)Y_{\kappa\ldots\ell m}(\Omega_{d-2}) \tag{12}$$

i.e., they are the product of two Laguerre polynomials and a (d-2)-dimensional harmonic.

The parabolic quantum numbers n_1 and n_2 are connected to the principal quantum number n by the relation:

$$n=n_1+n_2+\kappa+1. \tag{13}$$

We now proceed to establish the unitary transformation between the two basis sets given in (5) and (12). We represent the functions $U_{n_1n_2\kappa\ldots\ell m}$ as a linear combination of the functions $u_{n\lambda\kappa\ldots\ell m}$

$$U_{n_1n_2\kappa\ldots\ell m}=\sum_{\lambda=\kappa}^{n-1}T_{n\lambda}^{n_1n_2}u_{n\lambda\kappa\ldots\ell m}. \tag{14}$$

The orthonormal transformation coefficients $T_{n\lambda}^{n_1 n_2}$ can be evaluated with a procedure similar to that used by Suslov[47] for $d=3$ and derived by us in Ref. 33.

Letting ρ go to infinity in Eq. (14) we find:

$$\frac{(1+\cos\eta_{d-3})^{n_1+\frac{\kappa}{2}}(1-\cos\eta_{d-3})^{n_2+\frac{\kappa}{2}}}{[(n_1+\kappa+\frac{d-3}{2})!(n_2+\kappa+\frac{d-3}{2})!n_1!n_2!]^{\frac{1}{2}}} = \sum_{\lambda=\kappa}^{n-1} T_{n\lambda}^{n_1 n_2}(-)^{\kappa-\lambda} \times \frac{2^{n-2+\frac{d}{2}}\mathcal{N}_{\lambda\kappa}(1+\cos\eta_{d-3})^{\frac{\kappa}{2}}(1-\cos\eta_{d-3})^{\frac{\kappa}{2}}}{[(n+\lambda+d-3)!(n-\lambda-1)!]^{\frac{1}{2}}} C_{\lambda-\kappa}^{\kappa+\frac{d}{2}-1}(\cos\eta_{d-3}) \tag{15}$$

where a factor $(p_0\rho)^{n-1}\exp(-p_0\rho)Y_{\kappa\ldots\ell m}$ has been canceled in both sides. In order to obtain the above equation the asymptotic expression of Laguerre polynomials for $\rho \to \infty$[46] and the definition of parabolic coordinates, λ_1 and λ_2, in terms of the hyperradius ρ and the hyperangle η_{d-3}, Eq. (9), have been used.

Multiplying both the left-hand and right-hand sides of Eq. (15) by $\mathcal{N}_{\lambda\kappa}(1+\cos\eta_{d-3})^{\frac{\kappa}{2}}(1-\cos\eta_{d-3})^{\frac{\kappa}{2}}C_{\lambda-\kappa}^{\kappa+\frac{d}{2}-1}$ and integrating over the volume element $\sin^{d-3}\eta_{d-3}d\cos\eta_{d-3}$ we have:

$$T_{n\lambda}^{n_1 n_2} = (-)^{\lambda-\kappa}2^{-n-\frac{d}{2}+2}\mathcal{N}_{\lambda\kappa} \times \left[\frac{(n+\lambda+d-3)!(n-\lambda-1)!}{(n_1+\kappa+\frac{d-3}{2})!(n_2+\kappa+\frac{d-3}{2})!n_1!n_2!}\right]^{\frac{1}{2}} \times \int_{-1}^{1}(1+\cos\eta_{d-3})^{n_1+\kappa+\frac{d-3}{2}} (1-\cos\eta_{d-3})^{n_2+\kappa+\frac{d-3}{2}} C_{\lambda-\kappa}^{\kappa+\frac{d}{2}-1}(\cos\eta_{d-3})d\cos\eta_{d-3} \tag{16}$$

where use has been made of the orthonormality of Gegenbauer polynomials.

This integral is given in p. 826 of Ref. 48. After some algebra we obtain the following expression for the transformation coefficients:

$$\begin{aligned} T_{n\lambda}^{n_1 n_2} &= (-)^{\lambda-\kappa} \frac{(n-\kappa-1)!}{(\kappa+\frac{d-3}{2})!} \\ &\times \left[\frac{(2\lambda+d-2)(\lambda+\kappa+d-3)!(n_1+\kappa+\frac{d-3}{2})!(n_2+\kappa+\frac{d-3}{2})!}{(\lambda-\kappa)!(n+\lambda+d-3)!(n-\lambda-1)!n_1!n_2!} \right]^{\frac{1}{2}} \\ &\times {}_3F_2 \begin{pmatrix} \kappa-\lambda, & \lambda+\kappa+d-2, & -n_2 \\ \kappa+\frac{d-1}{2}, & -n+\kappa+1, & 1 \end{pmatrix}, \end{aligned} \tag{17}$$

where ${}_3F_2(\cdots;1)$ is a generalized hypergeometric function of unit argument. Hahn polynomials (see Fig. 5.1) can also be expressed in terms of such functions.[49] In particular, the transformation coefficients in the equation above are:

$$T_{n\lambda}^{n_1 n_2} = (-)^{\lambda-\kappa} Q_{\lambda-\kappa,n_2}^{\kappa+\frac{d-3}{2},\kappa+\frac{d-3}{2},n-\kappa-1} \tag{18}$$

here $Q_{\lambda-\kappa,n_2}^{\kappa+\frac{d-3}{2},\kappa+\frac{d-3}{2},n-\kappa}$ is the Hahn coefficient[50] of degree $\lambda-\kappa$ and n_2 is the discrete variable. These coefficients allow a representation in terms of generalized $3j$ symbols[50]:

$$\begin{aligned} T_{n\lambda}^{n_1 n_2} &= (-)^{\lambda-\kappa} \exp\left[\iota\pi\left(n_2-\kappa-\frac{d-3}{2}\right)\right](2\lambda+d-2)^{\frac{1}{2}} \\ &\begin{pmatrix} \lambda+\frac{d-3}{2} & \frac{n-1}{2}+\frac{d-3}{4} & \frac{n-1}{2}+\frac{d-3}{4} \\ \kappa+\frac{d-3}{2} & \frac{n_1-n_2}{2}-\frac{\kappa}{2}-\frac{d-3}{4} & -\frac{n_1-n_2}{2}-\frac{\kappa}{2}-\frac{d-3}{4} \end{pmatrix}. \end{aligned} \tag{19}$$

2.3 Two-dimensional Coulomb Sturmians

The investigation of two-dimensional hydrogenic systems seems academic but useful to gain insight into the deeper nature of matter[51,52]; however, the existence of hydrogenic systems with dimensionality smaller than three has been observed[53] or suggested as an important tool in nanotechnology (for instance in low dimensional semiconductor nanostructures[53]) or in quantum computation (e.g., one- and two-dimensional hydrogenic atoms have been proposed as qubits.[54,55])

The two-dimensional hydrogen atom was first investigated in the 1960s and 1970s from different perspectives, using hydrogenic orbitals in polar[56] and parabolic[57] coordinates. In Ref. 58 the connection between the Coulomb problem in a plane in parabolic coordinates and the two-dimensional harmonic oscillator in Cartesian coordinates has been established.

In Ref. 33 we introduced two-dimensional Coulomb Sturmians in polar and parabolic coordinates, explicitly deriving the connection between them.

The solutions to the Schrödinger equation for an electron in a two-dimensional attractive Coulomb field in polar (ρ, ϕ)[56] and parabolic[57] (λ_1, λ_2) coordinates (Fig. 5.2), which are defined through the following equations, can be obtained by specializing Eqs. (4) and (8):

$$\begin{aligned} x_1 &= \rho \sin\phi = \lambda_1\lambda_2, \\ x_2 &= \rho \cos\phi = \frac{1}{2}(\lambda_1^2 - \lambda_2^2), \end{aligned} \tag{20}$$

where x_1 and x_2 are the Cartesian coordinates in the plane and η_1 in Eq. (4) was replaced by the more common ϕ. The parabolic coordinates λ_1 and λ_2 are allowed to range from 0 to ∞ and from $-\infty$ to ∞, respectively. Note that the parabolic coordinates used here are squares of those used in the *d*-dimensional case (see previous section): this choice is most convenient for problems in the plane, making more evident the two-valued nature of the corresponding wavefunction, as shown below.

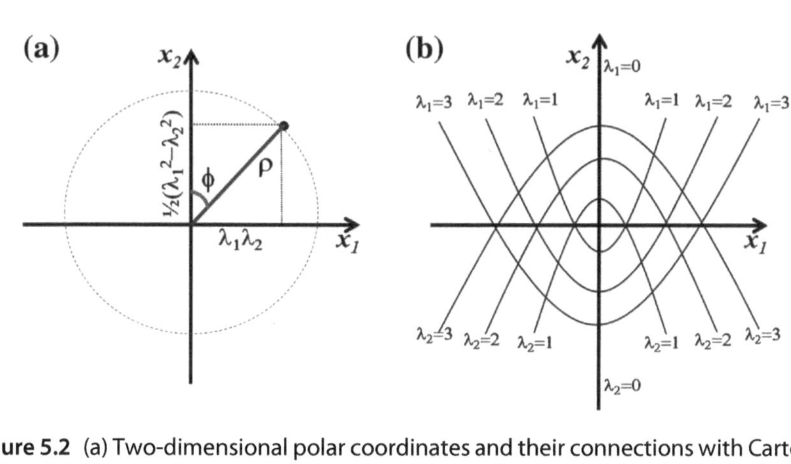

Figure 5.2 (a) Two-dimensional polar coordinates and their connections with Cartesian and parabolic coordinates and (b) parabolic coordinates: fixed values of λ_1 and λ_2 are represented on the Cartesian plane. For color version of this figure, the reader is referred to the online version of this chapter.

In polar coordinates, the Sturmian wavefunctions[56] corresponding to a fixed value of the energy E – or formally introducing $p_0 = \sqrt{-2E}$ – take the form:

$$u_{nm}(\rho, \phi) = \mathcal{N}_{nm}(2p_0\rho)^{|m|} e^{-p_0\rho} L_{n-|m|-1}^{2|m|}(2p_0\rho) e^{\imath m\phi}, \tag{21}$$

where p_0 is:

$$p_0 = \frac{2Z}{2n-1} \tag{22}$$

and Z is the nuclear charge. Atomic units are used throughout. The principal quantum number n takes the values $1, 2, \ldots$ and $m = 0, \pm 1, \ldots, \pm(n-1)$. The eigenfunction in Eq. (21) is the product of a radial part and an angular part: a Laguerre polynomial,[46] $L_{n-|m|-1}^{2|m|}$, and a harmonic of the circle S^1, $e^{\imath m\phi}$, respectively. The normalization factor $\mathcal{N}_{nm}$ is:

$$\mathcal{N}_{nm} = 2p_0 \left[\frac{(n-\mid m \mid -1)!}{2\pi(2n-1)[(n+\mid m \mid -1)!]^3} \right]^{\frac{1}{2}}$$

as appropriate for a Sturmian set.

In the parabolic coordinate system, the two-dimensional hydrogen atom problem was treated in Refs. [33,57]. The wave functions is expressed as:

$$U_{n_1 n_2}(\lambda_1, \lambda_2) = \mathcal{N}_{n_1 n_2} e^{-p_0 \frac{\lambda_1^2+\lambda_2^2}{2}} H_{n_1}\left(\sqrt{p_0 \lambda_1^2}\right) H_{n_2}\left(\sqrt{p_0 \lambda_2^2}\right), \tag{23}$$

where

$$p_0 = \frac{2Z}{(n_1 + n_2 + 1)} \tag{24}$$

n_1 and n_2 are the parabolic quantum numbers and H_{n_1} and H_{n_2} are Hermite polynomials (i.e., the first step in the Askey scheme, Figure 5.1). The normalization factor $\mathcal{N}_{n_1 n_2}$ is:

$$\mathcal{N}_{n_1 n_2} = 2p_0 \left(\frac{1}{2\pi(n_1 + n_2 + 1)\, 2^{n_1+n_2}\, n_1! n_2!} \right)^{\frac{1}{2}}.$$

By comparison of Eqs. (22) and (24) the relation of n_1 and n_2 with the quantum number n can be worked out:

$$n_1 + n_2 + 1 = 2n - 1. \tag{25}$$

Equation (25) serves to establish the range of allowed values for n_1 and n_2: most importantly it enforces that, as emphasized in Ref. 57, only values of n_1 and n_2 with the same parity give the correct energy levels. The acceptable wave functions are those given in Eq. (23) with the constraint that n_1 and n_2 are both even or both odd, i.e., the allowed wavefunctions must satisfy $U_{n_1 n_2}(\lambda_1, \lambda_2) = U_{n_1 n_2}(-\lambda_1, -\lambda_2)$ since (λ_1, λ_2) and $(-\lambda_1, -\lambda_2)$ represent the same point in the plane. In order to facilitate comparison with the well-known three-dimensional case, recall that Hermite polynomials are limiting cases of Laguerre polynomials [46] (Figure 5.1): for example, when n_1 is even or odd the Laguerre polynomial is respectively of the type $L_{\frac{n_1}{2}}^{-\frac{1}{2}}$ or $L_{\frac{n_1-1}{2}}^{\frac{1}{2}}$ (p. 786 in Ref. 46).

Let us now consider the transformation between the solutions u_{nm} and $U_{n_1 n_2}$:

$$U_{n_1 n_2}(\lambda_1, \lambda_2) = \sum_m T_{nm}^{n_1 n_2} u_{nm}(\rho, \phi), \tag{26}$$

where $T_{nm}^{n_1 n_2}$ are the transformation coefficients which are unitary matrix elements. (The inverse of (26) involves summation over allowed values for $n_1 - n_2$). In Ref. 33 we proved that these coefficients are Wigner *d*-functions by adopting the same procedure followed for the *d*-dimensional case.

Letting ρ go to infinity from Eq. (26) we find:

$$\sum_m T_{nm}^{n_1 n_2} \frac{(-)^{n-|m|-1} e^{\imath m\phi}}{[(n-|m|-1)!(n+|m|-1)!]^{\frac{1}{2}}} = \left(\frac{2^{n_1+n_2}}{n_1! n_2!}\right)^{\frac{1}{2}} \cos^{n_1}\frac{\phi}{2} \sin^{n_2}\frac{\phi}{2}, \tag{27}$$

where use has been made of the relationship between parabolic and polar coordinates:

$$\begin{aligned} \lambda_1^2 &= 2\rho \cos^2\frac{\phi}{2}, \\ \lambda_2^2 &= 2\rho \sin^2\frac{\phi}{2} \end{aligned} \tag{28}$$

and of the expression of Laguerre and Hermite polynomials for $\rho \to \infty$ (p. 785 in Ref. 46). Using the orthogonality of the functions $e^{\imath m\phi}$ we obtain the following integral representation for the transformation coefficients:

$$T_{nm}^{n_1 n_2} = (-)^{n-|m|-1} \left[\frac{2^{(n_1+n_2)}(n-|m|-1)!(n+|m|-1)!}{4\pi^2 n_1! n_2!}\right]^{\frac{1}{2}} \times \int_0^{2\pi} \cos^{n_1}\frac{\phi}{2} \sin^{n_2}\frac{\phi}{2} e^{-\imath m\phi} \, d\phi. \tag{29}$$

The integral in the above expression is given on p. 477 in Ref. 48 in terms of the hypergeometric series. We can express these coefficients explicitly in terms of Wigner reduced rotation matrix elements (d-functions) [59]:

$$T_{nm}^{n_1 n_2} = (-)^{\frac{n_1-n_2}{2}} e^{-\imath\pi\frac{n_1}{2}} d^{\frac{n_1+n_2}{2}}_{\frac{n_1-n_2}{2},m}\left(\frac{\pi}{2}\right). \tag{30}$$

Thus, for the two-dimensional hydrogen atom, i.e., when the motion of the electron around the proton is constrained to a plane, the transformation coefficients between wavefunctions obtained by variable separation in polar and parabolic coordinates are simply Wigner d-functions. Their involvement and argument $\frac{\pi}{2}$ imply that the symmetry group for the two-dimensional hydrogen atom is O(3) as stressed in Ref. 57. This conclusion will be more transparent from the momentum space analysis carried out in Section 3.

2.4 Three-dimensional Coulomb Sturmians

It is well known that Schrödinger's equation for the hydrogen atom is separable in configuration space for different sets of coordinates. Polar coordinates (ρ, θ, ϕ) are by far the most commonly encountered, because of the evident spherical symmetry of the hydrogen atom, and are defined as (Fig. 5.3a)

$$\begin{aligned} x &= \rho \sin\theta \cos\phi, \\ y &= \rho \sin\theta \sin\phi, \\ z &= \rho \cos\theta, \end{aligned} \tag{31}$$

where we have replaced x_1, x_2, and x_3 with the more familiar y, x, and z and the angles η_1 and η_2 with ϕ and θ.

Three-dimensional Coulomb Sturmians are therefore given by

$$\begin{aligned} u_{nlm}(r,\theta,\phi) &= (2p_0)^{3/2}\left[\frac{(n-l-1)!}{2n[(n+l)!]^3}\right]^{1/2} \\ &\quad \times e^{-p_0 r}(2p_0 r)^l L_{n-l-1}^{2l+1}(2p_0 r)\ Y_{lm}(\theta,\phi), \end{aligned} \tag{32}$$

where $L_{n-l-1}^{2l+1}(2p_0 r)$ is a Laguerre polynomial and $Y_{lm}(\theta, \phi)$ is a spherical harmonic and p_0 is now given by

$$p_0 = \frac{Z}{n}. \tag{33}$$

From the relation $p_0 = \sqrt{-2E}$ the familiar hydrogenic wavefunctions are easily recovered, corresponding to the energy spectrum $E_n = -\frac{Z^2}{n^2}$.

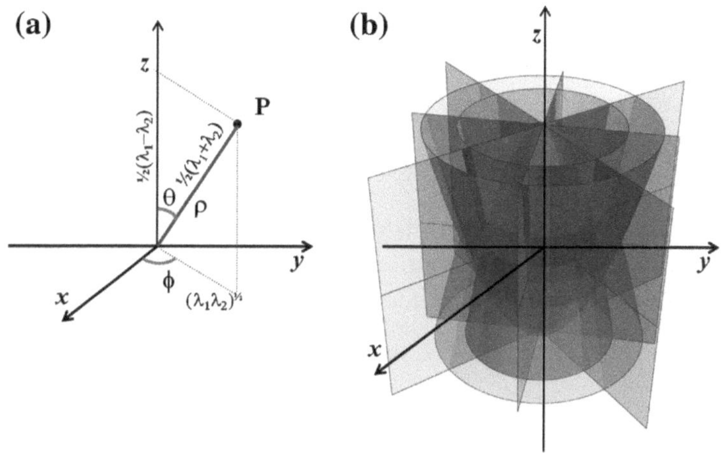

Figure 5.3 Three-dimensional polar and parabolic coordinates. Relations of polar coordinates with Cartesian and parabolic coordinates are given in (a); (b) the position of a point in parabolic coordinates is defined by the intersection of two paraboloids (the blue and the red surfaces, corresponding to fixed values of λ_1 and λ_2, respectively, see Figure. 5.2) and the green plane corresponding to a fixed value of the angle ϕ. (For interpretation of the references to color in this figure legend, the reader is referred to the web version of this book.)

The separation of variables in differential equations can be associated with algebraic structure described by continuous group theory.[60–62] Therefore Schrödinger's equation can be discussed with reference to its dynamical symmetries.

Polar Sturmians can be seen as satisfying the following eigenvalue equations, in which $\hat{\mathbf{L}}$ is the orbital angular momentum of the electron, and $\hat{L}_z$ its projection along the z-axis:

$$\left(-\frac{1}{2}\nabla^2 + \frac{p_0^2}{2} - \frac{np_0}{r}\right) u_{nlm}(r, \theta, \phi) = 0, \tag{34}$$

$$[\hat{\mathbf{L}}^2 - l(l+1)] u_{nlm}(r, \theta, \phi) = 0, \tag{35}$$

$$(\hat{L}_z - m) u_{nlm}(r, \theta, \phi) = 0. \tag{36}$$

Parabolic coordinates can be obtained from Eq. (10):

$$\begin{aligned} x &= \sqrt{\lambda_1\lambda_2}\cos\phi, \\ y &= \sqrt{\lambda_1\lambda_2}\sin\phi, \\ z &= \frac{1}{2}(\lambda_1 - \lambda_2). \end{aligned} \tag{37}$$

These coordinates have two points of reference: one is on the nucleus and the other is at infinite distance along the z-axis (see Fig. 5.3b).

Parabolic coordinates are related to polar coordinates through

$$\begin{aligned} \lambda_1 &= \rho(1 + \cos\theta), \\ \lambda_2 &= \rho(1 - \cos\theta). \end{aligned} \tag{38}$$

The corresponding parabolic Coulomb Sturmians can then be written as:

$$U_{n_1 n_2 m}(\lambda_1, \lambda_2, \phi) = \sqrt{\frac{p_0^3}{\pi}\frac{n_1!n_2!}{(n_1 + n_2 + m + 1)[(n_1 + m)!(n_2 + m)!]^3}} \times (p_0\sqrt{\lambda_1\lambda_2})^m e^{\left[-p_0\left(\frac{\lambda_1+\lambda_2}{2}\right)\right]} L_{n_1}^m(p_0\lambda_1) L_{n_2}^m(p_0\lambda_2) e^{im\phi}.$$

Note that different notations for parabolic quantum numbers are often employed. Coherently with Eq. (12) we use here n_1, n_2, and m, with $n = n_1 + n_2 + m + 1$. However, for symmetry reasons, which will become clearer below, use of the parabolic quantum number $\mu = n_1 - n_2$ will also be made. In this case, the explicit expression of parabolic Sturmians becomes:

$$U_{n\mu m}(\lambda_1\lambda_2\phi) = \sqrt{\frac{p_0^3}{\pi n}\frac{[(\mu + n - m - 1)/2]![(n - \mu - m - 1)/2]!}{\{[(\mu + n + m - 1)/2]![(n - \mu + m - 1)/2]!\}^3}} \times \left(p_0\sqrt{\lambda_1\lambda_2}\right)^m e^{-p_0\frac{\lambda_1+\lambda_2}{2}} \times L^m_{\frac{1}{2}(n+\mu-m-1)}\left(p_0\lambda_1\right) L^m_{\frac{1}{2}(n-\mu-m-1)}\left(p_0\lambda_2\right) e^{im\phi}, \tag{39}$$

where the integer μ takes the negative and positive values: $\mu = -(n - m - 1), -(n - m - 3), \ldots, (n - m - 3), (n - m - 1)$. Note that μ has the parity of $n - m - 1$.

A summary of other notation for parabolic quantum numbers and their relationships can be found in Ref. 16.

Parabolic orbitals satisfy the following eigenvalue equations:

$$\left(-\frac{1}{2}\nabla^2 + \frac{p_0^2}{2} - \frac{np_0}{r}\right)U_{n\mu m}(\lambda_1, \lambda_2, \phi) = 0, \tag{40}$$

$$(\hat{K}_z - \mu)U_{n\mu m}(\lambda_1, \lambda_2, \phi) = 0, \tag{41}$$

$$(\hat{L}_z - m)U_{n\mu m}(\lambda_1, \lambda_2, \phi) = 0. \tag{42}$$

In Eq. (41) the operator $\hat{K}_z$ corresponds to the component of the reduced Runge–Lenz vector[61,63,64] along the z-axis. This vector is classically related to the direction of the major axis of an elliptic orbit: it is a constant of motion and therefore commutes with the Hamiltonian. Thus, the quantum number μ appears to be the eigenvalue of the $\hat{K}_z$ operator.

The reduced Runge–Lenz vector operator is defined as:

$$\hat{\mathbf{K}} = \left[\frac{1}{2}(\hat{\mathbf{L}} \times \hat{\mathbf{p}} - \hat{\mathbf{p}} \times \hat{\mathbf{L}}) + \frac{z\mathbf{r}}{r}\right] / p_0. \tag{43}$$

The vectors $\hat{\mathbf{K}}$ and $\hat{\mathbf{L}}$ are orthogonal in classical mechanics, and for the quantum mechanical operators $\hat{\mathbf{L}} \cdot \hat{\mathbf{K}} = 0$.

The transformation between polar and parabolic Sturmians was given for the first time in configuration space by Park,[65] who did not specify the explicit form of the eigenfunctions and so did not define the phase convention. Hughes[66] and Tarter[67] calculated the elements by direct integration and the correct phase difference wss given by Englefield.[58] Alternatively, by applying the Suslov[47] approach one has:

$$U_{n_1 n_2 m}(\lambda_1, \lambda_2, \phi) = \sum_l T_{nl}^{n_1 n_2} u_{nm}(\rho, \theta, \phi) \tag{44}$$

and

$$T_{nl}^{n_1 n_2} = (-)^{n_2+m}\langle \frac{1}{2}(n-1), \frac{1}{2}(n-1); \frac{1}{2}(m+n_2-n_1), \frac{1}{2}(m+n_1-n_2)|lm\rangle. \tag{45}$$

We proposed[16,68] an alternative view for these matrix elements making use of Regge symmetry[69] for Clebsch–Gordan coefficients:

$$T_{nl}^{n_1 n_2} = (-)^{n_2+m}\langle\frac{1}{2}(n+m-1), \frac{1}{2}(n-m-1); \frac{n_2-n_1}{2}, \frac{n_1-n_2}{2}|l0\rangle. \tag{46}$$

The Clebsch–Gordan coefficients obtained in this way, having at least one zero element, suggest that $n_1 - n_2$ (that is the quantum number μ, eigenvalue of $\hat{K}_z$) can be interpreted as a *helicity* quantum number. Indeed, this interpretation becomes much clearer when analyzed from the momentum space point of view, where the symmetry group for the three-dimensional hydrogen atom manifestly appears to be $O(4)$.

2.5 Non-standard Coulomb Sturmians in three dimensions: spheroidal and Sphero-elliptic Sturmians

A basic study of the hydrogen atom in spheroidal coordinates (Fig. 5.4) is due to Coulson and collaborators.[70,71] This two-center coordinate system is suitable for studying the motion of the electron in H_2^+ within the Born–Oppenheimer approximation because it renders the Schrödinger equation separable, although not analytically solvable. In this way, diatomic orbitals, i.e., orbitals for one electron between two nuclei, can be computed and exploited for quantum calculations on two-electron problems.[72] For the hydrogen atom, one of the two centers coincides with the nucleus and the other is set at a distance ρ along a reference axis. The *spheroidal orbitals* are the best zeroth-order functions to treat the hydrogen atom perturbed by a point charge. However, analytical solutions for an electron interacting with two fixed nuclei are known only for special values of the internuclear distance and of the ratio of nuclear charges.[71]

To solve the Coulomb problem in spheroidal coordinates one writes the spheroidal Sturmians $T(\xi, \zeta, \phi)$ as a product $\Xi(\xi)N(\zeta)\Phi(\phi)$. This leads to the introduction of two separation constants, $-m^2$ and f.

By defining

$$\alpha \equiv \rho\left(-\frac{E_0}{2}\right)^{1/2} = \rho\frac{p_0}{2} = \rho\frac{Z_n}{2n} \tag{47}$$

the following separated equations can be written

$$\frac{d^2\Phi}{d^2\phi} = -m^2\Phi, \tag{48}$$

$$\frac{d}{d\xi}\left[(1-\xi^2)\frac{d\Xi}{d\xi}\right] + \left[f + \alpha^2\xi^2 - 2n\alpha\xi - \frac{m^2}{1-\xi^2}\right]\Xi = 0. \tag{49}$$

The equation in ζ is formally identical to the one in ξ, but the variable definition interval is different (Fig. 5.4). Equation (48) also occurs in the separation in the polar coordinates θ and ϕ: its normalized solution is $\frac{1}{\sqrt{2\pi}}e^{im\phi}$ and the $T(\xi, \zeta, \phi)$ are eigenfunctions of $\hat{L}_z$. The solutions of the other two equations can, however, only be written as an expansion in terms of associated Legendre polynomials,[73] as a consequence, neither the spheroidal eigenfunctions nor the eigenvalues f can be written as a closed expression.

It is possible, nevertheless, to write down the eigenvalue equations that completely characterize the spheroidal basis:

$$\left(-\frac{1}{2}\nabla^2 + \frac{p_0^2}{2} - \frac{np_0}{r}\right)T(\mathbf{r}) = 0, \tag{50}$$

$$[\hat{\mathbf{L}}^2 + 2\alpha\hat{K}_z - (f + \alpha^2)]T(\mathbf{r}) = 0, \tag{51}$$

$$(\hat{L}_z - m)T(\mathbf{r}) = 0. \tag{52}$$

Therefore, to obtain spheroidal orbitals without solving Eq. (49) one can calculate the expansion coefficients of the unknown functions in an appropriate basis by diagonalizing the matrix of the operator $\hat{\mathbf{L}}^2 + 2\alpha\hat{K}_z$ in such a basis. Looking at this operator, one understands how spheroidal orbitals are related to those we have encountered before. When the distance ρ tends to zero, α does too; therefore, the operator becomes $\hat{\mathbf{L}}^2$, which is associated with the polar Sturmian basis: As $\rho \to 0$ the $|n, f, m\rangle$ set

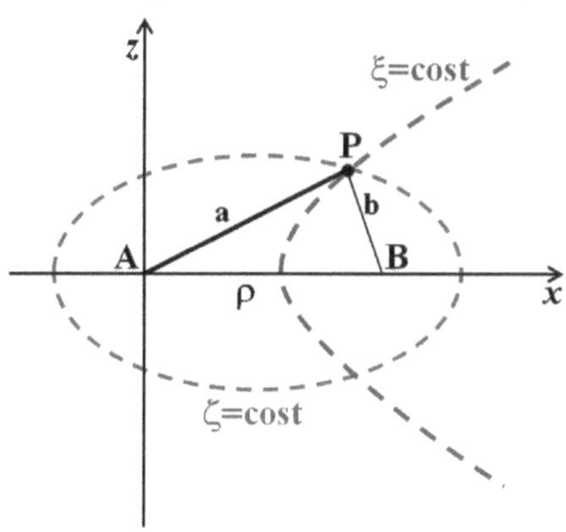

Figure 5.4 Definition of spheroidal coordinates: $\xi = \frac{a+b}{\rho}$; $\zeta = \frac{a-b}{\rho}$. The curves of constant ζ are oblate spheroids, those of constant ξ are the hyperboloids of revolution. The coordinate ζ is restricted by $0 \le \zeta < \infty$ and ξ is restricted by $-1 \le \xi < 1$. The angle φ (not represented in the figure) is the angle between the plane ABP and the xy plane (the y axis points down) as for polar coordinates (see Fig. 5.3). For color version of this figure, the reader is referred to the online version of this chapter.

coincides with the $|n, l, m\rangle$ set. On the other hand, as the distance between the two centers goes to infinity the contribution from $\hat{K}_z$ becomes dominant; thus, the $|n, f, m\rangle$ coincides with the $|n, \mu, m\rangle$ set.[73] This confirms the role of spheroidal coordinates as a bridge between the parabolic and the polar coordinates.[74]

The operator $\hat{\mathbf{L}}^2 + 2\alpha\hat{K}_z$ can be seen, in the neighborhood of its asymptotes, as a main part with a small perturbation added: For small α the perturbation is $2\alpha\hat{K}_z$ and spheroidal functions stem from the effect of this perturbation on the zero-order functions $|n, l, m\rangle$; on the other hand, when α is large the perturbation is $\hat{\mathbf{L}}^2$ and the $|n, f, m\rangle$ differ little from the zero-order functions $|n, \mu, m\rangle$.

This enables us to write down the spheroidal Sturmians in the proximity of the asymptotes in terms of polar or parabolic Sturmians.[73]

For example, for small α the first-order expression for spheroidal Sturmians, stemming from the perturbation of a level $|n, l, m\rangle$, is:

$$
\begin{aligned}
(T_{nfm})^{[1]} = u_{nlm} &- \frac{\alpha}{l+1}\sqrt{\frac{[(n-1)^2 - l^2](l^2 - m^2)}{(2l+3)(2l+1)}}\,u_{n,l+1,m} \\
&+ \frac{\alpha}{l}\sqrt{\frac{[n^2 - l^2](l^2 - m^2)}{(2l-1)(2l+1)}}\,u_{n,l-1,m}.
\end{aligned}
\tag{53}
$$

In the case of large α, spheroidal eigenstates come from corrections on the parabolic eigenstates with first-order eigenfunctions:

$$
\begin{aligned}
(T_{nfm})^{[1]} = U_{n,\mu,m} &- \frac{1}{8}\sqrt{[n^2 - (m+\mu+1)^2][n^2 - (m-\mu-1)^2]}\,U_{n,\mu+2,m} \\
&+ \frac{1}{8}\sqrt{[n^2 - (m-\mu+1)^2][n^2 - (m+\mu-1)^2]}\,U_{n,\mu-2,m}.
\end{aligned}
\tag{54}
$$

Sphero-elliptic Sturmians derive from the existence of *elliptic harmonics*, which arise from the parametrization of the sphere by elliptic coordinates (also called conical, e.g., in Ref. 46), an alternative kind of harmonic besides the familiar spherical harmonics $Y_{l,m}(\theta, \phi)$.

Elliptic harmonics can be used for building up a new type of hydrogenic orbitals, which we will call sphero-elliptic[61] (their relation with elliptic coordinates is reported in Figure 5.5). They are given as the product of the radial part of Eq. (32) and an elliptic harmonic. There exist

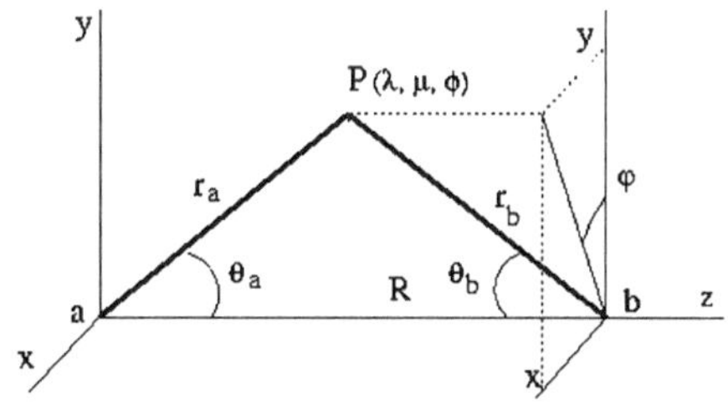

Figure 5.5 Relation between elliptic and polar coordinates for the point **P** in the three-dimensional space:
$r_a = \frac{R}{2}(\lambda + \mu)$; $r_b = \frac{R}{2}(\lambda - \mu)$;
$\cos\theta_a = \frac{1+\lambda\mu}{\lambda+\mu}$; $\sin\theta_a = \frac{[(\lambda^2-1)(1-\mu^2)]^{\frac{1}{2}}}{\lambda+\mu}$; $\cos\theta_b = \frac{1-\lambda\mu}{\lambda-\mu}$; $\sin\theta_b = \frac{[(\lambda^2-1)(1-\mu^2)]^{\frac{1}{2}}}{\lambda-\mu}$.
The figure has been adapted from Figure 3 of Ref. 76.

alternative equivalent parameterizations that render the motion on the sphere separable and lead to elliptic harmonics. Two of them are:

$$\begin{aligned} x &= \mathrm{sn}(\alpha,k)\mathrm{dn}(\beta,k') = \cos\psi\sqrt{1-k'^2\cos^2\omega} \quad && k^2+k'^2=1, \\ y &= \mathrm{cn}(\alpha,k)\mathrm{cn}(\beta,k') = \sin\psi\sin\omega && 0\le k\le 1, \\ z &= \mathrm{dn}(\alpha,k)\mathrm{sn}(\beta,k') = \cos\omega\sqrt{1-k^2\cos^2\psi} && 0\le k'\le 1. \end{aligned} \tag{55}$$

We see that Jacobi elliptic functions cn, sn, and dn[75] appear. They contain a continuous parameter (k or k'), called *modulus*, which can assume any value between zero and one. The parametrization in Eq. (55) leads to the elliptic harmonics:

$$\mathcal{Y}^{\,k}_{l,\eta}(\alpha,\beta) = M^k_{l,\eta}(\alpha)N^k_{l,\eta}(\beta), \tag{56}$$

where $M^k_{l,\eta}(\alpha)$ and $N^k_{l,\eta}(\beta)$ solve Lamé equations in Jacobian form[60]; and η are the quantum numbers that characterize these harmonics. $\mathcal{Y}^{\,k}_{l,\eta}(\alpha,\beta)$ depends parametrically on k, they thus have one degree of freedom more than the spherical harmonics.

This harmonic is the eigenfunction of the following operators:

$$[\hat{\mathbf{L}}^2 - l(l+1)]\mathcal{Y}^{\,k}_{l,\eta}(\alpha,\beta) = 0, \tag{57}$$

$$(\hat{E} - \eta)\mathcal{Y}^{\,k}_{l,\eta}(\alpha,\beta) = 0 \quad \hat{E} = \hat{L}_x^2 + k'^2\hat{L}_y^2. \tag{58}$$

We can expand the harmonics $\mathcal{Y}^{\,k}_{l,\eta}(\alpha,\beta)$ in terms of spherical harmonics $Y_{l,m}(\theta,\phi)$, because they are solutions of the same physical problem and are both eigenfunctions of $\hat{\mathbf{L}}^2$. In the transformation between the two basis sets the quantum number l is conserved; thus, the transformation matrix is factorized in blocks of dimension $2l+1$:

$$\mathcal{Y}^{k}_{l,\eta}(\alpha,\beta) = \sum_{m=-l}^{l} x^{l}_{\eta,m} Y_{l,m}(\theta,\phi). \tag{59}$$

By computing explicitly the matrix elements of the operator $\hat{E}$ on $Y_{l,m}(\theta,\phi)$, one sees that the transformation coefficients obey the following three-term recurrence [60]:

$$A_{m-2}x^{l}_{\eta,m-2} + (B_m - \eta)x^{l}_{\eta,m} + A_m x_{\eta,m+2} = 0, \tag{60}$$

where

$$\begin{aligned} A_m &= \frac{1}{4}(1-k'^2)\sqrt{(l-m)[l^2-(m+1)^2](l+m+2)}, \\ B_m &= \frac{1}{2}(1+k'^2)[l(l+1)-m^2]. \end{aligned}$$

When k' tends to its limiting values, zero or one, the elliptic harmonics coincide with simpler functions. As $k' \to 0, \hat{E} \to \hat{L}^2_x$, that is the $\mathcal{Y}^{k}_{l,\eta}(\alpha,\beta)$ tend to a combination of "rotated" spherical harmonics where x is the projection axis, instead of z. Thus the quantum number η tends to m', i.e., the quantum number angular momentum projection on the x axis. On the other hand, as $k' \to 1, \hat{E} \to \hat{L}^2_x + \hat{L}^2_y$, so that by linearly combining $\hat{E}$ and $\hat{\mathbf{L}}^2$ one sees that the $\mathcal{Y}^{k}_{l,\eta}(\alpha,\beta)$ become eigenfuctions of $\hat{L}^2_z$; they coincide with spherical harmonics quantized on the z axis, $Y_{l,m}(\theta,\phi)$.

3. A MOMENTUM SPACE PERSPECTIVE

One of the major advantages in the use of d-dimensional Coulomb Sturmian basis sets is that their counterparts in momentum space are hyperspherical harmonics of $(d+1)$ dimension. Because these functions are the eigenfunctions of the Beltrami–Laplace operator on a S^d unit sphere embedded into R^{d+1} Euclidean space, we can get useful information on the nature and symmetry properties of Coulomb Sturmian eigenfunctions. Furthermore, the integrals on two or three hyperspherical harmonics of any dimension, which are usually involved in the solution of atomic and molecular problems, can be usefully expressed as discrete summations over elements of angular momentum algebra, namely $3j$, $6j$, and $9j$ coefficients and their generalizations, greatly simplifying their mathematical solution, also in the light of the significant progresses achieved in the computation and storage of such tools.[77–79]

Though the solution of the Fourier transform of hydrogenic eigenfunctions had been explicitly worked out as soon as 1929 by Podolski and Pauling,[80] Fock in 1935,[81] by using *ante litteram* Sturmian functions and by exploiting the "inverse" stereographic projection of the three-dimensional Euclidean space onto the surface of the S^3 sphere, could identify the eigenfunctions of the hydrogen atom in momentum space with four-dimensional hyperspherical harmonics (apart from a phase factor). This identification, among other things, allowed the correct interpretation of the principal quantum number n as a hyperangular momentum quantum number and of the related *hidden* symmetry and *accidental* degeneracy, as manifestations of a greater symmetry than that appearing in configuration space.

This procedure, as will be shown in the next sections, can readily be extended to any dimension, giving momentum space Coulomb Sturmians. We will also show that the variety of momentum space Coulomb Sturmians is enriched by new elements which do not have an explicit counterpart in configuration space,[10] but can nonetheless be exploited to speed up convergence for problems with a specific kind of symmetry. As a further bonus, the connections between any kind of hyperspherical harmonic with different symmetry properties are fully known and can always be written in terms of generalized $6j$ coefficients, in the most complicated case, but often reduce to simpler ordinary $3j$ coefficients or even to Wigner d functions.

3.1 *d*-Dimensional momentum space Coulomb Sturmians

By a generalization of Fock's treatment[82] of hydrogenic atoms it is possible to show (see e.g., Ref. 45) that in a d-dimensional momentum space the hydrogen wavefunctions in hyperspherical coordinates are directly related to $(d+1)$-dimensional hyperspherical harmonics $Y_{n-1,\lambda\kappa\ldots lm}$.

Fock's procedure in d-dimensional space consists of projecting a point in the related Euclidean space onto the surface of a unit radius S^d sphere in the $(d+1)$-dimensional Euclidean space. In polar coordinates the position of the momentum of the electron $\mathbf{p}$ is given by:

$$\begin{aligned} p_1 &= p \sin \chi_{d-1} \cdots \sin \chi_3 \sin \chi_2 \sin \chi_1, \\ p_2 &= p \sin \chi_{d-1} \cdots \sin \chi_3 \sin \chi_2 \cos \chi_1, \\ p_3 &= p \sin \chi_{d-1} \cdots \sin \chi_3 \cos \chi_2, \\ &\vdots \\ p_{d-1} &= p \sin \chi_{d-1} \cos \chi_{d-2}, \\ p_d &= p \cos \chi_{d-1}, \end{aligned} \tag{61}$$

where the modulus p and the angles $\chi_{d-1}, \ldots, \chi_2, \chi_1$ define the orientation of $\mathbf{p}$ in the d-dimensional momentum space.

The generic point $\mathbf{u}$ on the surface of the $(d+1)$-dimensional hypersphere is given by:

$$\begin{aligned} u_1 &= p_1 \sin \chi_d = \sin \chi_d \sin \chi_{d-1} \cdots \sin \chi_3 \sin \chi_2 \sin \chi_1, \\ u_2 &= p_2 \sin \chi_d = \sin \chi_d \sin \chi_{d-1} \cdots \sin \chi_3 \sin \chi_2 \cos \chi_1, \\ u_3 &= p_3 \sin \chi_d = \sin \chi_d \sin \chi_{d-1} \cdots \sin \chi_3 \cos \chi_2, \\ &\vdots \\ u_{d-1} &= p_{d-1} \sin \chi_d = \sin \chi_d \sin \chi_{d-1} \cos \chi_{d-2}, \\ u_d &= p_d \sin \chi_d = \sin \chi_d \cos \chi_{d-1}, \\ u_{d+1} &= \cos \chi_d, \end{aligned} \tag{62}$$

where the angle χ_d is defined by $\cos \chi_d = \frac{p_0^2 - p^2}{p_0^2 + p^2}$. Such a set of angles corresponds to an angular parametrization of S^d in spherical coordinates.[83]

Momentum space spherical Sturmians can now be obtained either by solving the integral Schrödinger equation in the d-dimensional momentum space making use of Fock's projection, or by Fourier transforming $u_{n\lambda\kappa\ldots\ell m}(\rho, \eta_{d-1}, \ldots, \eta_1)$ in Eq. (5) as in Ref. 33. In both cases one gets:

$$\begin{aligned} \psi_{n\lambda\kappa\ldots\ell m}(p, \chi_{d-1}, \ldots, \chi_2, \chi_1) = {} & (-)^{n-1-|m|} (\imath)^{\lambda+m-|m|} \frac{(2p_0)^{\frac{d}{2}+1}}{\sqrt{2}\,(p_0^2 + p^2)^{\frac{d+1}{2}}} \\ & \times Y_{n-1\lambda\kappa\ldots\ell m}(\chi_d, \ldots, \chi_2, \chi_1). \end{aligned} \tag{63}$$

In this space the wavefunctions $\Psi_{n_1 n_2 \kappa\ldots\ell m}$, i.e., the Fourier transforms of wavefunctions in parabolic coordinates, can be obtained[33] as a linear combination (through the transformation coefficients given in Eq. (18), expressed, for instance, as Hahn coefficents) of the $\psi_{n\lambda\kappa\ldots\ell m}$ of Eq. (63):

$$\Psi_{n_1 n_2 \kappa\ldots\ell m}(\Theta,\Phi,\chi_{d-2},\ldots,\chi_2,\chi_1) = (-)^{n-1-|m|}(\imath)^{\kappa+m-|m|}\frac{(2p_0)^{\frac{d}{2}+1}}{\sqrt{2}\,(p_0^2+p^2)^{\frac{d+1}{2}}}$$
$$\sum_{\lambda=\kappa}^{n-1}(-\imath)^{\lambda-\kappa}\,Q_{\lambda-\kappa,n_2}^{\kappa+\frac{d-3}{2},\kappa+\frac{d-3}{2},n-\kappa}\,Y_{n-1\lambda\kappa\ldots\ell m}(\chi_d,\ldots,\chi_2,\chi_1). \tag{64}$$

The summation appearing in the above equation yields[87,33] a hyperspherical harmonic $Y_{n-1,n_1-n_2,\kappa\ldots\ell m}(\Theta,\Phi,\chi_{d-2},\ldots,\chi_2,\chi_1)$, alternative to that in (63). Thus, for the momentum space wavefunctions in parabolic coordinates one finally obtains:

$$\Psi_{n_1 n_2 \kappa\ldots\ell m}(\Theta,\Phi,\chi_{d-2},\ldots,\chi_2,\chi_1) = (-)^{n-1-|m|}(\imath)^{\kappa+m-|m|}\frac{(2p_0)^{\frac{d}{2}+1}}{\sqrt{2}\,(p_0^2+p^2)^{\frac{d+1}{2}}}$$
$$\times\, Y_{n-1,n_1-n_2,\kappa\ldots\ell m}(\Theta,\Phi,\chi_{d-2},\ldots,\chi_2,\chi_1), \tag{65}$$

where the angles $\Theta,\Phi,\chi_{d-2},\ldots,\chi_2,\chi_1$ are related to the angles $\chi_d,\ldots,\chi_2,\chi_1$ through the equations:

$$\begin{aligned}
u_1 &= \sin\chi_d\sin\chi_{d-1}\cdots\sin\chi_3\sin\chi_2\sin\chi_1 = \sin\Theta\sin\chi_{d-2}\cdots\sin\chi_3\sin\chi_2\sin\chi_1,\\
u_2 &= \sin\chi_d\sin\chi_{d-1}\cdots\sin\chi_3\sin\chi_2\cos\chi_1 = \sin\Theta\sin\chi_{d-2}\cdots\sin\chi_3\sin\chi_2\cos\chi_1,\\
u_3 &= \sin\chi_d\sin\chi_{d-1}\cdots\sin\chi_3\cos\chi_2 = \sin\Theta\sin\chi_{d-2}\cdots\sin\chi_3\cos\chi_2,\\
&\vdots\\
u_{d-1} &= \sin\chi_d\sin\chi_{d-1}\cos\chi_{d-2} = \sin\Theta\cos\chi_{d-2},\\
u_d &= \sin\chi_d\cos\chi_{d-1} = \cos\Theta\sin\Phi,\\
u_{d+1} &= \cos\chi_d = \cos\Theta\cos\Phi.
\end{aligned} \tag{66}$$

This set of angles is usually identified with an angular parametrization in cylindrical coordinates.[83]

The same hyperspherical harmonic could also be obtained by the Fourier transform of the parabolic Sturmians of Eq. (12). The complete reciprocity between configuration and momentum space is depicted in Figure 5.6.

The hyperspherical harmonics of Eqs. (63) and (65) are both eigenfunctions of the same $(d+1)$-dimensional Beltrami–Laplace operator with eigenvalue $n-1$, but they are eigenstates for Beltrami–Laplace operators of

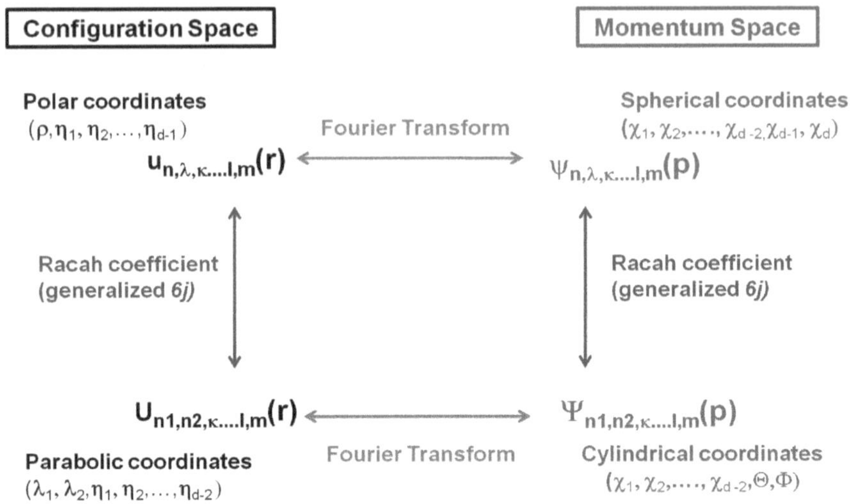

Figure 5.6 Summarizing scheme illustrating the reciprocity between transformations in configuration and momentum space: we can pass from *d*-dimensional configuration space to *d*-dimensional momentum space (and vice versa) through Fourier transforms and their inverses, and from the *d*-dimensional spherical to the *d*-dimensional cylindrical set (and vice versa) through a generalized 6*j* coefficient. Special cases are $d=2$ when the generalized 6*j* coefficient becomes a Wigner *d* function and $d=3$ when it becomes a 3*j* coefficient. For color version of this figure, the reader is referred to the online version of this chapter.

different subgroups, with eigenvalues λ and $n_1 - n_2$ (or μ, as introduced in Section 2.4), respectively.

Indeed, such a result can be directly related to different reduction schemes for the continuous rotation group in $(d+1)$-dimensional space $O(d+1)$: The hyperspherical harmonic in spherical coordinates corresponds to the *canonical* reduction $O(d+1) \supset O(d) \supset O(d-1) \supset \cdots \supset O(2)$, whereas $\Psi_{n_1 n_2 \kappa \ldots \ell m}(\Theta, \Phi, \chi_{d-2}, \ldots, \chi_2, \chi_1)$ in cylindrical coordinates corresponds to $O(d+1) \supset O(2) \times [O(d-2) \supset \cdots \supset O(2)]$.

These are only two of the many distinct hyperspherical harmonics which can be written for the S^d eigenfunctions of various Beltrami–Laplace operators for different subspaces of $O(d+1)$, thus enjoying different symmetry properties. The number of such hyperspherical harmonics amounts to [22,90]:

$$\mathcal{N} = \frac{(2d)!}{d!}. \tag{67}$$

However, this number can be reduced[68,88] by a factor 2^d, because some of these harmonics differ only by the definition of an angle (i.e., they differ for a phase factor) and are thus eigenfunctions of the same rotation operator, or refer to right-handed or left-handed coordinate systems.

A full classification of these functions in the general *d*-dimensional case, though possible, is beyond the scope of this work (and quite useless unless specific applications are needed, like in Ref. 9), but it can readily be performed for the two- and the physical three-dimensional spaces, allowing to gain an insight into the symmetry properties of both momentum and configuration space Coulomb Sturmians, as will be shown in the following.

It might be difficult to get the configuration space counterparts of these basis sets analytically; however, by recalling the reciprocity between configuration and momentum space, they can be recovered by applying the same superposition coefficients between alternative hyperspherical harmonics. These superposition coefficients are well known and can be classified according to the Askey and Nikiforov schemes (Fig. 5.1),[25,22,23] because they coincide, in the most general case, with Racah orthonormalized coefficients (generalized 6*j* coefficient) and can be easily computed.The explicit form of these coefficients can be found in Ref. 22,23 and, because of their importance for applications, is summarized in the Appendix.

3.2 Two-dimensional momentum space Coulomb Sturmians

In this section we consider the wave functions of the two-dimensional Coulomb problem in momentum space. The momentum space eigenfunctions are solutions of the related Schrödinger equation obtained by Fourier-transforming the position space Schrödinger equation,[84] that is

$$(p_o^2 + p^2)\psi_{n,m}(p,\varphi) = \frac{2}{\pi}\int \frac{\psi_{n,m}(p',\varphi')}{|\mathbf{p}-\mathbf{p}'|} p' \mathrm{d}p' \,\mathrm{d},\varphi' \tag{68}$$

where

$$\psi_{n,m}(p,\varphi) = \frac{1}{2\pi}\int e^{-\imath pr\cos(\phi-\varphi)} u_{nm}(r,\phi) r \mathrm{d}r\,\mathrm{d}\phi \tag{69}$$

and $|\mathbf{p}-\mathbf{p}'|$ is the distance between two points identified by the two momentum vectors $\mathbf{p}$ and $\mathbf{p}'$.

In Ref. 84 applying Fock's projection to the two-dimensional hydrogen atom in momentum space, it is shown that the eigenfunctions are ordinary spherical harmonics [85]:

$$\psi_{n,m}(p,\varphi) = \imath^m (-)^{n-1} \frac{\sqrt{8} p_0^2}{(p_0^2 + p^2)^{\frac{3}{2}}} Y_{n-1,m}(\vartheta,\varphi) \tag{70}$$

where p is the modulus of the momentum of the electron $\mathbf{p}$, φ is its orientation in the two-dimensional momentum space, with Cartesian coordinates p_1 and p_2:

$$\begin{aligned} p_1 &= p \sin\varphi, \\ p_2 &= p \cos\varphi \end{aligned} \tag{71}$$

with

$$\begin{aligned} p &= p_0 \tan\frac{\vartheta}{2}, \\ \cos\vartheta &= \frac{p_0^2 - p^2}{p_0^2 + p^2}. \end{aligned}$$

The two-dimensional Fock stereographic projection has been depicted in Figure 5.7, since only the two-dimensional case allows its visualization. [1] It consists of projecting a point in the plane onto the surface of a unit radius sphere. As shown in the figure, each point P of the sphere, with coordinates u_1, u_2, u_3 [with $(u_1^2 + u_2^2 + u_3^2 = 1)$] corresponds to the point $\mathbf{p}$ in the plane with coordinates p_1 and p_2. According to such projection, u_1, u_2, u_3 in spherical coordinates are given by [84]:

$$\begin{aligned} u_1 &= \sin\vartheta \sin\varphi, \\ u_2 &= \sin\vartheta \cos\varphi, \\ u_3 &= \cos\vartheta. \end{aligned} \tag{72}$$

In order to obtain the wavefunctions $\Psi_{n_1 n_2}$ in parabolic coordinates, which are the momentum space counterpart of the orbitals $U_{n_1 n_2}$ in Eq. (23), we use the transformation given in Eq. (30) of the previous section:

$$\Psi_{n_1 n_2}(\Theta, \Phi) = \sum_m (-)^{\frac{n_1 - n_2}{2}} e^{-\imath\pi\frac{n_1}{2}} d^{\frac{n_1+n_2}{2}}_{\frac{n_1-n_2}{2},m}\left(\frac{\pi}{2}\right) \psi_{nm}(p, \varphi) \tag{73}$$

[1] "Only one or two people have ever attained the ability to visualize hyper-solids as simply and naturally as we ordinary mortals visualize solids; but a certain facility in that direction may be acquired by contemplating the analogy between one and two dimensions, then two and three, and so (by a kind of extrapolation) three and four. [...] However, there is some danger of being led astray unless we check our results with the aid of one of the other two procedures." [in Coxeter, H.S.M. *Regular Polytopes*; Dover Publications: New York, 1973; p 119]. An example of how this could be misleading in the passage from three to two dimensions can be found in *The Planiverse*, Dewdney, A. K. Pan Books (1984) (p. 244) which, concerning the two-dimensional hydrogen atom, reads: "the principal quantum number summarizes the total energy of an atom and this energy can take on only the values of $n = 0, 1, 2, 3, \ldots$, and so on. Likewise, the orbital angular momentum number, which describes the atom's energy of rotation, also has only $l = 0, 1, 2, 3, \ldots$, as its possible values." As shown here, the hidden spherical symmetry of the two-dimensional hydrogen atom leads to different conclusions: in particular, the orbital quantum number m ranges from $-(n-1)$ to $n-1$.

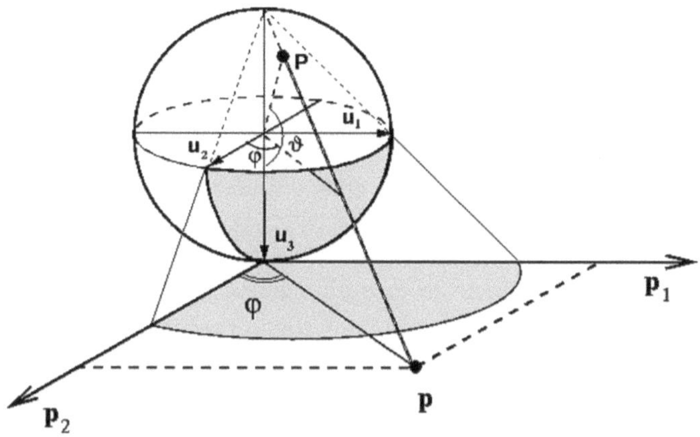

Figure 5.7 Stereographic projection of a point **P** in a plane with coordinates $p_1 = p \sin \varphi$ and $p_2 = p \cos \varphi$ onto the surface of a unit sphere with coordinates u_1, u_2, and u_3. The shaded area represents the projection of the points of the plane onto the part of the spherical bowl comprised between the equator and the axes u_1 and u_2. The points fulfilling the condition $p = p_0$ are projected onto the equator. For color version of this figure, the reader is referred to the online version of this chapter.

or, inserting Eq. (70) into Eq. (73):

$$\Psi_{n_1 n_2}(\Theta, \Phi) = e^{\imath \pi \frac{n_1}{2}} \frac{\sqrt{8} p_0^2}{(p_0^2 + p^2)^{\frac{3}{2}}} \sum_m D^{n-1}_{m, \frac{n_1 - n_2}{2}} \left(\frac{\pi}{2}, \frac{\pi}{2}, \pi \right) Y_{n-1,m}(\vartheta, \varphi), \tag{74}$$

where $D^{n-1}_{m, \frac{n_1 - n_2}{2}}$ is a Wigner's rotation matrix.[85] Using a known relation between spherical harmonics in two different angular coordinates[85] one can write:

$$\Psi_{n_1 n_2}(\Theta, \Phi) = e^{\imath \pi \frac{n_1}{2}} \frac{\sqrt{8} p_0^2}{(p_0^2 + p^2)^{\frac{3}{2}}} Y_{n-1, \frac{n_1 - n_2}{2}} \left(\frac{\pi}{2} - \Theta, \Phi \right) \tag{75}$$

where Θ and Φ, as suggested in Ref. 58, are bipolar coordinates:

$$p_1 = p_0 \frac{\cos \Theta \sin \Phi}{1 + \cos \Theta \cos \Phi} \quad p_2 = p_0 \frac{\sin \Theta}{1 + \cos \Theta \cos \Phi}.$$

The wavefunction in Eq. (75) corresponds to a spherical harmonic relative to a rotated polar axis with respect to that in (70). Two-dimensional Sturmians in momentum space and spherical harmonics are in one-to-one fashion correspondence via the two-dimensional analog of Fock's stereographic projection,[84] as represented, in the more general *d*-dimensional case, in Figure 5.6. Indeed, the momentum space Sturmians of Eq. (75) can also be obtained by performing Fock's transformation and by establishing the relationship between the variables ϑ and φ and the alternatives Θ and Φ, which parameterize the same point on a unit sphere:

$$\begin{aligned} u_1 &= \sin\vartheta \sin\varphi = \sin\Theta, \\ u_2 &= \sin\vartheta \cos\varphi = \cos\Theta \sin\Phi, \\ u_3 &= \cos\vartheta = \cos\Theta \cos\Phi. \end{aligned} \tag{76}$$

These relationships illustrate two alternative angular parameterizations of the rotation group $O(3)$,[44] which is the symmetry group of the two-dimensional hydrogen atom.[57]

Thus, all the different spherical harmonics of $O(3)$ are, apart from a normalization factor, solutions to the Schrödinger equation for the two-dimensional Coulomb problem in momentum space. Their symmetry properties can be better understood by recalling that in quantum mechanics spherical harmonics $Y_{\ell m}(\Omega)$ are eigenfunctions of the angular momentum operator $\mathbf{L}^2$ with eigenvalues $\ell(\ell+1)$. [In the specific application of the two-dimensional hydrogen atom the quantum number $n-1$ can be identified with ℓ]. The quantum number m is the projection of ℓ along a quantization axis (generally z), which, for the spherical harmonic of Eq. (70), is u_3. The application of a transformation coefficient, as that in Eq. (30), leads to a spherical harmonic with a different projection axis, e.g., the harmonic of Eq. (75) has u_1 as the projection axis. Quantum numbers for this harmonic are $n-1$ and m', which can be obtained by identifying $n_1+n_2+1=2n-1$ (Eq. 25) and $m'=(n_1-n_2)/2$.

In $O(3)$ another spherical harmonic, with different symmetry properties, can be written as solution of Eq. (68), with u_2 as the projection axis, connected to the previous two by a transformation coefficient also related to a Wigner d function. This gives a total number of three spherical harmonics (as can be obtained using Eq. (67) with $d=2$) in eigenstate for $\mathbf{L}^2$ and one of its Cartesian components L_x, L_y, or L_z with $\mathbf{L}^2 = L_x^2 + L_y^2 + L_z^2$.

Another view of this is related to the construction of the S^2 sphere (the physical three-dimensional sphere) starting from the circle S^1, lying on the (u_1, u_2) plane, and rotating it by an angle θ around the additional u_3 axis (see Fig. 5.8). Since we can start from three different planes [(u_1, u_2) or (u_2, u_3) or (u_1, u_3)] with as many additional axes, we are led to three different spherical harmonics.

3.3 Three-dimensional momentum space Coulomb Sturmians

In 1935 Fock obtained the wavefunctions for the hydrogen atom $\psi_{n,l,m}(p,\vartheta,\varphi)$ in momentum space by solving the integral Schrödinger equation related by Fourier transform to the one in configuration space:

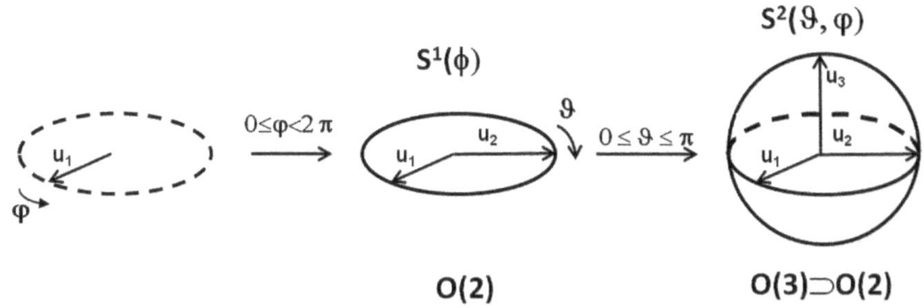

Figure 5.8 Construction of the S^2 sphere: it is obtained by rotating the circle S^1, lying on the (u_1, u_2) plane, by the angle θ with respect of u_3. A different choice of the plane (i.e., of the projection axis) leads to different spherical harmonics.

$$(p_o^2 + p^2)\psi_{n,l,m}(p, \vartheta, \varphi) = \frac{1}{\pi^2} \int \frac{\psi_{n,l,m}(p', \vartheta', \varphi')}{|\mathbf{p} - \mathbf{p}'|^2} p'^2 \sin\vartheta' \mathrm{d}p' \mathrm{d}\vartheta' \,\mathrm{d}\varphi', \tag{77}$$

where

$$\psi_{n,l,m}(p, \vartheta, \varphi) = \frac{1}{(2\pi)^{3/2}} \int e^{-\imath \mathbf{pr}} u_{nlm}(r, \theta, \phi)\mathrm{d}\mathbf{r} \tag{78}$$

p_0 corresponds to the energy $E_0 = -p_0^2/2$. The wavefunctions $\psi_{n,l,m}(p, \vartheta, \varphi)$ are connected to four-dimensional spherical harmonics, i.e., they are related to the eigenfunctions of the Beltrami–Laplace operator on S^3. Indeed, the momentum $\mathbf{p}$ with components:

$$\begin{aligned} p_1 &= p \sin\vartheta \cos\varphi, \\ p_2 &= p \sin\vartheta \sin\varphi, \\ p_3 &= p \cos\vartheta \end{aligned} \tag{79}$$

is projected onto a tetra-dimensional hypersphere of unit radius:

$$\begin{aligned} u_1 &= \frac{2p_0 p_y}{p_0^2 + p^2} = \sin\chi \sin\vartheta \sin\varphi, \\ u_2 &= \frac{2p_0 p_x}{p_0^2 + p^2} = \sin\chi \sin\vartheta \cos\varphi, \\ u_3 &= \frac{2p_0 p_z}{p_0^2 + p^2} = \sin\chi \cos\vartheta, \\ u_4 &= \frac{p_0^2 - p^2}{p_0^2 + p^2} = \cos\chi. \end{aligned} \tag{80}$$

Equations (80) show the relationship between the Euclidean space (p_1, p_2, p_3) and the surface S^3 of a four-dimensional sphere $(u_1^2 + u_2^2 + u_3^2 + u_4^2 = 1)$, and give an explicit parameterization in polar coordinates $(\chi, \vartheta, \varphi)$.

The eigenfunction for the hydrogen atom in momentum space is thus [81]:

$$\psi_{nlm}(\mathbf{p}) = (-)^{n-1-|m|}(\imath)^{l+m-|m|} \frac{4p_0^{5/2}}{(p_0^2 + p^2)^2} \cdot Y_{n-1,l,m}(\chi, \vartheta, \varphi), \quad (81)$$

where $Y_{n-1,l,m}(\chi, \vartheta, \varphi)$ is a four-dimensional hyperspherical harmonic, the wavefunction for the free motion of a particle on the S^3 surface. Because of this identification, the principal quantum number n (which labels the energy spectrum) can appropriately be interpreted as a hyperangular momentum quantum number, manifesting that the hidden symmetry giving rise to the accidental degeneracy emerging in the three-dimensional configuration space treatment is actually a four-dimensional symmetry, which has been analyzed and discussed in various articles,[61] reviews,[86] and books.[58] The invariance group for the Hamiltonian of the hydrogen atom is thus $O(4)$, the continuous group of rotations in the four-dimensional Euclidean space. The actual symmetry of the hydrogen atom Schrödinger equation has therefore become apparent only in the momentum space approach: The degeneracy existing for a given n leads to n^2 Sturmians belonging to the same irreducible representation of O(4).

Under this perspective the nature of the eigenvalue Eqs. (34)–(36) obtained in the configuration space treatment can be better understood. Indeed, the hydrogenic Hamiltonian can be expressed as a function of the orbital angular momentum $\hat{\mathbf{L}}$ and the reduced Runge–Lenz vector $\hat{\mathbf{K}}$[89] introduced in Section 2.4, as

$$\hat{H} = -\frac{1}{2}(\hat{\mathbf{L}}^2 + \hat{\mathbf{K}}^2 + 1)^{-1}. \quad (82)$$

The Cartesian components of $\hat{\mathbf{L}}$ and $\hat{\mathbf{K}}$ can be identified with the six generators $\hat{J}_{ij}$ of two-dimensional rotations in a four-dimensional space. A rotation of an angle α on the (ij) plane is performed by applying the operator $e^{\imath\alpha\hat{J}_{ij}}$.

We can associate the generators of $O(4)$ to the components of $\hat{\mathbf{L}}$ and $\hat{\mathbf{K}}$ on the basis of their commutation relations [58,63]:

$$\begin{aligned} \hat{L}_x &= \hat{J}_{23}\ \hat{L}_y = \hat{J}_{31}\ \hat{L}_z = \hat{J}_{12}, \\ \hat{K}_x &= \hat{J}_{14}\ \hat{K}_y = \hat{J}_{24}\ \hat{K}_z = \hat{J}_{34}. \end{aligned} \quad (83)$$

The Beltrami–Laplace operator on the four-dimensional sphere can be written in terms of the generators of $O(4)$:

$$(\hat{J}_{12}^2 + \hat{J}_{23}^2 + \hat{J}_{31}^2 + \hat{J}_{14}^2 + \hat{J}_{24}^2 + \hat{J}_{34}^2) Y_{j,\alpha,\beta}(\Omega) = j(j+2) Y_{j,\alpha,\beta}(\Omega), \quad (84)$$

where α and β are two quantum numbers. Use of Eqs. (83) shows that the operator in Eq. (84) coincides with $(\hat{\mathbf{L}}^2 + \hat{\mathbf{K}}^2)$, and therefore momentum space Sturmians, which are essentially hyperspherical harmonics, can be identified with basis sets for irreducible representations of the rotation group $O(4)$. The hyperangular momentum $j = n - 1$ labels the energy levels.

The hyperspherical harmonic in polar coordinates in Eq. (81) is also eigenfunction of $\hat{\mathbf{L}}^2 = \hat{L}_x^2 + \hat{L}_y^2 + \hat{L}_z^2 = \hat{J}_{23}^2 + \hat{J}_{31}^2 + \hat{J}_{12}^2$, which is related to rotations on the tridimensional subspace x, y, z (or u_1, u_2, u_3 in the present case) of the four-dimensional space, and of $\hat{L}_z = \hat{J}_{12}$ related to rotation on the bidimensional subspace (u_1, u_2). This corresponds to a reduction of $O(4)$ into its subgroups of the kind: $O(4) \supset O(3) \supset O(2)$ and to the construction of the four-dimensional sphere represented in Figure 5.9a.

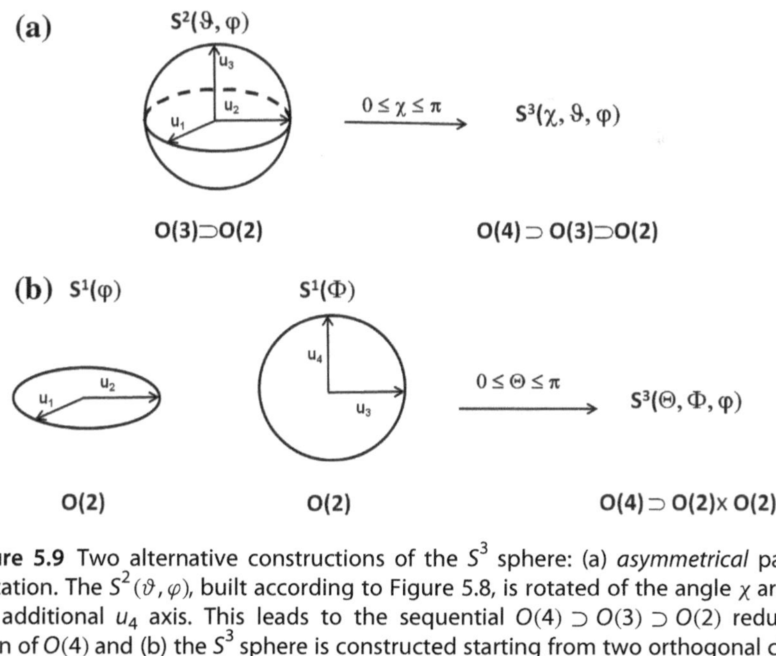

Figure 5.9 Two alternative constructions of the S^3 sphere: (a) *asymmetrical* parametrization. The $S^2(\vartheta, \varphi)$, built according to Figure 5.8, is rotated of the angle χ around the additional u_4 axis. This leads to the sequential $O(4) \supset O(3) \supset O(2)$ reduction chain of $O(4)$ and (b) the S^3 sphere is constructed starting from two orthogonal circles $S^1(\varphi)$ and $S^1(\phi)$ which are rotated by the angle Θ. This is related to the *symmetrical* $O(4) \supset O(2) \times O(2)$ reduction chain.

Table 5.1 The 15 momentum space Sturmians classified according to their symmetry properties

Asymmetrical parametrization

3-dim subspace	*O(3) operator*	*2-dim subspace*	*O(2) operator*
$[x, y, z]$	$\hat{L}_x^2 + \hat{L}_y^2 + \hat{L}_z^2$	$[x, y]$	$\hat{L}_z$
$[x, y, z]$	$\hat{L}_x^2 + \hat{L}_y^2 + \hat{L}_z^2$	$[y, z]$	$\hat{L}_x$
$[x, y, z]$	$\hat{L}_x^2 + \hat{L}_y^2 + \hat{L}_z^2$	$[z, y]$	$\hat{L}_y$
$[x, y, w]$	$\hat{K}_x^2 + \hat{K}_y^2 + \hat{L}_z^2$	$[x, y]$	$\hat{L}_z$
$[x, y, w]$	$\hat{K}_x^2 + \hat{K}_y^2 + \hat{L}_z^2$	$[y, w]$	$\hat{K}_y$
$[x, y, w]$	$\hat{K}_x^2 + \hat{K}_y^2 + \hat{L}_z^2$	$[w, x]$	$\hat{K}_x$
$[x, z, w]$	$\hat{K}_x^2 + \hat{K}_z^2 + \hat{L}_z^2$	$[z, x]$	$\hat{L}_y$
$[x, z, w]$	$\hat{K}_x^2 + \hat{K}_z^2 + \hat{L}_z^2$	$[z, w]$	$\hat{K}_z$
$[x, z, w]$	$\hat{K}_x^2 + \hat{K}_z^2 + \hat{L}_z^2$	$[w, x]$	$\hat{K}_x$
$[y, z, w]$	$\hat{K}_y^2 + \hat{K}_z^2 + \hat{L}_x^2$	$[y, z]$	$\hat{L}_x$
$[y, z, w]$	$\hat{K}_y^2 + \hat{K}_z^2 + \hat{L}_x^2$	$[y, w]$	$\hat{K}_y$
$[y, z, w]$	$\hat{K}_y^2 + \hat{K}_z^2 + \hat{L}_x^2$	$[w, z]$	$\hat{K}_z$

Symmetrical parametrization

2-dim subspace	*O(2) operator*	*2-dim subspace*	*O(2) operator*
$[y, z]$	$\hat{L}_x$	$[x, w]$	$\hat{K}_x$
$[x, z]$	$\hat{L}_y$	$[y, w]$	$\hat{K}_y$
$[x, y]$	$\hat{L}_z$	$[z, w]$	$\hat{K}_z$

Because of the sequential fashion of building the hypersphere (S^1 circle $\rightarrow S^2$ sphere $\rightarrow S^3$ hypersphere) this parametrization has also been referred to as *asymmetrical*.[44]

From the equivalence of the generators in Eqs. (84), it follows that this is only one of the many ways the parametrization of S^3 can be achieved, leading to hyperspherical harmonics different, in terms of symmetry properties, from that of Eq. (81), as shown below.

From the previous sections we can gather that the momentum space Sturmian in parabolic coordinates can be found by performing the Fourier transform of the $U_{n\mu m}(\lambda_1, \lambda_2, \phi)$ of Eq. (39), or by directly solving the momentum space integral, Eq. (77), in parabolic coordinates by making use of Fock's projection with an alternative angular parametrization, or by using the transformation coefficient given in Eq. (45) upon Eq. (81).

In the latter case we have:

$$\begin{aligned}\Psi_{n\mu m}(\mathbf{p}) &= \sum_l (-)^{\frac{n-1+m-\mu}{2}} \langle \frac{1}{2}(n-1), \frac{m-\mu}{2}; \frac{1}{2}(n-1), \frac{m+\mu}{2} | lm \rangle \cdot \psi_{nlm}(\mathbf{p}) \\ &= \frac{4p_0^{5/2}}{(p^2+p_0^2)^2} (-)^{n-1} (\imath)^m \left(\frac{n}{2\pi^2}\right)^{\frac{1}{2}} D^{\frac{1}{2}(n-1)}_{\frac{\mu+m}{2}, \frac{\mu-m}{2}} (-\Phi - \varphi, 2\Theta, \varphi - \Phi).\end{aligned} \tag{85}$$

To obtain (85) use has been made of the following relationship between harmonics [44,63]:

$$\begin{aligned}&\sum_l (-)^{\frac{n-1+m-\mu}{2}} (\imath)^{l+m} \langle \frac{1}{2}(n-1), \frac{m-\mu}{2}; \frac{1}{2}(n-1), \frac{m+\mu}{2} | lm \rangle Y_{nlm}(\chi, \vartheta, \varphi) \\ &= \left(\frac{n}{2\pi^2}\right)^{\frac{1}{2}} D^{\frac{1}{2}(n-1)}_{\frac{\mu+m}{2}, \frac{\mu-m}{2}} (-\Phi - \varphi, 2\Theta, \varphi - \Phi).\end{aligned} \tag{86}$$

The Wigner D function in Eq. (85) is the well-known non-standard hyperspherical harmonic corresponding to a parametrization of the four-dimensional unit sphere in terms of the cylindrical angles Θ, Φ, and φ.[63] A point u_1, u_2, u_3, u_4 on the sphere is now given by:

$$\begin{aligned} u_1 &= \sin\chi \sin\vartheta \sin\varphi = \sin\Theta \sin\varphi, \\ u_2 &= \sin\chi \sin\vartheta \cos\varphi = \sin\Theta \cos\varphi, \\ u_3 &= \sin\chi \cos\vartheta = \cos\Theta \sin\Phi, \\ u_4 &= \cos\chi = \cos\Theta \cos\Phi. \end{aligned} \tag{87}$$

This kind of parametrization is conceptually very different from that of Eqs. (80), because it is related to a different subgroup chain reduction of $O(4)$: $O(4) \supset O(2) \times O(2)$, that is to the construction of the S^3 hypersphere starting from two different S^1 circles, each described by an angle (Φ and ϕ in the present case), and introducing a further angle Θ to give S^3, as depicted in Figure 5.9.[2] Because we start from two equivalent (though orthogonal) bidimensional spaces, this parametrization is also referred to as *symmetrical*.[44]

The corresponding four-dimensional hyperspherical harmonic of Eq. (85) is an eigenfunction of the Beltrami–Laplace operator (84) and of the

[2] This second construction of the tetradimensional sphere may recall a similarity with Dante's universe, which M.A. Peterson in his article "Dante and the 3-sphere", *Am. J. Phys.* **1979** (47), 1031, places into tetradimensional Euclidean space.

two bidimensional operators $\hat{L}_z = \hat{J}_{12}$ and $\hat{A}_z = \hat{J}_{34}$. Such a basis function thus gives the z axis the role of a privileged direction in the physical space.

Other Coulomb Strumians can be written in momentum space, for both asymmetrical and symmetrical parameterizations, corresponding to different choices of the operators in three-dimensional and two-dimensional subspaces. By using the formula in Eq. (67) for $d=3$ we can calculate 120 hyperspherical harmonics, which can be reduced by the factor 2^d to 15 different basis sets. A full illustration of this classification, of the symmetry properties, of the transformation coefficients between the various bases, and of the useful sum rules which can be derived can be found in Ref. 88. Here we limit to a short description in terms of their correspondence to the $O(4)$ operators as summarized in Table 5.1.

Three of the 15 momentum space Coulomb Sturmians are symmetrical hyperspherical harmonics with a privileged z (Eq. 85), x (in eigenstate for $\hat{L}_z = \hat{J}_{23}$ and $\hat{A}_x = \hat{J}_{14}$) or y ($\hat{L}_y = \hat{J}_{13}$ or $\hat{A}_y = \hat{J}_{24}$).

The other 12 Sturmians correspond to the asymmetrical type, and they are diagonal for an operator acting on three dimensional ($\hat{L}_x^2 + \hat{L}_y^2 + \hat{L}_z^2 = \hat{\mathbf{L}}^2$, or $\hat{K}_x^2 + \hat{K}_y^2 + \hat{L}_z^2$, or $\hat{K}_x^2 + \hat{L}_y^2 + \hat{K}_z^2$, or $\hat{L}_x^2 + \hat{K}_y^2 + \hat{K}_z^2$) and for one acting on a two dimensional subspace. With the exception of the spherical basis of Eq. (81), the other bases are little known or employed. However, in the last years a growing interest has emerged toward the basis in eigenstates of $\hat{K}_x^2 + \hat{K}_y^2 + \hat{L}_z^2$ and $\hat{L}_z$, which was called λ basis,[91,92] because it can be labeled by the quantum number λ eigenvalue of $\hat{K}_x^2 + \hat{K}_y^2 + \hat{L}_z^2$, or *Zeeman*.[9,10,68] Though its counterpart in configuration space is not known, the superposition coefficients connecting this hyperspherical harmonic with those in Eq. (81) and (85) can be easily calculated (see Appendix) as $6j$ generalized coefficients or as sum over $3j$ coefficients,[23,68,88] the discrete elements placed in the upper part of Figure 5.1.

3.4 Non-standard Coulomb Sturmians in momentum space: Elliptic Sturmians

Besides the canonical bases described in previous sections, deriving from parameterizations of the hypersphere S^3 in terms of the well-known circular function sin and cos, there exist bases of O(4) (and thus three-dimensional momentum space Sturmians) resulting from elliptic parameterizations of S^3.[61,60,10] Some of these bases can be related by Fourier transform to their direct space counterparts described in Section 2.5. As emerged from the configuration space analysis, a crucial fact is that the value that one gets for such functions also depends on a number, k, called modulus, that can have

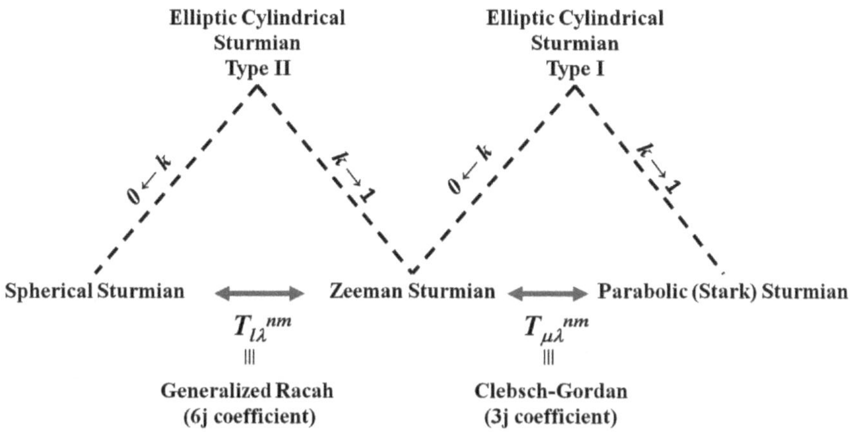

Figure 5.10 Limiting cases of elliptic Sturmians lead to canonical Spherical, Zeeman, and Stark Sturmians. For color version of this figure, the reader is referred to the online version of this chapter.

any value in the range [0, 1]. This gives the Sturmians from elliptic parameterizations an additional degree of freedom with respect to the canonical Sturmians: the former are more flexible but, as mentioned before, their manipulation is more complicated. For example, they cannot be expressed in closed form and in fact, as far as we know, they have never been used for the calculation of properties of atomic and molecular systems other than the hydrogen atom. In the following, the symmetry properties of elliptic Sturmians will be analyzed to clarify their relation with canonical Sturmians and situate them in the overall scheme of the alternative hydrogenic Sturmians.

The simplest kind of elliptic Sturmians can be derived from the parametrization of the hypersphere S^3 in sphero-elliptic coordinates:

$$\begin{aligned} u_1 &= \sin\chi\,\mathrm{sn}(\alpha,k)\mathrm{dn}(\beta,k'), \\ u_2 &= \sin\chi\,\mathrm{cn}(\alpha,k)\mathrm{cn}(\beta,k'), \\ u_3 &= \sin\chi\,\mathrm{dn}(\alpha,k)\mathrm{sn}(\beta,k') \\ u_4 &= \cos\chi. \end{aligned} \tag{88}$$

As in configuration space, the relationship between the modulus k and k' is $k + k' = 1$, and the constants of motion for the corresponding harmonics are $\hat{\mathbf{L}}^2 + \hat{\mathbf{K}}^2, \hat{\mathbf{L}}^2$ and $\hat{L}_x^2 + k'\hat{L}_y^2$.[61,60,10] The momentum space Sturmian set resulting from the parametrization (88) is the Fourier transform of the corresponding counterpart in configuration space, introduced in Ref. 60 and analyzed in Section 2.5, as can be seen by comparing operators and coordinates of the two reciprocal spaces.

There are additional elliptic parameterizations, deriving from elliptic cylindrical coordinates of types I and II, which do not have a counterpart in configuration space (the answer to the intriguing question of why some elliptic Sturmians in momentum space do not have counterparts in configuration space can be found in Ref. 10).

Elliptic cylindrical coordinates of type I parameterize S^3 as

$$\begin{aligned} u_1 &= \mathrm{sn}(\alpha, k)\mathrm{dn}(\beta, k') \cos\varphi, \\ u_2 &= \mathrm{sn}(\alpha, k)\mathrm{dn}(\beta, k') \sin\varphi, \\ u_3 &= \mathrm{sn}(\beta, k)\mathrm{dn}(\alpha, k'), \\ u_4 &= \mathrm{cn}(\alpha, k)\mathrm{cn}(\beta, k'). \end{aligned} \tag{89}$$

The corresponding hyperharmonics are given as $\mathscr{L}(\alpha)\mathscr{L}(\beta)\begin{Bmatrix}\cos m\varphi \\ \sin m\varphi\end{Bmatrix}$ where $\mathscr{L}(\alpha)$ and $\mathscr{L}(\beta)$ are solutions of equations related to the associated Lamé equation.[61] These solutions are not available in closed form but information on a certain basis set and on its connections with the others can be obtained by studying the corresponding commuting operators, which are[10] $\hat{\mathbf{L}}^2 + \hat{\mathbf{K}}^2, \hat{\mathbf{L}}^2 + \left(\frac{1}{1-k^2}\right)\hat{K}_z^2$ and $\hat{L}_z^2$. In the limit $k \to 0, \hat{\mathbf{L}}^2 + \left(\frac{1}{1-k^2}\right)\hat{K}_z^2$ tends to $\hat{L}_x^2 + \hat{L}_y^2 + \hat{L}_z^2 + \hat{K}_z^2$ which can be manipulated to give $\hat{K}_x^2 + \hat{K}_y^2 + \hat{L}_z^2$: This means that the elliptic cylindrical harmonics of type I tend, in this limit, to the Zeeman set. In the opposite limit $k \to 1$, the operator tends to $\hat{K}_z^2$; thus, the set tends to the Stark basis set.

The elliptic cylindrical coordinates of type II are:

$$\begin{aligned} u_1 &= \mathrm{cn}(\alpha, k)\mathrm{cn}(\beta, k') \cos\varphi, \\ u_2 &= \mathrm{cn}(\alpha, k)\mathrm{cn}(\beta, k') \sin\varphi, \\ u_3 &= \mathrm{sn}(\alpha, k)\mathrm{dn}(\beta, k'), \\ u_4 &= \mathrm{dn}(\alpha, k)\mathrm{sn}(\beta, k'). \end{aligned} \tag{90}$$

This basis can be characterized by the set of operators $\hat{\mathbf{L}}^2 + \hat{\mathbf{K}}^2, \hat{\mathbf{L}}^2 + k^2\hat{K}_z^2$ and $\hat{L}_z^2$. When $k \to 0$, the set coincides with the spherical Sturmian because $\hat{\mathbf{L}}^2 + k^2\hat{K}_z^2 \to \hat{\mathbf{L}}^2$, while as $k \to 1$ $\hat{\mathbf{L}}^2 + k^2\hat{K}_z^2$ tends to $\hat{\mathbf{L}}^2 + \hat{K}_z^2$ which leads to $\hat{K}_x^2 + \hat{K}_y^2 + \hat{L}_z^2$, i.e., the operator characterizing the Zeeman basis.

Figure 5.10 depicts a summarizing scheme, reporting connections and limiting relations between elliptic O(4) Sturmians and the canonical ones. As in Section 2.5, though their explicit expression is unknown, it is possible to expand the elliptic Sturmians in terms of their limiting canonical cases, exploiting three-term recursion relations to calculate the coefficients of the expansion. The appropriate formulae can be found in Ref. 10.

4. APPLICATIONS TO ATOMIC AND MOLECULAR PROBLEMS

All Coulomb Sturmians in configuration and momentum space that were examined in this paper are perfectly equivalent in a field with hyperspherical symmetry. This equivalence fails when the symmetry is broken, for example by introducing an electric and/or a magnetic field or, in any case, when a particular direction in the space is preferred (like the nuclear axis in H_2^+). By considering the symmetry properties of the perturbation, it is possible to build up a subgroup chain and thus a wavefunction basis set well adapted to the description of the anisotropy.

In the following, some applications of Sturmian basis sets will be sketched with a critical analysis of the results. However, not all the related mathematical apparatus will be reported (the interested reader is referred to the bibliography), with the exception of some selected formulae (in particular those involving useful overlap integrals) of more general interest.

4.1 Applications to atomic structure

Different Coulomb Sturmians are well suited for the investigation of the atomic behavior in the presence of electric and/or magnetic fields. In such cases it is possible to write down the perturbed Hamiltonian in terms of the operators examined in the previous sections.

It is well known, for instance, that when a weak electric field is applied along the z-axis for the hydrogen atom (Stark effect), the Schrödinger equation is separable in the parabolic basis[64] – whose generators are $\hat{L}_z$ and $\hat{K}_z$ – thus privileging the z-axis (the explicit expression for the perturbation is $V = F \cdot z$ where F is the intensity of the applied electric field). In a first order perturbation treatment, it is sufficient to consider degeneracy lifting for a level corresponding to a fixed value of the quantum number n. The perturbation operator can be written as $V_n = -\frac{3}{2}F\hat{K}_z$. Thus the parabolic basis (which, for this reason, is also known as *Stark* basis[10,68]) commutes with the perturbed Hamiltonian with the energy spectrum [64]: $E(n, \mu) = -\frac{1}{2}n^2 + \frac{3}{2}n\mu F$, where μ is the parabolic quantum number introduced in Section 2.4.

The treatment of the Zeeman effect, upon application of a magnetic field, can also benefit from this approach. The Hamiltonian for this system (after the separation of the center of mass motion) is $\hat{H} = \hat{H}_0 + \frac{B}{2}\hat{L}_z + \frac{B^2}{8}V$, where $\hat{H}_0$ is the unperturbed Hamiltonian of Eq. (82), $\frac{B}{2}\hat{L}_z$ is the first-order perturbation (paramagnetic term) and $\frac{B^2}{8}V$ is the second-order perturbation (diamagnetic term), with B the intensity of the magnetic field and $V = (x^2 + y^2)$.

In the first-order approximation, the spherical basis, the Stark basis and the Zeeman basis, all exact eigenfunctions of $\hat{L}_z$, commute with the Hamiltonian, and are therefore exact solutions of the related Schrödinger equation. On the other hand, the second-order term, responsible for the physics of the problem and for the effects in stronger magnetic fields, does not allow the separation of variables for this problem in any of the Sturmians that we introduced in Section 3.3,[3] nor, as already stressed, permits simple algebraic calculations. However, by noting that the position vector **r** and the reduced Runge–Lenz vector **K** are related by $\mathbf{r} = -3/2n\mathbf{K}$, one can hope that $\hat{K}_x^2 + \hat{K}_y^2$ might be a good approximation of $V = (x^2 + y^2)$, and thus that the Coulomb Sturmian having $\hat{K}_x$, $\hat{K}_y$, and $\hat{L}_z$ as generators might be a good basis set for this problem. Indeed, the use of such wavefunctions leads to a quasi-diagonal behavior,[91] making this kind of basis very well suited also for the treatment of quantum chaos in intense magnetic fields.[93,94] This is why this kind of momentum space Sturmian has also been referred to as *Zeeman* basis set.[68]

Moreover, Sturmian basis sets of *asymmetrical* type, including the Zeeman basis or its variations, are a valuable tool for the comprehension of circular states (single-electron states characterized by the highest allowed angular momentum for a given hyperangular momentum quantum number n) and Rydberg states (states with high hyperangular momentum quantum number n).[95,96,36]

The quantum mechanics of multi-electron atoms and molecules can also be discussed in terms of the breaking of the hyperspherical symmetry of a d-dimensional hydrogenic atom with $d=3(N-1)$ for N body Coulomb problems, due to the introduction of further charged particles (electrons and/or nuclei).[9] In this case, the generalization of Fock's treatment to

[3]Such basis does exist: it is the Coulomb Sturmian separable in elliptic cylindrical coordinates of type I of Section 3.4,[61,10] which, however, as already stressed, does not permit simple algebraic calculations.

spaces of mathematical dimensions higher than the physical one allows the study of atomic and molecular structure from the point of view of broken symmetry of hyperspheres. In configuration space, Sturmian basis functions can thus be used as expansion basis sets to build up atomic and molecular orbitals. One way to do this uses the hyperspherical approach, where the motion of a N body system is mapped to that of a single particle with reduced mass $M = (\prod m_i / \sum m_i)^{1/(N-1)}$ in a d-dimensional space. In this space the position of the particle is given by an hyperradius ρ and a set of (d-1) hyperangles, collectively denoted as ω.

This method has been long known[97,99,100,98] and successfully applied for the treatment of interactions among few particles (in particular for problems involving nuclear dynamics[101,102]). The formulation described in Ref.9, which will be sketched here, has the advantage of fully exploiting the benefits one can obtain from the use of Coulomb Sturmian basis sets. First of all, the secular equation can be very compactly formulated: All integrals can be written in closed form as matrix elements corresponding to coupling, recoupling, or transformation coefficients of hyperangular momenta algebra. Note that this is feasible also for the so-called 'radial' integrals over Laguerre polynomials (see Eq. (97) in the following). As a consequence, the structure of the relevant matrices and of their zeroes can be foreseen and exploited, sparseness being crucial in applications. Furthermore, one can resort to different basis sets, with different symmetry properties, to accelerate convergence. Indeed, since the counterparts of d-dimensional Sturmian functions in momentum space are (d+1)-dimensional hyperspherical harmonics, the possibility of exploiting different parametrizations of the (d+1)-dimensional sphere can be considered. As a matter of fact, though in Ref. [9] the wavefunction was expanded in configuration space Sturmians, alternative sets can be obtained from the momentum space perspective, by using the transformation coefficients given in the Appendix.

The method is general and thus can be applied to any N-body Coulomb problem, i.e., to arbitrary mass ratios: the central symmetry of the hydrogen atom can be broken either by introducing electrons, leading to a multi-electron atomic system, or by introducing further nuclei, leading to a molecular multi-center problem, to be treated without the Born-Oppenheimer approximation. Applications so far have concerned three-body problems, considering in particular the two-electron series and the H_2^+ molecular ion.

In the general case one can write the Schrödinger equation for a system of N particles interacting through Coulomb forces as:

$$\left[\frac{1}{r^{d-1}}\frac{\partial}{\partial r}r^{d-1}\frac{\partial}{\partial r}-\frac{\Delta(\Omega_{d-1})^2}{r^2}+2V-2E\right]\Psi(\mathbf{r})=0. \tag{91}$$

Here, $\Delta(\Omega_{d-1})$ is the Beltrami–Laplace operator on the d-dimensional sphere, introduced in Section 2, with eigenvalues $\lambda(\lambda+d-2)$. Thus, λ here plays the role of the grand orbital angular momentum quantum number and σ collectively represents the set of d-2 projections of λ (i.e., the quantum numbers σ are eigenvalues of the rotation operators of the subgroups related to the chosen chain reduction). The Coulomb potential in hyperspherical coordinates takes the simple factorized form [103]:

$$V=-\frac{Z(\Omega)}{r}, \tag{92}$$

where $Z(\Omega)$ can be seen as an anisotropic charge. Because of this term Eq. (91) is not separable: if $Z(\Omega)$ were a constant Eq. (91) would coincide with the Schrödinger equation of the multidimensional hydrogen atom, whose solutions can be obtained exactly as d-dimensional Sturmian basis functions, as shown in Section 2. Therefore, in configuration space the many-body Coulomb problem is isomorphic to that of a multidimensional hydrogen atom with an anisotropic charge. For this reason it seems appropriate to expand multi-electronic orbitals $\Psi(\mathbf{r})$ in terms of d-dimensional Sturmian basis sets:

$$\Psi(\mathbf{r})=\sum_{n\lambda\sigma}c_{n\lambda\sigma}u_{n\lambda\sigma}(\mathbf{r}), \tag{93}$$

where $u_{n\lambda\sigma}(\mathbf{r})$ are the polar Sturmians of Eq. (5).

If the expansion (93) is inserted into Eq. (91), after some manipulations,[9] one gets the following secular equation:

$$|\mathbf{A}-p_o\mathbf{I}|\mathbf{c}=0, \tag{94}$$

where $\mathbf{I}$ is the unit matrix, $\mathbf{c}$ is the eigenvector of the expansion coefficients and $\mathbf{A}$, the matrix to be diagonalized, is written as a product of a radial and an angular integral:

$$A_{nn'\lambda\lambda'mm'}=R_{nn'\lambda\lambda'}\Omega_{\lambda\lambda'mm'}, \tag{95}$$

where $\Omega_{\lambda\lambda'mm'}$ is the integral over the ω angular variables

$$\Omega_{\lambda'\sigma'\lambda\sigma}=\int Y^*_{\lambda'\sigma'}(\omega)Z(\Omega)Y_{\lambda\sigma}(\omega)\,d\omega \tag{96}$$

and $R_{nn'\lambda\lambda'}$ is the radial integral

$$R_{n\lambda n'\lambda'} = 2\sqrt{\frac{(n'-\lambda'-1)!(n-\lambda-1)!}{(2n+d-3)(2n'+d-3)[(n'+\lambda'+d-3)!(n+\lambda+d-3)!]^3}} \times \int \exp(2p_0r)(2p_0r)^{\lambda+\lambda'+d-2} L_{n'-\lambda'-1}^{2\lambda'+d-2}(2p_0r) L_{n-\lambda-1}^{2\lambda+d-2}(2p_0r)\mathrm{d}(2p_0r). \tag{97}$$

The result of the integration (96) depends on the selected parametrization for the angular variables and on the particular Coulomb system under analysis. In any case, it can always be expressed in terms of vector coupling and recoupling coefficients of hyperangular momenta, because Eq. (96) is an integral over hyperspherical harmonics.[4]

The explicit expression of the radial integral $R_{n\lambda n'\lambda'}$ depends only on the number of particles (i.e., on the dimension of configuration space) and not on the nature of the particles. Furthermore, $R_{n\lambda n'\lambda'}$ can be written as a linear combination of vector coupling coefficients.[9] In the case of the three-body problem, with $d=6$, its explicit expression is:

$$R_{nn'\lambda\lambda'} = \frac{1}{(n'+\lambda+3)^{\frac{1}{2}}}\left[(n+\lambda+3)^{\frac{1}{2}}F + (n-\lambda)^{\frac{1}{2}}B\right] \tag{98}$$

with

$$F = \langle\frac{n'+\lambda'+2}{2}, \frac{-n+\lambda'+2\lambda+4}{2}, \frac{n'-n+\lambda'-\lambda}{2}, \frac{n'-n+\lambda'-\lambda}{2} | \frac{n'+\lambda+2}{2}, \frac{-n'+2\lambda'+\lambda+4}{2}\rangle,$$

$$B = \langle\frac{n'+\lambda'+3}{2}, \frac{\lambda-n'+2\lambda+3}{2}, \frac{n'-n+\lambda'-\lambda-1}{2}, \frac{n'-n+\lambda'-\lambda+1}{2} | \frac{n'+\lambda+2}{2}, \frac{2\lambda'-n'+\lambda+4}{2}\rangle,$$

where F and B are Clebsch–Gordan coefficients, so $R_{nn'\lambda\lambda'}$ depends on the involved quantum numbers only and not on the nature of the system. Even

[4]In the application to the three-body problem, the involved six-dimensional harmonics simplify, for total angular momentum equal to zero, to four-dimensional ones. The integral $\Omega_{\lambda\lambda'}$ mm'can be expressed as a sum over a $9j$ coefficient if the harmonics of Eq. (81) are employed, or a sum over $3j$ coefficients if the alternative harmonics of Eq. (85) are used.

if this result might seem surprising from a configuration space point of view, if we look at the problem from the momentum space perspective the whole matrix $A_{nn'\lambda\lambda'mm'}$ can in fact be obtained from the integration over angular variables only, parameterizing the S^6 hypersphere embedded in a seven-dimensional Euclidean space. Thus the quantum numbers n and λ label the eigenvalues of the generators of the seven-dimensional hypersphere S^6 and of a six-dimensional subspace, respectively. In this perspective, the possibility of exploiting different parameterizations of S^6, whose symmetry properties are closer to those of the system, represents an extremely attractive feature. Such a possibility was exploited[9] to eliminate the λ quantum number (i.e., the cause of poor convergence for most systems) by applying a transformation coefficient like the one in Eq. (111) in the Appendix. This resulted in a more rapid convergence, in particular for the case of the ground state energy of the H_2^+ ion.

Though it can be accelerated by an appropriate choice of the Sturmian basis set, the Coulomb Sturmians used in Ref. 9 show in general a rather slow convergence, mainly because of the large number of functions needed to correctly deal with cusp conditions at Coulomb singularities. For this reason alternative Sturmian bases (mostly in configuration space) have been constructed, often as the product of one-electron Sturmians, and applied for extensive calculations of the energy spectrum of atoms and ions[105] also in strong external fields.[106,107] Convergence of these sets is rapid, although the spectrum of the generating eigenproblem is mixed, containing, apart from discrete eigenvalues, a continuum part: therefore, there is no guarantee that every well-behaved function of the same variables obeying the same boundary conditions can be expanded in terms of this basis.[104]

Remarkable performances for atomic bound states have been reached through the introduction of *optimal* Sturmian basis functions, possessing the correct physical behavior at short and long distances from the nucleus.[12]

It is also worth mentioning that Coulomb Sturmian basis sets are being increasingly employed in nuclear physics to describe systems of interacting protons and neutrons, thanks to their distinctive property of constituting a complete, discrete set of square-integrable functions, while also possessing a realistic exponential asymptotes appropriate to the nuclear problem.[108]

4.2 Applications to molecular structure

The hyperspherical approach described in the previous subsection, where electrons and nuclei are treated on the same footing, can also be used to deal with molecular systems beyond the Born Oppenheimer approximation. If we take into account approaches using this approximation, applications of Sturmian basis sets in momentum space to the description of molecular structure date back to 1965 when Shibuya and Wulfman[84] developed a treatment for the problem of one electron in the presence of many fixed nuclei by extending Fock's momentum space method: in particular they applied this analysis to the H_2^+ molecular ion. The basic ingredient of their treatment is the explicit calculation of a matrix, usually indicated with $S(\mathbf{R})$, representing the overlap between the basis eigenfunction relative to a certain nucleus and the eigenfunction basis set relative to another nucleus placed at an arbitrary distance. One-electron energies and orbitals can be obtained, solving a secular problem involving only the $\mathbf{S}$ matrix. Further progress was made by other authors[109,110] and most important in this track is the work of Koga and co-workers.[111] The present authors[16] have extended Shibuya-Wulfman analysis using, instead of conventional spherical orbitals, parabolic (or Stark) basis sets which provide a simpler expression for the $\mathbf{S}$ matrix and a better convergence when reduced basis sets are employed.

One of the crucial points in Shibuya–Wulfman treatment is the introduction of a remarkable plane wave expansion in terms of hyperspherical harmonics and Coulomb Sturmians, extremely useful for molecular calculations:

$$\exp(i\mathbf{p}\cdot\mathbf{R}) = (2\pi)^{3/2}\sum_{nlm} u^*_{nlm}(\mathbf{R})\cdot\frac{p_0^2+p^2}{2p_0^2}\psi_{nlm}(\mathbf{p}). \tag{99}$$

For a detailed mathematical analysis of such an expansion see Ref. 32. Alternatively we can use parabolic Sturmians and the non-standard hypespherical harmonics of Eqs. (39) and (85) to have [16]:

$$\exp(i\mathbf{p}\cdot\mathbf{R}) = (2\pi)^{3/2}\sum_{n\mu m} U^*_{n\mu m}(\mathbf{R})\cdot\frac{p_0^2+p^2}{2p_0^2}\Psi_{n\mu m}(\mathbf{p}). \tag{100}$$

Extensions to the *d*-dimensional space,[33] in view of multi-electron molecular applications, have also been introduced by the present authors.

In the multi-center Coulomb problem nuclear attraction integrals between orbitals $u_{nlm}(\mathbf{r})$ centered on one nucleus at the origin and those $u_{n'l'm'}(\mathbf{r}-\mathbf{R})$ centered on another nucleus at a position $\mathbf{R}$ have to be calculated. The vector $\mathbf{r}$ specifies the position of the electron. The matrix $\mathbf{S}(\mathbf{R})$, with elements

$$S^{nlm}_{n'l'm'}(\mathbf{R}) = \frac{n}{p_0}\int u^*_{n'l'm'}(\mathbf{r}-\mathbf{R})\,\frac{1}{r}\,u_{nlm}(\mathbf{r})d\mathbf{r} \tag{101}$$

physically corresponds to the potential energy matrix between $u(\mathbf{r}-\mathbf{R})$ and $u(\mathbf{r})$. This expression provides a useful expansion of the orbital on a nucleus at a distance $\mathbf{R}$ ($u_{nlm}(\mathbf{r}-\mathbf{R})$) in terms of the orbitals of a nucleus at the origin $u_{NLM}(\mathbf{r})$:

$$u_{nlm}(\mathbf{r}-\mathbf{R}) = \sum_{NLM} u_{NLM}(\mathbf{r}) S_{nlm}^{NLM}(\mathbf{R}). \tag{102}$$

Shibuya and Wulfman elegantly solved the integral (101) by working in momentum space, exploiting the inverse transform of (78):

$$u_{nlm}(\mathbf{r}-\mathbf{R}) = (2\pi)^{-3/2} \int e^{i\mathbf{p}(\mathbf{r}-\mathbf{R})} \psi_{nlm}(\mathbf{p}) d\mathbf{p} \tag{103}$$

which allows the separation of the dependence on the nuclear coordinate $\mathbf{R}$ from that of the electron coordinate $\mathbf{r}$. In momentum space, the integral is equivalent to one over the product of three four-dimensional spherical harmonics, and a closed form expression is obtained in terms of angular momentum coupling and recoupling coefficients (3-j, 6-j, and 9-j symbols).

By the use of plane wave expansion (99) and by working in momentum space it is possible to write down the overlap integral between two polar Sturmians centered on two different nuclei a and b, separated by a distance $\mathbf{R}$ [112]:

$$\begin{aligned}
&\int u^*_{a,n',l',m'}(\mathbf{x}) u^*_{b,n,l,m}(\mathbf{x}) d\mathbf{x} \\
&= \left(\frac{2\pi}{p_0}\right)^{3/2} \sum_{NLM} \left[\frac{nn'N(2l+1)(2l'+1)}{2\pi^2}\right] \langle l', m', l, m | L, M\rangle \\
&\begin{Bmatrix} \frac{n'-1}{2} & \frac{n-1}{2} & \frac{N-1}{2} \\ \frac{n'-1}{2} & \frac{n-1}{2} & \frac{N-1}{2} \\ l' & l & L \end{Bmatrix} \times \left\{ \left[1 + \frac{(N-L)(N+L+1)}{4N(N+1)} + \frac{(N+L)(N-L-1)}{4N(N-1)}(1-\delta_{N,1})\right] u_{NLM}(\mathbf{R}) \right. \\
&- \sqrt{\frac{(N-L)(N+L+1)}{N(N+1)}} u_{N+1,LM}(\mathbf{R}) - \sqrt{\frac{(N+L)(N-L-1)}{N(N-1)}} u_{N-1,LM}(\mathbf{R}) \\
&+ \frac{1}{4(N+1)} \sqrt{\frac{(N-L)(N+L+2)[(N+1)^2-L^2]}{N(N+2)}} u_{N+2,LM}(\mathbf{R}) \\
&\left. + \frac{1}{4(N-1)} \sqrt{\frac{(N+L)(N-L-2)[(N-1)^2-L^2]}{N(N-2)}} u_{N-2,LM}(\mathbf{R})(1-\delta_{N,1})(1-\delta_{N,2}) \right\}.
\end{aligned} \tag{104}$$

Though this method seems very promising, up to date only some of the integrals involved in the calculations can be fully expressed within this approach. Alternative paths have thus been followed, either by expanding the Sturmians in the required integrals in terms of Slater type orbitals[112] or by using generalized Sturmians discussed by Avery and coworkers.[113,14,15]

5. ADDITIONAL AND CONCLUDING REMARKS

As also recently discussed elsewhere,[114] the research developed in the present paper is part of current efforts devoted to exploiting the connection between the search and use of orbitals as basis sets in applied quantum mechanics and those advances in the mathematics of special functions and orthogonal polynomials, which have been motivated by the theory of angular momentum, and can also be comprehended within the common label of "spin network" algorithms. Ref. 114 gives also an update of numerical applications beyond those reviewed in Ref. 9.

In the course of this paper, the ingredients that have been presented (the special functions and the orthogonal polynomials) mostly belong to the scheme in Figure 5.1. Reference 24 provides additional relevant details, specific attention being paid to those aspects related to functions appearing at the upper floors of the ladder, i.e., above the Jacobi polynomials. In a sense, ascending this *Jacobi ladder* we encounter functions defined on discrete grids rather than on variables spanning continuous ranges. Their use is basic to the hyperquantization algorithm,[102] so far exploited in molecular dynamics but of perspective relevance in quantum chemistry. In discrete approaches, the Sturm-Liouville theory for differential equations has its counterpart in the theory of Jacobi matrices. The recursion relationships that define orthogonal polynomials play the role of difference equations. We have encountered such relationships in the previous sections. In some cases, e.g., spheroidal and sphero-elliptic Sturmians, they cannot be given explicit solutions and their properties and uses are object of current investigations.

ACKNOWLEDGMENTS

We thank John Avery and James Avery for useful discussions and suggestions. We are grateful to Maria Belen Ruiz for discussions and for the use of Figure 5.5.

APPENDIX A. CALCULATION OF SUPERPOSITION COEFFICIENTS BETWEEN HYPERSPHERICAL HARMONICS

The transformation coefficients between alternative d-dimensional Coulomb Sturmians can be calculated as superposition integrals between sets of hyperspherical harmonics pertaining to alternative parameterizations of the $(d+1)$-dimensional sphere S^d.

Let us consider two hyperspherical harmonics on the S^d sphere[44] (i.e., eigenfunctions of the $(d+1)$-dimensional Beltrami–Laplace operator) $Y_{t,t_{12},t_1,t_2,t_3}(\Omega)$ and $Y_{t,t_{23},t_1,t_2,t_3}(\Omega')$, where Ω and Ω' collectively indicate two sets of d angles defining two different parameterizations of the S^d sphere, that is $O(d+1) \supset O(d_3) \times O(d_1+d_2)$ and $O(d) \supset O(d_1) \times O(d_2+d_3)$ (with $d_1+d_2+d_3 = d+1$). In general $Y_{t,t_{12},t_1,t_2,t_3}(\Omega)$ can be written as a product of Jacobi polynomials (see Fig. 5.1) each of them function of a different angle χ_i belonging to the set Ω and defining rotations on a subspace of S^d of dimension d_i. In special cases, Jacobi polynomials simplify to Legendre, or Gegenbauer polynomials. A full treatment can be found in Refs. 44, 22, 23.

The superposition between the two harmonics is essentially a $(d-1)$ -dimensional integral over four Jacobi polynomials[87,22] and, apart from a phase factor, can be identified with a Racah orthonormalized coefficient, related (see below) to Racah polynomials (at the upper level of the Askey scheme of Figure 5.1):

$$
\begin{aligned}
T_{t_{12},t_{23}} &= \int Y_{t,t_{12},t_1,t_2,t_3}(\Omega)\, Y_{t,t_{23},t_1,t_2,t_3}(\Omega')\, d\Omega \\
&= (-)^{\frac{t_{12}-t_1-t_2}{2}} W_{\frac{t_{12}-t_1-t_2}{2},\frac{t_{23}-2}{2}+\frac{d_2+d_3}{4}} \\
&\quad \times \left(\frac{t_2+t_3-2}{2}+\frac{d_2+d_3}{4}, \frac{t-t_1}{2}+\frac{d_2+d_3}{4}, t_1+\frac{d_1-2}{2}, t_2+\frac{d_2-2}{2}\right).
\end{aligned} \tag{105}
$$

Racah orthonormalized coefficients $W_{n,s}(a,b,\alpha,\beta)$ are defined starting from Racah polynomials $u_n^{(\alpha,\beta)}(s,a,b)$[22,23] as

$$
W_{n,s}(a,b,\alpha,\beta) = \frac{\sqrt{\rho(s)(2s+1)}}{d_n} u_n^{(\alpha,\beta)}(s,a,b), \tag{106}
$$

where n (positive integer, $0 \le n \le b-a-1$) is the degree of the polynomial and $s(s+1)$ is the discrete variable. The grid is therefore quadratic in the parameters defined in the interval:

$$a < s \leq b - 1, \quad \text{with} \quad -\frac{1}{2} \leq a \leq b - 1. \tag{107}$$

The parameters α and β are restricted to the values:

$$\alpha > -1 \quad \text{and} \quad -1 < \beta < 2a + 1. \tag{108}$$

Racah orthonormalized coefficients can also be written as an extended or generalized $6j$ coefficient[23] whose entries can be integers, half integers or proportional to $\frac{1}{4}$ depending on the values of d_1, d_2, d_3. From the identifications:

$$\begin{array}{llll} n = \frac{t_{12}-t_1-t_2}{2} & s = \frac{t_{23}}{2} + \frac{d_2+d_3}{4} - 1 & a = \frac{t_2+t_3}{2} + \frac{d_2+d_3}{4} - 1 \\ b = \frac{t-t_1}{2} + \frac{d_2+d_3}{4} & \alpha = t_1 + \frac{d_1}{2} - 1 & \beta = t_2 + \frac{d_2}{2} - 1 \end{array} \tag{109}$$

and from the range relations (107) and (108): the well-known rules for hyperangular momentum quantum numbers can be recovered [22,23]:

$$t \geq t_{12} + t_3, \quad t \geq t_{23} + t_1, \quad t_{12} \geq t_1 + t_2, \quad t_{23} \geq t_2 + t_3. \tag{110}$$

When d_1, d_2 and d_3 are all even, the coefficient in Eq. (105) reduces to an ordinary $6j$ symbol with integer and/or half integer entries only. An alternative expression for the superposition coefficient in Eq. (105) can be obtained making use of Regge symmetry[69] for $6j$ coefficients by taking care that the selection rules (110) for hyperspherical harmonics still hold valid. In this case in each entry – i.e., every (hyper-)angular momentum- it appears only a specific quantum number t_i and the corresponding numbers counting the dimensions of the subgroup reduction chain:

$$\begin{aligned} T_{t_{12},t_{23}} = \; & (-)^{\frac{t+t_{12}+t_{23}-t_2+d_1+d_2+d_3}{2}} \left[(t_{12} + \frac{d_1+d_2}{2} - 1)(t_{23} + \frac{d_2+d_3}{2} - 1) \right]^{1/2} \\ & \times \begin{Bmatrix} -\frac{t_1}{2} - \frac{d_1}{4} & \frac{t_2}{2} + \frac{d_2}{4} & \frac{t_{12}}{2} + \frac{d_1+d_2}{4} - 1 \\ -\frac{t_3}{2} - \frac{d_3}{4} & \frac{t}{2} + \frac{d_1+d_2+d_3}{4} - 1 & \frac{t_{23}}{2} + \frac{d_2+d_3}{4} - 1 \end{Bmatrix}. \end{aligned} \tag{111}$$

For spaces of low dimensions, hyperspherical harmonics (and the expression of the connecting $6j$ coefficient) may simplify: when one of the d_i is one, in (105) or (111) t_i is conveniently taken to be 0 or 1 according to parity (t_i must be even when $t_{ij} - t_j$ is even and odd when $t_{ij} - t_j$ is odd, for the Jacobi polynomials involved to be different from zero,[22]) thus the corresponding entry in coefficient (111) reduces to a fixed value $-\frac{1}{4}$ or $-\frac{3}{4}$, respectively.

For the d-dimensional hydrogen atom, the overlaps between polar and parabolic wavefunctions are generalized $6j$ symbols, which simplify either to a $3j$ coefficient or to a generalized $3j$ coefficient,[50] depending on the even or odd value of d. This simplification is connected to Racah polynomials properties (see Eq. (3.7.23), Ref. 22): the Racah orthonormal coefficients involved, $W_{n,s}(\pm\frac{1}{2}, b, \alpha, \pm\frac{1}{2})$ can be expressed in terms of Hahn coefficients $Q^{\alpha,\alpha,2b+1}_{2n+\frac{1}{2}\pm\frac{1}{2},b+s}$ $(Q^{m,m,\nu-m}_{l+\frac{1}{2}\pm\frac{1}{2},\frac{\nu-m+\mu}{2}})$, i.e., $3j$ and generalized $3j$ coefficients,[50] as shown in Eq. (19).

In the three-dimensional case, momentum space Coulomb Sturmians corresponding to polar and parabolic coordinates are essentially the hyperspherical harmonics $Y_{n-1lm}(\Omega)$ of Eq. (81) and $Y_{n-1\mu m}(\Omega')$ of Eq. (85). The superposition between the two harmonics[16] is identical (possibly within a normalization dependent phase difference) to the overlap between the two Sturmian basis functions in configuration space (Eq. (32) and Eq. (39)) and can be expressed as a generalized $6j$ coefficient by appropriately substituting the quantum numbers $n-1, l, \mu, m$ to t, t_{12}, t_{23}, t_1 in Eq. (111) $(d_1 = 2,\ d_2 = 1,\ d_3 = 1)$ -thus giving $n-1 \geq l,\ l \geq m$, and $n-1 \geq \mu + m$, and introducing a normalization factor $\frac{1}{\sqrt{2}}$ which takes into account the fact that the harmonics $Y_{n-1lm}(\Omega)$, since d_3=1 can now be written in terms of Gegenbauer polynomials, instead of Jacobi polynomials [22,44]:

$$\begin{aligned}
T_{l,\mu} &= \int Y_{n-1lm}(\Omega)Y_{n-1\mu m}(\Omega')\mathrm{d}\Omega \\
&= (-1)^{\frac{n-2-m-\mu}{2}}\frac{1}{\sqrt{2}}\left[(l+\frac{1}{2})\mu\right]^{1/2}\begin{Bmatrix} -\frac{m+1}{2} & -\frac{1}{4}(-\frac{3}{4}) & \frac{l-1}{2} \\ -\frac{1}{4}(-\frac{3}{4}) & \frac{n-1}{2} & \frac{\mu-1}{2} \end{Bmatrix} \\
&= (-)^{\frac{n-2+m-\mu}{2}}(2l+1)^{1/2}\begin{pmatrix} \frac{n-2}{2} & \frac{n-2}{2} & l \\ \frac{m-\mu}{2} & \frac{m+\mu}{2} & -m \end{pmatrix}.
\end{aligned} \tag{112}$$

Equation (112) shows that, for particular values of its arguments, the $6j$ generalized coefficient reduces to the well-known $3j$ symbol representing the superposition between the hydrogen orbitals in polar and parabolic coordinates.

REFERENCES

1. Bouferguene, A.; Fares, M.; Hoggan, P. E. STOP: A Slater-type Orbital Package for Molecular Electronic Structure Determination. *Int. J. Quantum Chem.* **1996,** *57,* 801–810.
2. Safouhi, H.; Hoggan, P. E. New Methods for Accelerating the Convergence of Molecular Electronic Integrals Over Exponential Type Orbitals. *Mol. Phys.* **2003,** *101,* 19.
3. Hoggan, P. E. General Two-electron Exponential Type Orbital Integrals in Polyatomics Without Orbital Translations. *Int. J. Quantum Chem.* **2009,** *109,* 2926.
4. Fernández Rico, J.; López, R.; Ema, I.; Ramírez, G. Reference Program for Molecular Calculations with Slater-type Orbitals. *J. Comput. Chem.* **1998,** *19,* 1284.
5. Fernández Rico, J.; López, R.; Ema, I.; Ramírez, G. Electrostatic Potentials and Fields from Density Expansions of Deformed Atoms in Molecules. *J. Comput. Chem.* **2004,** *25,* 1347.
6. Ruiz, M. B.; Peuker, K. Analytical Expressions of Exchange and Three-center Nuclear Attraction Integrals Over 1s and 2s Slater Orbitals with Different Exponents. *Chem. Phys. Lett.* **2005,** *412,* 244.
7. Shull, H.; Löwdin, P.-O. Superposition of Configurations and Natural Spin Orbitals. Applications to the He Problem. *J. Chem. Phys.* **1959,** *30,* 617.
8. Avery, J. *Hyperspherical Harmonics and Generalized Sturmians*; Kluwer Academic Publishers: Dordrecht, 2000.
9. Aquilanti, V.; Cavalli, S.; Coletti, C.; Di Domenico, D.; Grossi, G. Hyperspherical Harmonics as Sturmian Orbitals in Momentum Space: A Systematic Approach to the Few-body Coulomb Problem. *Int. Rev. Phys. Chem.* **2001,** *20,* 673.
10. Aquilanti, V.; Caligiana, A.; Cavalli, S.; Coletti, C. Hydrogenic Orbitals in Momentum Space and Hyperspherical Harmonics. Elliptic Sturmian Basis Sets. *Int. J. Quantum Chem.* **2003,** *92,* 212.
11. Randazzo, J. M.; Frappicini, A. L.; Colavecchia, F. D.; Gasaneo, G. Discrete Sets of Sturmian Functions Applied to Two-electron Atoms. *Phys. Rev. A* **2009,** *81,* 022507.
12. Randazzo, J. M.; Ancarani, L. U.; Gasaneo, G.; Frappicini, A. L.; Colavecchia, F. D. Optimal Sturmian Basis Functions for Atomic three-body Systems. *Phys. Rev. A* **2010,** *81,* 042520.
13. Mitnik, D. M.; Colavecchia, F. D.; Gasaneo, G.; Randazzo, J. M. Computational Methods for Generalized Sturmans Basis. *Comput. Phys. Commun.* **2011,** *182,* 1145.
14. Avery, J.; Avery, J. Generalized Sturmian Solutions for Many-particle Schrödinger Equations. *J. Phys. Chem. A* **2004,** *108,* 8848.
15. Avery, J.; Avery, J. Coulomb Sturmians as a Basis for Molecular Calculations. *Mol. Phys.* **2012,** *110,* 1593.
16. Aquilanti, V.; Cavalli, S.; Coletti, C.; Grossi, G. Alternative Sturmian Bases and Momentum Space Orbitals: An Application to the Hydrogen Molecular Ion. *Chem. Phys.* **1996,** *209,* 405.
17. Guseinov, I. I.; Sahin, E. Evaluation of One-electron Molecular Integrals Over Complete Orthonormal Sets of Ψ^{α}-ETO Using Auxiliary Functions. *Int. J. Quantum Chem.* **2010,** *110,* 1803–1808.
18. Avery, J. Many-particle Sturmians. *J. Math. Chem.* **1997,** *21,* 285–304.
19. Hoggan, P. E.; Bouferguene, A. In *Advances in The Theory of Quantum Systems in Chemistry and Physics, Why Specific ETOs are Advantageous for NMR and Molecular Interactions*; Hoggan, P. E., Brändas, E. J., Maruani, J., Piecuch, P. Delgado-Barrio, G., Eds. Advances in The Theory of Quantum Systems in Chemistry and Physics, Why Specific ETOs are Advantageous for NMR and Molecular Interactions; Springer, pp 83–102.
20. Guseinov, I.I. New Complete Orthonormal Sets of Exponential-type Orbitals and Their Application to Translation of Slater Orbitals. *Int. J. Quantum Chem.* **2002,** *90,* 114–118.

21. Calderini, D.; Cavalli, S.; Coletti, C.; Grossi, G.; Aquilanti, V. Hydrogenoid Orbitals Revisited: From Slater Orbitals to Coulomb Sturmians. *J. Chem. Sci.* **2012,** *124*, 187.
22. Nikiforov, A.; Suslov, S.; Uvarov, V. *Classical Orthogonal Polynomials of a Discrete Variable*; Springer-Verlag: Berlin, 1991.
23. Aquilanti, V.; Cavalli, S.; Coletti, C. Angular and Hyperangular Momentum Recoupling, Harmonic Superposition and Racah Polynomials: A Recursive Algorithm. *Chem. Phys. Lett.* **2001,** *344*, 587.
24. Ragni, M.; Bitencourt, A. C. P.; Da S Ferreira, C.; Aquilanti, V.; Anderson, R. W.; Littlejohn, R. G. Exact Computation and Asymptotic Approximations of 6j Symbols: Illustration of Their Semiclassical Limits. *Int. J. Quantum Chem.* **2010,** *110*, 731.
25. Koekoek, R.; Swarttouw, R. The Askey-scheme of Hypergeometric Orthogonal Polynomials and its q-Analogue, TU Delft, The Netherlands (1998). Available via anonymous <ftpsite:ftp.twi.tudelft.nl,directory:/pub/publications/tech-reports>.
26. Askey, R. Duality for Classical Orthogonal Polynomials. *J. Comput. Appl. Math.* **2005,** *178*, 37–43.
27. Thakkar, A. J. The Momentum Density Perspective of the Electronic Structure of Atoms and Molecules. *Adv. Chem. Phys.* **2004,** *128*, 303.
28. Öztekin, E.; Özcan, S. Overlap Integrals Between Irregular Solid Harmonics and STOs Via the Fourier Transform Methods. *J. Math. Chem.* **2007,** *42*, 337.Öztekin, E.; Özcan, S. Analytical Evaluation for Two-center Nuclear Attraction Integrals Over Slater Type Orbitals by Using Fourier Transform Method. *J. Math. Chem.* **2009,** *45*, 1153–1164.
29. Yavuz, M.; Yükçu, N.; Öztekin, E.; Yilmaz, H.; Döndür, S. On the Evaluation Overlap Integrals with the Same and Different Screening Parameters Over Slater Type Orbitals via the Fourier-transform Method. *Commun. Theor. Phys.* **2005,** *43*, 151.
30. Rotenberg, M. Application of Sturmian Functions to the Schroedinger Three-body Problem: Elastic e^+-H Scattering. *Ann. Phys. (N.Y.)* **1962,** *19*, 262–278.
31. Kalnins, E. G.; Miller, W., Jr. Structure Theory for Extended Kepler–Coulomb 3D Classical Superintegrable Systems in Symmetry, Integrability and Geometry: Methods and Applications. *SIGMA* **2012**, *8*, 034.
32. Weniger, E. J. Weakly Convergent Expansions of a Plane Wave and Their Use in Fourier Integrals. *J. Math. Phys.* **1985,** *26*, 276.
33. Aquilanti, V.; Cavalli, S.; Coletti, C. The d-Dimensional Hydrogen Atom: Hyperspherical Harmonics as Momentum Space Orbitals and Alternative Sturmian Basis Sets. *Chem. Phys.* **1997,** *214*, 1–13.
34. Van Gorder, R. A. Wave Functions and Energy Spectra for the Hydrogenic Atom in $R^3 \times M$. *J. Math. Phys.* **2010,** *51*, 122104.
35. Szmytkowski, R. Alternative Approach to the Solution of the Momentum-space Schrödinger Equation for Bound States of the N-dimensional Coulomb Problem. *Ann. Phys. (Berlin)* **2012,** *524*, 345.
36. Dehesa, J. S.; López-Rosa, S.; Martínez-Finkelshtein, A.; Yañez, R. J. Information Theory of D-dimensional Hydrogenic Systems: Application to Circular and Rydberg States. *Int. J. Quantum Chem.* **2010,** *110*, 1529.
37. López-Rosa, S.; Manzano, D.; Dehesa, J. S. Complexity of d-dimensional Hydrogenic Systems in Position and Momentum Spaces. *Physica A* **2009,** *388*, 3273.
38. Slen, L.; Nepstad, R.; Hansen, J. P.; Madsen, L. B. The N-dimensional Coulomb Problem: Stark Effect in Hyperparabolic and Hyperspherical Coordinates. *J. Phys. A: Math. Theor.* **2007,** *40*, 1097–1104.
39. Burgbacher, F.; Lämmerzhal, C.; Macias, A. Is there a Stable Hydrogen Atom in Higher Dimensions? *J. Math. Phys.* **1999,** *40*, 625.
40. Herschbach, D. R., Avery, J., Goscinski, O., Eds.; *Dimensional Scaling in Chemical Physics*; Kluwer: Dordrecht, 1993.

41. Dong, S. H.; Ma, Z. Q. Nonrelativistic Levinsons Theorem in D Dimensions. *Phys. Rev. A* **2002,** *65*, 42717.
42. Itzykson, C.; Zuber, J. B. *Quantum Field Theory*; Dover: New York, 2006.
43. Amelino-Camelia, G., Kowalski-Glikman, J., Eds.; *Planck Scale Effects in Astrophysics and Cosmology*; Springer: Berlin, 2005.
44. Aquilanti, V.; Cavalli, S.; Grossi, G. Hyperspherical Coordinates for Molecular Dynamics by the Method of Trees and the Mapping of Potential Energy Surfaces for Triatomic Systems. *J. Chem. Phys.* **1986,** *85*, 1362.
45. Avery, J. *Hyperspherical Harmonics, Applications in Quantum Theory*; Kluwer Academic Publishers: Dordrecht, 1989.
46. Morse, P. M.; Feshbach, H. *Methods of Theoretical Physics*; McGraw-Hill: New York, 1953.
47. Suslov, S. K. The Hahn Polynomials in the Coulomb Problem. *Sov. J. Nucl. Phys.* **1984,** *15*, 79–82.
48. Gradshteyn, I. S.; Ryzhik, I. M. *Tables of Integrals, Series and Products*; Academic Press: New York, 1980.
49. Karlin, S.; McGregor, J. L. The Hahn Polynomials, Formulas and Applications. *Scr. Math.* **1961,** *26*, 33.
50. Aquilanti, V.; Cavalli, S.; De Fazio, D. Angular and Hyperangular Momentum Coupling Coefficients as Hahn Polynomials. *J. Phys. Chem.* **1995,** *99*, 15694.
51. Amore, P.; Fernandez, F. M. Bound States for the Quantum Dipole Moment in Two Dimensions. *J. Phys. B: At. Mol. Opt. Phys.* **2012,** *45*, 235004.
52. Yang, X. L.; Guo, S. H.; Chan, F. T.; Wong, K. W.; Ching, W. Y. Analytic Solution of a Two-dimensional Hydrogen Atom. I. Nonrelativistic Theory. *Phys. Rev. A* **1991,** *43*, 1186.
53. Li, S. S.; Xia, J. B. Electronic States of a Hydrogenic Donor Impurity in Semiconductor Nano-structures. *Phys. Lett. A* **2007,** *366*, 120.
54. Nieto, M. M. Electrons Above a Helium Surface and the One-dimensional Rydberg Atom. *Phys. Rev. A* **2000,** *61*, 034901.
55. Dykman, M. I.; Platzman, P. M.; Seddigard, P. Qubits with Electrons on Liquid Helium. *Phys. Rev. B* **2003,** *67*, 155402.
56. Zaslow, B.; Zandler, M. E. Two-Dimensional Analog to the Hydrogen Atom. *Am. J. Phys.* **1967,** *35*, 1118.
57. Cisneros, A.; McIntosh, H. V. Symmetry of the Two-dimensional Hydrogen Atom. *J. Math. Phys.* **1969,** *10*, 277.
58. Englefield, M. J. *Group Theory and The Coulomb Problem*; Wiley-Interscience: New York, 1972.
59. Varshalovich, D. A.; Moskalev, A. N.; Khersonskii, V. K. *Quantum Theory of Angular Momentum*; World Scientific: Singapore, 1988.
60. Patera, J.; Winternitz, J. A New Basis for the Representations of the Rotation Group. Lamé and Heun Polynomials. *J. Math. Phys.* **1973,** *14*, 1130.
61. Kalnins, E. G.; Miller, W.; Winternitz, P. The Group *O*(4), Separation of Variables and the Hydrogen Atom. *S.I.A.M J. Appl. Math.* **1976,** *30*, 630.
62. Miller, W., Jr.; *Symmetry and Separation of Variables*; Addison-Wesley Publishing Company: Reading, MA, 1977.
63. Judd, B. R. *Angular Momentum Theory for Diatomic Molecules*; Academic Press: New York, 1975.
64. Landau, L. D.; Lifshitz, E. M. *Quantum Mechanics. Nonrelativistic Theory*; Addison-Wesley: Reading, MA, 1958.
65. Park, D. Relation Between the Parabolic and Spherical Eigenfunctions of hydrogen. *Z. Phys.* **1960,** *159*, 155.
66. Hughes, J. W. B. Stark States and O(4) Symmetry of Hydrogenic Atoms. *Proc. Phys. Soc.* **1967,** *91*, 810.

67. Tarter, C. B. Coefficients Connecting the Stark and Field-free Wavefunctions for Hydrogen. *J. Math. Phys.* **1970,** *11*, 3192.
68. Aquilanti, V.; Cavalli, S.; Coletti, C. Hyperspherical Symmetry of Hydrogenic Orbitals and Recoupling Coefficients among Alternative Bases. *Phys. Rev. Lett.* **1998,** *80*, 3209.
69. Regge, T. Symmetry Properties of Clebsch–Gordan Coefficients. *Nuovo Cimento* **1958,** *10*, 544.
70. Coulson, C. A.; Joseph, A. Spheroidal Wave Functions for the Hydrogen Atom. *Proc. Phys. Soc. London* **1967,** *90*, 887.
71. Coulson, C. A.; Robinson, P. D. Wave Functions for the Hydrogen Atom in Spheroidal Coordinates I: The Derivation and Properties of the Functions. *Proc. Phys. Soc. London* **1958,** *71*, 815.
72. Auber, M.; Bessis, N.; Bessis, G. Prolate-spheroidal Orbitals for Homonuclear and Heteronuclear Diatomic Molecules. I. Basic Procedure. *Phys. Rev. A* **1974,** *10*, 51.
73. Aquilanti, V.; Caligiana, A.; Cavalli, S. Hydrogenic Elliptic Orbitals, Coulomb Sturmian Sets, and Recoupling Coefficients Among Alternative Bases. *Int. J. Quantum Chem.* **2003,** *92*, 99.
74. Sung, S. M.; Herschbach, D. R. Hidden Symmetry and Explicit Spheroidal Eigenfunctions of the Hydrogen Atom. *J. Chem. Phys.* **1991,** *95*, 7437.
75. Abramowitz, M.; Stegun, I. A. *Handbook of Mathematical Function*; Dover: New York, 1964.
76. Hoggan, P. E.; Ruiz, M. B.; Özdoğan, T. Molecular Integrals over Slater-type Orbitals. From Pioneers to Recent Progress. In *Quantum Frontiers of Atoms and Molecules*; Putz, M.V., Ed.;; Nova Science Publishers, pp 63–90.
77. Anderson, R. W.; Aquilanti, V.; da Silva Ferreira, C. Exact Computation and Large Angular Momentum Asymptotics of 3nj Symbols: Semiclassical Disentangling of Spin Networks. *J. Chem. Phys.* **2008,** *129*, 161101.
78. Rasch, J.; Yu, A. C. H. Efficient Storage Scheme for Precalculated Wigner 3j, 6j and Gaunt Coefficients. *SIAM J. Sci. Comput.* **2003,** *25*, 1416.
79. Pinchon, D.; Hoggan, P. E. New Index Functions for Storing Gaunt Coefficients. *Int. J. Quantum Chem.* **2007,** *107*, 2186–2196.
80. Podolsky, B.; Pauling, L. The Momentum Distribution in Hydrogen-like Atoms. *Phys. Rev.* **1929,** *34*, 109.
81. Fock, V. Zur Theorie des Wasserstoffatoms. *Z. Phys.* **1935,** *98*, 145.
82. Alliluev, S. P. On the Relation Between Accidental Degeneracy and Hidden Symmetry of a System. *Sov. Phys. JETP* **1958,** *6*, 156–159.
83. Biedernharn, L. C.; Louck, J. D. *Angular Momentum in Quantum Physics. Theory and Applications*; Addison-Wesley Publishing Co.: Reading, MA, 1981.
84. Shibuya, I.; Wulfman, C. E. Molecular Orbitals inl Momentum Space. *Proc. Roy. Soc. A* **1965,** *286*, 376.
85. Brink, D. M.; Satchler, G. R. *Angular Momentum*; Clarendon: Oxford, 1968.
86. Bander, M.; Itzykson, C. Group Theory and the Hydrogen Atom (I). *Rev. Mod. Phys.* **1996,** *38*, 330.
87. Kil'dyushov, M. S. Hyperspherical Functions of Tree Type in the N-body Problem. *Sov. J. Nucl. Phys.* **1972,** *15*, 113.
88. Aquilanti, V.; Coletti, C. 3*nj*-Symbols and Harmonic Superposition Coefficients: An Icosahedral Abacus. *Chem. Phys. Lett.* **2001,** *344*, 601.
89. Valent, G. The Hydrogen Atom in Electric and Magnetic Fields: Pauli's 1926 Article. *Am. J. Phys.* **2003,** *71*, 171.
90. Coletti, C. Struttura Atomica e Molecolare come Rottura della Simmetria Ipersferica, Ph.D. Thesis, Universitá di Perugia, 1997.
91. Labarthe, J. J. A Basis for the Study of the Quadratic Zeeman Effect in Atomic Hydrogen. *J. Phys. B* **1981,** *14*, L-467.
92. Herrick, D. R. Symmetry of the Quadratic Zeeman Effect for Hydrogen. *Phys. Rev. A* **1982,** *26*, 323.

93. Delande, D.; Gay, J. C. Scars of Symmetries in Quantum Chaos. *Phys. Rev. Lett.* **1987,** *59*, 1809.
94. Delande, D.; Gay, J. C. Quantum Chaos and Statistical Properties of Energy Levels: Numerical Study of the Hydrogen Atom in a Magnetic Field. *Phys. Rev. Lett.* **1986,** *57*, 2006.
95. Delande, D.; Gay, J.C. A New Method for Producing Circular Rydberg States. *Europhys. Lett.* **1988,** *5*, 303.
96. Penant, F.; Delande, D.; Gay, J. C. Rydberg States of Rubidium in Crossed Electric and Magnetic Fields. *Phys. Rev. A* **1988**, *37*, 4707.
97. Macek, J. Properties of Autoionizing States of He. *J. Phys. B: At. Mol. Phys.* **1968,** *1*, 831.
98. Ballot, L.; Fabre de la Ripelle, M. Application of the Hyperspherical Formalism to the Trinucleon Bound State Problems. *Ann. Phys.* **1980,** *127*, 62.
99. Smith, F. T. Generalized Angular Momentum in Many-body Collisions. *Phys. Rev.* **1960,** *120*, 1058.
100. Smith, F. T. A Symmetric Representation for Three-Body Problems. I. Motion in a Plane. *J. Math. Phys.* **1962,** *3*, 735.
101. Aquilanti, V.; Cavalli, S.; Coletti, C.; De Fazio, D.; Grossi, G. Hyperangular Momentum: Applications to Atomic and Molecular Science. In *New Methods in Quantum Theory*, ; Tsipis, C. A., Popov, V. S., Herschbach, D. R., Eds.; Kluwer, pp 233–250.
102. Aquilanti, V.; Cavalli, S.; De Fazio, D. Hyperquantization Algorithm. II. Implementation for the F + H_2 Reaction Dynamics Including Open-Shell and Spin-Orbit Interactions. *J. Chem. Phys* **1998**, *109*, 3805.
103. Klar, H.; Klar, M. An Accurate Treatment of Two-electron Systems Using Hyperspherical Coordinates. *J. Phys. B* **1980,** *13*, 1057.
104. Szmytkowski, R. The Dirac–Coulomb Sturmian Functions in the $Z = 0$ Limit: Properties and Applications to Series Expansions of the Dirac Green Function and the Dirac Plane Wave. *J. Phys. A* **2000,** *33*, 427.
105. Avery, J.; Avery, J. The Generalized Sturmian Method for Calculating Spectra of Atoms and Ions. *J. Math. Chem.* **2003,** *33*, 145.
106. Avery, J.; Coletti, C. Generalized Sturmians Applied to Atoms in Strong External Fields. *J. Math. Chem.* **2000,** *27*, 43.
107. Avery, J.; Coletti, C. Many-electron Sturmians Applied to Atom and Ions in Strong External Fields. In *New Trends in Quantum Systems in Chemistry and Physics*; Maruani, J., Minot, C., McWeeny, R., Smeyers, Y. G., Wilson S, S., Eds.; Kluwer: Dordrecht, 2001; Vol. I, pp 77–93.
108. Caprio, M. A.; Maris, P.; Vary, J. P. Coulomb-Sturmian Basis for the Nuclear Many-body Problem. *Phys. Rev. C* **2012,** *86*, 034312.
109. Novosadov, B. K. Hydrogen-like Atomic Orbitals: Addition and Expansion Theorems, Integrals. *Int. J. Quantum Chem.* **1983,** *24*, 1.
110. Monkhorst, H.; Jeziorski, B. No Linear Dependence oF Many-center Integral Problems in Momentum Space Quantum Chemistry. *J. Chem. Phys.* **1979,** *71*, 5268.
111. Koga, T. Direct Solution of the $H(1s)H^+$ Long-range Interaction Problem in Momentum Space. *J. Chem. Phys.* **1985,** *82*, 2022. Koga, T.; Matsuhashi, T. One-electron Diatomics in Momentum Space. V. Nonvariational LCAO Approach. *J. Chem. Phys.* **1988,** *89*, 983.
112. Aquilanti, V.; Caligiana, A. *J. Mol. Struct. (Theochem)* **2004,** *709*, 15.
113. Avery, J. Many-center Coulomb Sturmians and ShibuyaWulfman Integrals. *Int. J. Quantum Chem.* **2004,** *100*, 121.
114. Calderini, D.; Coletti, C.; Grossi, G.; Aquilanti, V. Continuous Discrete Algorithms in Quantum Chemistry: Polynomial Sets, Spin Networks and Sturmian Orbitals. In *Lecture Notes in Computer Science;* Murgante, B., Ed.; Springer-Verlag, pp 32–45.

CHAPTER SIX

Fast Electron Repulsion Integrals for Molecular Coulomb Sturmians

J.E. Avery
Niels Bohr Institute, University of Copenhagen, Copenhagen, Denmark

Contents

Abstract

A new method is presented for calculating interelectron repulsion integrals for molecular Coulomb Sturmian basis sets. This makes use of an expansion of densities in terms of 2*k*-Sturmians, and the interelectron repulsion integrals are then calculated by a method based on the theory of hyperspherical harmonics.

A rudimentary software library has been implemented and preliminary benchmarks indicate very good performance: On average 40 ns, or approximately 80 clock cycles, per electron repulsion integral. This makes molecular Coulomb Sturmians competitive

Advances in Quantum Chemistry, Volume 67
ISSN 0065-3276, http://dx.doi.org/10.1016/B978-0-12-411544-6.00006-6

with Gaussian type orbitals in terms of speed, and is three to four orders of magnitude faster than methods based on expanding the exponential type orbitals in Gaussians. A full software library will be made available during autumn 2013.

1. INTRODUCTION

Historically, when quantum theorists approached the problem of describing electrons in atoms and molecules, it was thought that hydrogen-like orbitals could be used as basis functions. However, it was soon realized that these functions are not complete without including the continuum. In order to overcome this problem, Shull and Löwdin[21] introduced orbitals which came to be called *Coulomb Sturmians*. These are solutions to a one-electron wave equation identical with the hydrogenlike wave equation, except that Z/n is replaced everywhere by a constant $k > 0$. Coulomb Sturmians have the desired completeness and rapid convergence properties and they are now widely used in atomic physics and scattering calculations.

However, the mathematics required for calculation of electron repulsion integrals (ERI) proved to be very difficult, not only for Coulomb Sturmians but for all exponential type orbitals. Pioneering work to overcome these difficulties was done by Harris, Michels, Steinborn, Weniger, Weatherford, et al.[13,22,24,25] The mathematical obstacles to evaluating multicenter ERIs led S.F. Boys and Roy McWeeny in 1950[5] to introduce Cartesian Gaussian basis functions of the form $x^{n_x}y^{n_y}z^{n_z}e^{-\alpha r^2}$. Because of the great ease with which many-center ERIs can be evaluated in a Gaussian basis, exponential type orbitals were virtually abandoned over the past decades and main-stream quantum chemistry now relies almost exclusively on Gaussians, despite their serious drawbacks.

While exponential type orbitals (ETOs) have the correct behavior for synthesizing electronic wave functions in a screened Coulomb field, both at long and short ranges, Gaussians cannot reproduce the sharp features of wave functions and densities close to the nucleus and fall off too fast at long ranges, no matter how many are used. ETOs provide much more compact and accurate descriptions of wave functions.

Recent years have seen great progress in tackling the difficult mathematics involved.[11,12,14,15,19] However, evaluating multicenter electron repulsion integrals for ETOs still remains the bottleneck for quantum chemical calculations using ETOs.

In this paper, we focus on a particular type of ETO, the Coulomb Sturmians. While they are less general than the Slater type orbitals, they have a number of mathematical properties that make their use particularly attractive: They provide accurate, compact representation of electrons in screened Coulomb fields; they form a complete basis; their Fourier transform can be expressed simply in terms of hyperspherical harmonics; near-optimal exponents are determined automatically by solving the Sturmian secular equation; and all required integrals can be calculated independently of the scaling factor k. For these reasons, Coulomb Sturmians have found wide applications in atomic physics, scattering, nuclear magnetic resonance, and other applications where multicenter interelectron repulsion is not required.[6,10,16,20]

I show that the mathematical properties of Coulomb Sturmians make it possible to simplify electron repulsion integral evaluation greatly and confine all the heavy computations to a precomputation step. A special feature of this calculation is that two-center densities are not expanded about an intermediate center but about both of the two atoms involved. This leads to very accurate representations of two-center densities. The method is implemented in a software library, `libsturm2k`, which will be made publically available on

`http://sturmian.kvante.org/coulomb/`

during autumn 2013. The software will be described in a separate publication.

An extension to the method is in preparation for evaluating electron repulsion integrals for the more general Slater type orbitals. This will allow STO-based software to directly use our library to achieve substantial speedups. We are currently investigating the possibility of applying a similar method to certain classes of numerical orbitals, for example the type of generalized Sturmian functions described in Ref. 17.

2. THEORY

2.1 Coulomb Sturmians

Sturmians are isoenergetic solutions to a wave equation with a weighted potential. For Coulomb Sturmians, this equation is:

$$\left[-\frac{1}{2}\nabla^2 - \beta_n\frac{Z}{r} - E\right]\chi_{nlm}(\mathbf{x}) = 0. \tag{1}$$

The weighting factors β_n are chosen to be $\beta_n = \frac{kn}{Z}$, and all of the solutions correspond to the same energy,

$$E = -\frac{k^2}{2}. \tag{2}$$

Coulomb Sturmians are identical in form with the familiar hydrogenlike orbitals except that Z/n is everywhere replaced by a constant k, which is the same for all members of the basis set:

$$\chi_{nlm}(k, \mathbf{x}) = k^{3/2} N_{nl} (2kr)^l L_{n-l-1}^{2l+1}(2kr) e^{-kr} Y_{lm}(\hat{\mathbf{x}}), \tag{3}$$

where L_{n-l-1}^{2l+1} is an associated Laguerre polynomial. We will omit the k in $\chi_{nlm}(k, \mathbf{x})$ below when unambiguous. The Coulomb Sturmians obey a potential-weighted orthogonality relation:

$$\int d^3x \chi_{\mu'}^*(\mathbf{x}) \frac{1}{r} \chi_\mu(\mathbf{x}) = \frac{k}{n} \delta_{\mu'\mu} \tag{4}$$

and for every $k > 0$, they form a complete basis for the Hilbert space

$$H^1(\mathbb{R}^3) \equiv \left\{ \psi \in L^2(\mathbb{R}^3) | \nabla\psi \in L^2(\mathbb{R}^3) \right\}. \tag{5}$$

Importantly, by way of the Fock projection[8,9]

$$\hat{\mathbf{u}} = \left(\frac{2kp_1}{k^2 + p^2}, \frac{2kp_2}{k^2 + p^2}, \frac{2kp_3}{k^2 + p^2}, \frac{k^2 - p^2}{k^2 + p^2} \right) \tag{6}$$

the Fourier transform of Coulomb Sturmians can be written simply in terms of four-dimensional hyperspherical harmonics:

$$\begin{aligned} \chi_{nlm}^t(\mathbf{p}) &= M(p) Y_{n-1,l,m}(\hat{\mathbf{u}}), \\ M(p) &\equiv \frac{4k^{5/2}}{(k^2 + p^2)^2}. \end{aligned} \tag{7}$$

Because of this identity, Coulomb Sturmians inherit many beautiful mathematical properties from hyperspherical harmonics; this is explored in great depth by Vincenzo Aquilanti and his group.[1–4,7]

2.2 Many-center Sturmians

Let

$$\tau \equiv (a, l, m, n) \tag{8}$$

and

$$\chi_\tau(\mathbf{x}) \equiv \chi_{n,l,m}(\mathbf{x} - \mathbf{X}_a), \tag{9}$$

where $\mathbf{X}_a$ is a nuclear position, and $\chi_{n,l,m}(\mathbf{x})$ is a Coulomb Sturmian. We express a molecular orbital $\varphi_\zeta(\mathbf{x})$ as a linear combination of Coulomb Sturmians $\chi_\tau(\mathbf{x})$ centered on the atoms of the molecule:

$$\varphi_\zeta(\mathbf{x}) = \sum_{a,n,l,m} \chi_{n,l,m}(\mathbf{x} - \mathbf{X}_a) C_{(a,n,l,m),\zeta} \equiv \sum_\tau \chi_\tau(\mathbf{x}) C_{\tau,\zeta}. \tag{10}$$

2.3 Exact results: one-center densities in terms of 2*k*-Sturmians

We would like to use the relationship

$$\begin{aligned} J_{\tau_1,\tau_2,\tau_3,\tau_4} &\equiv \int \mathrm{d}^3x \int \mathrm{d}^3x' \rho_{\tau_1,\tau_2}(\mathbf{x}) \frac{1}{|\mathbf{x}-\mathbf{x}'|} \rho_{\tau_3,\tau_4}(\mathbf{x}') \\ &= 4\pi \int_0^\infty \mathrm{d}p \int \mathrm{d}\Omega_p \rho^t_{\tau_1,\tau_2}(\mathbf{p}) \rho^t_{\tau_3,\tau_4}(-\mathbf{p}) \end{aligned} \tag{11}$$

to evaluate the interelectron repulsion integral $J_{\tau_1,\tau_2,\tau_3,\tau_4}$. To do this, we must find the Fourier transform of the density

$$\rho_{\tau_1,\tau_2}(\mathbf{x}) \equiv \chi^*_{\tau_1}(\mathbf{x})\chi_{\tau_2}(\mathbf{x}) \equiv \chi^*_{\mu_1}(\mathbf{x}-\mathbf{X}_{a_1})\chi_{\mu_2}(\mathbf{x}-\mathbf{X}_{a_2}) \quad \mu \equiv (n,l,m), \tag{12}$$

where $\mathbf{X}_{a_1}$ and $\mathbf{X}_{a_2}$ are nuclear positions. In the special case where $\mathbf{X}_{a_2} = \mathbf{X}_{a_1}$, this Fourier transform is very easy to evaluate: We expand the product of Coulomb Sturmians in terms of single Sturmians with double the *k*-value:

$$\begin{aligned} \rho_{\mu_1,\mu_2}(\mathbf{x}) &= \chi^*_{\mu_1}(\mathbf{x})\chi_{\mu_2}(\mathbf{x}) \\ &= \sum_\mu \mathcal{C}_{\mu;\mu_1,\mu_2} \chi_\mu(2k,\mathbf{x}), \end{aligned} \tag{13}$$

where, using the potential-weighted orthonormality relation, we find that

$$\begin{aligned} \mathcal{C}_{\mu;\mu_1,\mu_2} &= \frac{n}{2} \int_0^\infty \mathrm{d}r\, r R_{n,l}(2r) R_{n_1,l_1}(r) R_{n_2,l_2}(r) \\ &\quad \times \int \mathrm{d}\Omega_3\, Y^*_{l,m}(\hat{\mathbf{x}}) Y^*_{l_1,m_1}(\hat{\mathbf{x}}) Y_{l_2,m_2}(\hat{\mathbf{x}}). \end{aligned} \tag{14}$$

The selection rules from the radial- and angular-integral in Eq. (14) limit the sum in (13):

$$\rho_{\mu_1,\mu_2} = \sum_{n=1}^{n_1+n_2-1} \sum_{l=|l_1-l_2|}^{l_1+l_2} C_{\mu;\mu_1,\mu_2} \chi_\mu(2k,\mathbf{x}). \tag{15}$$

The relationship shown in Eq. (15) is not an approximation. It is exact.

Evaluating (14) is simple and efficient. If we write

$$\mathfrak{c}^{lm}_{l_1 m_1; l_2 m_2} \equiv \int d\Omega\, Y^*_{lm}(\hat{\mathbf{x}}) Y^*_{l_1 m_1}(\hat{\mathbf{x}}) Y_{l_2 m_2}(\hat{\mathbf{x}}) \tag{16}$$

for the spherical harmonic coupling coefficients, and define the polynomial $p(r)$ by

$$R_{nl}(2r) R_{n_1 l_1}(r) R_{n_2 l_2}(r) = p(r) e^{-4r} \tag{17}$$

then

$$C_{\mu;\mu_1,\mu_2} = \frac{n}{2} \mathfrak{c}^{lm}_{l_1 m_1; l_2 m_2} \sum_{j=0}^{deg(p)} p_j \frac{(n+2)!}{4^{3+j}}. \tag{18}$$

By defining the vector $f_j = 4^{-(3+j)}(j+2)!$, we can compute the expansion coefficients as simple dot products of the polynomial coefficient vectors $\mathbf{p}$ and $\mathbf{f}$:

$$C_{\mu;\mu_1,\mu_2} = \frac{n}{2} \mathfrak{c}^{lm}_{l_1 m_1; l_2 m_2} \mathbf{p} \cdot \mathbf{f}, \tag{19}$$

which is extremely efficient to compute. The angular coupling coefficients $\mathfrak{c}^{lm}_{l_1 m_1; l_2 m_2}$ form a sparse tensor and are precalculated and stored, since they occur very often.

2.4 Fourier transformed densities and interelectron repulsion integrals in terms of hyperspherical harmonics

We next make use of the form of Fock's relationship that is appropriate for the $2k$ Sturmian basis:

$$\begin{aligned} \chi^t_\mu(2k,\mathbf{p}) &= \tilde{M}(p) Y_{n-1,l,m}(\hat{\mathbf{w}}) \equiv \tilde{M}(p) Y_\mu(\hat{\mathbf{w}}), \\ \hat{\mathbf{w}} &= \left(\frac{4kp_1}{4k^2+p^2}, \frac{4kp_2}{4k^2+p^2}, \frac{4kp_3}{4k^2+p^2}, \frac{4k^2-p^2}{4k^2+p^2} \right), \\ \tilde{M}(p) &\equiv \frac{4(2k)^{5/2}}{(4k^2+p^2)^2}. \end{aligned} \tag{20}$$

Here $Y_\mu(\hat{\mathbf{w}})$ is a four-dimensional hyperspherical harmonic. Thus we obtain the desired Fourier transform in terms of hyperspherical harmonics:

$$\begin{aligned}\rho^t_{\tau_1,\tau_2}(\mathbf{p}) &= \frac{1}{(2\pi)^{3/2}}\int \mathrm{d}^3x\, e^{-i\mathbf{p}\cdot\mathbf{x}}\rho_{\tau_1,\tau_2}(\mathbf{x}-\mathbf{X}_{a_1})\\ &= \frac{e^{-i\mathbf{p}\cdot\mathbf{X}_{a_1}}}{(2\pi)^{3/2}}\int \mathrm{d}^3x\, e^{-i\mathbf{p}\cdot(\mathbf{x}-\mathbf{X}_{a_1})}\rho_{\tau_1,\tau_2}(\mathbf{x}-\mathbf{X}_{a_1})\\ &= e^{-i\mathbf{p}\cdot\mathbf{X}_{a_1}}\sum_\mu \tilde{M}(p)Y_\mu(\hat{\mathbf{w}})\mathcal{C}_{\mu;\mu_1,\mu_2}.\end{aligned} \tag{21}$$

In this way we obtain the interelectron repulsion integral between two 1-center densities, located respectively at positions $\mathbf{X}_{a_1}$ and $\mathbf{X}_{a_3}$:

$$J_{\tau_1,\tau_2,\tau_3,\tau_4} = \sum_{\mu'}\sum_\mu J_{\mu',\mu}\mathcal{C}_{\mu';\mu_1,\mu_2}\mathcal{C}_{\mu;\mu_1,\mu_2}, \tag{22}$$

where

$$J_{\mu',\mu} = 4\pi\int_0^\infty \mathrm{d}p\,\tilde{M}^2(p)\int \mathrm{d}\Omega_p\, e^{i\mathbf{p}\cdot\mathbf{R}}(-1)^l Y_{\mu'}(\hat{\mathbf{w}})Y_\mu(\hat{\mathbf{w}}). \tag{23}$$

The integral $J_{\mu',\mu}$ can be further simplified:

$$\begin{aligned}J_{\mu',\mu} &= \sum_{\mu''} J_{\mu''}\tilde{c}_{\mu'';\mu',\mu},\\ \tilde{\mathfrak{c}}_{\mu'';\mu',\mu} &\equiv (-1)^l\int \mathrm{d}\Omega_4 Y^*_{\mu''}(\hat{\mathbf{w}})Y_{\mu'}(\hat{\mathbf{w}})Y_\mu(\hat{\mathbf{w}}),\\ J_{\mu''} &= 4\pi\int_0^\infty \mathrm{d}p\,\tilde{M}^2(p)\int \mathrm{d}\Omega_p\, e^{i\mathbf{p}\cdot\mathbf{R}}Y_{\mu''}(\hat{\mathbf{w}})\\ &= kY_{l,m}(\hat{\mathbf{S}})I_{nl}(S),\end{aligned} \tag{24}$$

where $\mathbf{S} = k\mathbf{R}$ and where $\tilde{\mathfrak{c}}_{\mu'';\mu',\mu}$ is a very sparse and highly symmetric tensor that can be precalculated and stored efficiently, for example using the method described in Ref. 18. Similarly, $I_{nl}(S)$ is a small matrix of radial functions that can easily be precalculated, and requires only a lookup at runtime. Equation (22) can be written as a matrix-vector product followed by a dot product:

$$J_{\tau_1,\tau_2,\tau_3,\tau_4} = \mathbf{c}_{12}\cdot(\mathbf{J}\mathbf{c}_{34}). \tag{25}$$

This gives us an exact and very rapid method for calculating interelectron repulsion integrals involving one-center densities.

2.5 Two-center densities in terms of 2*k*-Sturmians

Two-center densities are more difficult, but they too can be expanded accurately in terms of a 2*k*-Sturmian basis. Let

$$\rho_{\tau_1,\tau_2}(\mathbf{x}) = \chi^*_{\tau_1}(\mathbf{x})\chi_{\tau_2}(\mathbf{x}) \equiv \chi^*_{\mu_1}(\mathbf{x}-\mathbf{X}_1)\chi_{\mu_2}(\mathbf{x}-\mathbf{X}_2) \qquad \mu \equiv (n,l,m) \tag{26}$$

be a two-center density. We would like to represent the density as a series of the form

$$\begin{aligned} \rho_{\tau_1,\tau_2}(\mathbf{x}) &= \sum_\tau \chi_\tau(2k,\mathbf{x})c_{\tau_1,\tau_2,\tau} \\ &= \sum_\mu \chi_\mu(2k,\mathbf{x}-\mathbf{X}_1)c_{\tau_1,\tau_2;1,\mu} + \chi_\mu(2k,\mathbf{x}-\mathbf{X}_2)c_{\tau_1,\tau_2;2,\mu}, \end{aligned} \tag{27}$$

where $\mathbf{X}_1$ and $\mathbf{X}_2$ are the positions of the two centers. Multiplying from the left by a conjugate 2*k* basis function and integrating, we have

$$\begin{aligned} \int d^3x\, \chi^*_{\tau'}(2k,\mathbf{x})\rho_{\tau_1,\tau_2}(\mathbf{x}) &= \sum_\tau \int d^3x\, \chi^*_{\tau'}(2k,\mathbf{x})\chi_\tau(2k,\mathbf{x})c_{\tau_1,\tau_2,\tau} \\ &= \sum_\tau m_{\tau',\tau}(2k,\mathbf{X}_2-\mathbf{X}_1)c_{\tau_1,\tau_2,\tau}, \end{aligned} \tag{28}$$

which we can write in matrix form as

$$\mathbf{M}\mathbf{c}_{\tau_1\tau_2} = \mathbf{y}_{\tau_1\tau_2}. \tag{29}$$

A problem occurs when we try to invert the overlap matrix $m_{\tau',\tau}(2k,\mathbf{X}_2-\mathbf{X}_1)$: As internuclear separation decreases, this matrix becomes progressively ill conditioned due to overcompleteness of the basis set. Appendix B describes how the overcompleteness is removed.

Before we are able to calculate the coefficients $c_{\tau_1,\tau_2,\tau''}$, we must solve integrals of the form

$$\begin{aligned} &\int d^3x\, \chi^*_{\tau'}(2k,\mathbf{x})\rho_{\tau_1,\tau_2}(\mathbf{x}) \\ &= \int d^3x\, \chi^*_{\tau'}(2k,\mathbf{x})\chi^*_{\mu_1}(\mathbf{x}-\mathbf{X}_1)\chi_{\mu_2}(\mathbf{x}-\mathbf{X}_2). \end{aligned} \tag{30}$$

Two terms are needed, one of which is

$$
\begin{aligned}
I_1 &= \int \mathrm{d}^3x\, \chi^*_{\mu'}(2k, \mathbf{x} - \mathbf{X}_1)\chi^*_{\mu_1}(\mathbf{x} - \mathbf{X}_1)\chi_{\mu_2}(\mathbf{x} - \mathbf{X}_2) \\
&\equiv \int \mathrm{d}^3x f^*_{\mu',\mu_1}(\mathbf{x} - \mathbf{X}_1)\chi_{\mu_2}(\mathbf{x} - \mathbf{X}_2) \\
&= \int \mathrm{d}^3p\, e^{i\mathbf{p}\cdot\mathbf{R}} f^{t^*}_{\mu',\mu_1}(\mathbf{p})\chi^t_{\mu_2}(\mathbf{p}).
\end{aligned}
\tag{31}
$$

The second term, I_2, is identical up to a phase factor and renumbering $\mu_1 \leftrightarrow \mu_2$.

The problem of evaluating I_1 separates into two easily performed angular integrations and a more difficult radial p integration:

$$
\begin{aligned}
I_1 &= 4k^3\pi\, Y_{l'',m_2-m_1-m}(\hat{\mathbf{S}}) \sum_{l'=|l-l_1|}^{l+l_1} \sum_{l''=|l'-l_2|}^{l'+l_2} (-1)^{\frac{3l'+l''+l_2}{2}} \\
&\times \int \mathrm{d}\,\Omega Y^*_{l,m}(\hat{\mathbf{x}})\, Y^*_{l_1,m_1}(\hat{\mathbf{x}})\, Y_{l',m+m_1}(\hat{\mathbf{x}}) \\
&\times \int \mathrm{d}\Omega_q Y^*_{l',m+m_1}(\hat{\mathbf{q}})\, Y^*_{l'',m_2-m_1-m}(\hat{\mathbf{q}})\, Y_{l_2,m_2}(\hat{\mathbf{q}}) \\
&\times \int_0^\infty \mathrm{d}q\, q^2 \tilde{f}^t_{nl;n_1l_1;l'}(q)\tilde{g}^t_{n_2l_2}(q) j_{l''}(qS).
\end{aligned}
\tag{32}
$$

Noticing that the radial integral is a spherical Bessel transform (SBT), we can write the full integral as

$$
\begin{aligned}
I_1 &= 4k^3\pi\, Y_{l'',m_2-m_1-m}(\hat{\mathbf{S}}) \sum_{l'=|l-l_1|}^{l+l_1} \sum_{l''=|l'-l_2|}^{l'+l_2} (-1)^{\frac{3l'+l''+l_2}{2}} \\
&\times \mathbf{c}^{lm}_{l_1m_1;l',m+m_1}\, \mathbf{c}^{l',m+m_1}_{l'',m_2-m_1-m;l_2m_2} \\
&\times SBT^{q\to S}_{l''}\left(\tilde{f}^t_{nl;n_1l_1;l'}(q)\tilde{g}^t_{n_2l_2}(q)\right).
\end{aligned}
\tag{33}
$$

In the above two equations, $q \equiv p/k$, and

$$
\begin{aligned}
\mathbf{S} &\equiv k(\mathbf{X}_2 - \mathbf{X}_1), \\
\tilde{f}^t_{nl;n_1l_1;l'}(q) &\equiv \sqrt{\frac{2}{\pi}} \int_0^\infty \mathrm{d}s\, s^2 R_{nl}(2s) R_{n_2l_2}(s) j_{l'}(qs), \\
\tilde{g}^t_{n_1l_1}(q) &\equiv \sqrt{\frac{2}{\pi}} \int_0^\infty \mathrm{d}s\, s^2 R_{n_1l_1}(s) j_{l_1}(qs).
\end{aligned}
\tag{34}
$$

The integrals over ds are easily evaluated in closed form, for example by Mathematica, and since they are few in numbers, they are precalculated and stored. The final dq integration in the expression for I_1 can be expressed in terms of Harris W-functions,[13] or alternatively it can be evaluated numerically for the needed values of the scaled internuclear distance. There exist a number of publically available numerical SBT codes that are both accurate and fast; we have used J.D. Talman's implementation from Ref. 23 for testing. The computational impact of online evaluation of the radial integrals is manageable, since the number of these integrals is negligible compared to the number of repulsion integrals. However, for basis sets with a moderate number of atomic orbitals, this factor can be completely eliminated by storing the full radial integrals, since they are pure functions of the scaled internuclear distance S.

Having pre-calculated the coefficients in the expansion

$$\rho_{\tau_1,\tau_2}(\mathbf{x}) \approx \sum_{\tau} \chi_{\tau}(2k, \mathbf{x}) c_{\tau_1,\tau_2,\tau} \tag{35}$$

we can use them to obtain interelectron repulsion integrals involving two-center densities in a manner similar to our previous calculation for one-center densities:

$$J_{\tau_1,\tau_2,\tau_3,\tau_4} \approx \sum_{\tau'} \sum_{\tau} J_{\tau',\tau} c_{\tau_1,\tau_2,\tau'} c_{\tau_3,\tau_4,\tau} \tag{36}$$

and analogous to Eq. (25), we get

$$J_{\tau_1,\tau_2,\tau_3,\tau_4} \approx \mathbf{c}_{12} \cdot (\mathbf{J}\mathbf{c}_{34}). \tag{37}$$

However, the matrices $c_{\tau_1,\tau_2,\tau'}$ and $c_{\tau_3,\tau_4,\tau}$ are not sparse, and the result is an approximation rather than an exact equality.

2.6 Exact solution in special cases

In order to check the accuracy of the approximation by $2k$-Sturmians, we can compare the results to cases where we can obtain exact solutions. While the case where only one-center densities are involved is exact, it is useful mostly as a check of the implementation, not of the approximation, because in this case the $2k$-Sturmian approximation method becomes exact. However, there is a second class of cases where exact solutions can be found.

Using the relationship

$$F(|\mathbf{x}-\mathbf{R}|)=\sum_{l=0}^{\infty} f_l(r,R)P_l(\hat{\mathbf{x}}\cdot\hat{\mathbf{R}}), \tag{38}$$

where P_l is a Legendre polynomial and where $f_l(r,R)$ given by Sommer-Larsen's integral

$$f_l(r,R)=\frac{2l+1}{2rR}\int_{|r-R|}^{r+R} dr' r' F(r')P_l\left(\frac{r^2+R^2-r'^2}{2rR}\right) \tag{39}$$

we can obtain exact results for some cases of interelectron repulsion integrals where one-center densities interact with two-center densities. These give us useful checks on our approximate methods for treating two-center densities. If a one-center density has a Fourier transform of the form

$$\rho_1^t(\mathbf{p})=\sum_{l=0}^{l_{max}} \rho_{1,l}^t(p)Y_{l,m}(\hat{\mathbf{p}}) \tag{40}$$

and if the two-center density has a Fourier transform of the form

$$\rho_2^t(-\mathbf{p})=\sum_{l'=0}^{\infty} \rho_{2,l'}^t(p)P_{l'}(-\hat{\mathbf{p}}\cdot\hat{\mathbf{R}})$$

then the interelectron repulsion integral between the two densities is given by

$$\begin{aligned} J &= 4\pi\int_0^{\infty} dp\int d\Omega_p \sum_{l=0}^{l_{max}} \rho_{1,l}^t(p)Y_{l,m}(\hat{\mathbf{p}})\sum_{l'=0}^{\infty} \rho_{2,l'}^t(p)P_{l'}(-\hat{\mathbf{p}}\cdot\hat{\mathbf{R}}) \\ &= (4\pi)^2\sum_{l=0}^{l_{max}} Y_{l,m}(-\hat{\mathbf{R}})\int_0^{\infty} dp\rho_{1,l}^t(p)\rho_{2,l}^t(p). \end{aligned} \tag{41}$$

This yields an exact result, which can be obtained in closed form using Mathematica. The infinite sum over l' has disappeared because of the integration over $d\Omega_p$.

2.7 The integral transformation to m.o. repulsion integrals

Now let a molecular orbital be represented by a series of many-center Coulomb Sturmians:

$$\varphi_{\zeta_1}(\mathbf{x}) = \sum_{\tau_1} \chi_{\tau_1}(\mathbf{x}) C_{\tau_1,\zeta_1}.$$

Then the molecular orbital densities are obtained by linear transformation of the atomic orbital densities

$$\rho_{\zeta_1,\zeta_2}(\mathbf{x}) = \sum_{\tau_1} \sum_{\tau_2} C^{\dagger}_{\zeta_1,\tau_1} \rho_{\tau_1,\tau_2}(\mathbf{x}) C_{\tau_2,\zeta_2} \tag{42}$$

and the molecular orbital repulsion integrals are then

$$J_{\zeta_1,\zeta_2,\zeta_3,\zeta_4} \approx \sum_{\tau'} \sum_{\tau} J_{\tau',\tau}\, w_{\zeta_1,\zeta_2,\tau'} w_{\zeta_3,\zeta_4,\tau}, \tag{43}$$

where the m.o. density coefficients $w_{\zeta_1 \zeta_2;\tau}$ are

$$w_{\zeta_1,\zeta_2,\tau'} = \sum_{\tau_1} \sum_{\tau_2} C^{\dagger}_{\zeta_1,\tau_1} c_{\tau_1,\tau_2,\tau'} C_{\tau_2,\zeta_2}. \tag{44}$$

Once again, the full electron repulsion integral is computed by combining the $\tau_1\tau_2$-coefficients and the $\tau_3\tau_4$-coefficients with a dot product:

$$J_{\zeta_1,\zeta_2,\zeta_3,\zeta_4} \approx \mathbf{w}_{\tau_1\tau_2} \cdot (\mathbf{J}\mathbf{w}_{\tau_3\tau_4}). \tag{45}$$

In many large-scale quantum chemical calculations, also when using Gaussian type orbitals, the heaviest computational factor is due to storing, retrieving, and sometimes sorting billions of molecular ERIs on disk, since there are too many to fit in main memory, yet the m.o.-transformation is too heavy to perform every time an integral is needed. However, there are quadratically fewer $\mathbf{w}_{\tau_1\tau_2}$ and $\mathbf{J}\mathbf{w}_{\tau_3\tau_4}$ than there are molecular ERIs and combining them using (45) just takes a dot product.This means that even large-scale calculations can be performed in-core, avoiding disk access during computation altogether.

3. PRELIMINARY RESULTS

The following calculations were performed for a small Coulomb Sturmian basis consisting of the $m=0$ orbitals 1s,..., 3d, expanded in a $2k$-Sturmian basis consisting of the $m = 0$ orbitals 1s,..., 6g.

3.1 Accuracy

To the extent that the expansions in $2k$-sturmians are able to reproduce the exact orbital densities, we can expect electron repulsion integrals and other calculated properties, to be accurate. Figure 6. 1 shows a selection of

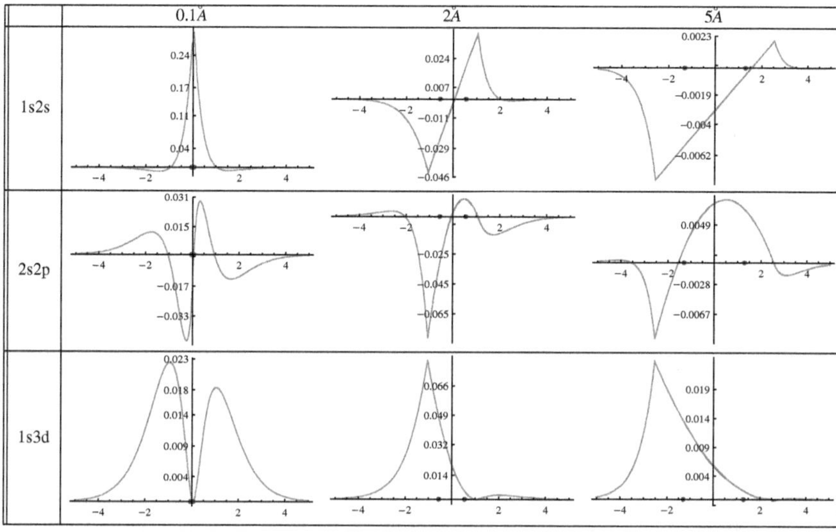

Figure 6. 1 Reproduction of electronic orbital densities from computed expansion coefficients shown together with the exact densities. The densities are plotted as functions of kz through the line $x = 0$, $y = 0$. We see that the $2k$-Sturmian approximation faithfully reproduces both soft and sharp features to the point where approximated and exact densities are visually indistinguishable. Figure 6. 2 shows the total error integrated over $\mathbb{R}^3$ as a function of nuclear separation. For color version of this figure, the reader is referred to the online version of this chapter.

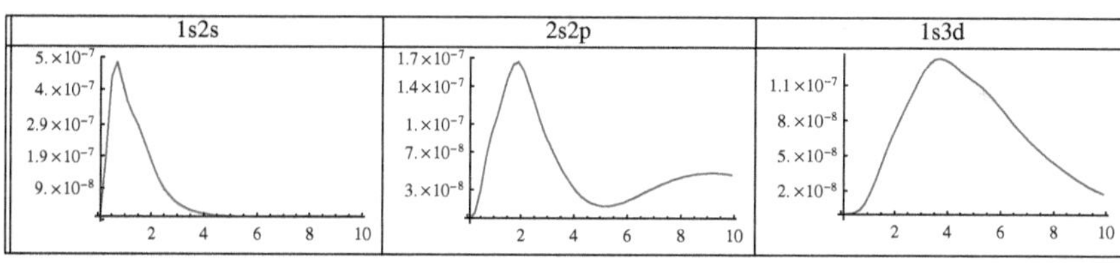

Figure 6. 2 L^2-error $\sqrt{(\int d^3x(\rho_{\text{exact}}(\mathbf{x}) - \rho_{approx}(\mathbf{x}))^2}$ as a function of scaled nuclear separation distance S/Å. For color version of this figure, the reader is referred to the online version of this chapter.

reproduced densities plotted together with the exact ones at various scaled nuclear separation distances. Since they a re visually indistinguishable, we also show the total errors integrated over space as functions of S, defined by

$$\|\rho_{\text{exact}} - \rho_{\text{approx}}\|^2 = \int \mathrm{d}^3x(\rho_{\text{exact}}(\mathbf{x}) - \rho_{\text{approx}}(\mathbf{x}))^2 \tag{46}$$

in Figure 6. 2. At $S = 0$, the approximation becomes exact, and at very large distances, the relative error increases but the integrals become very small and

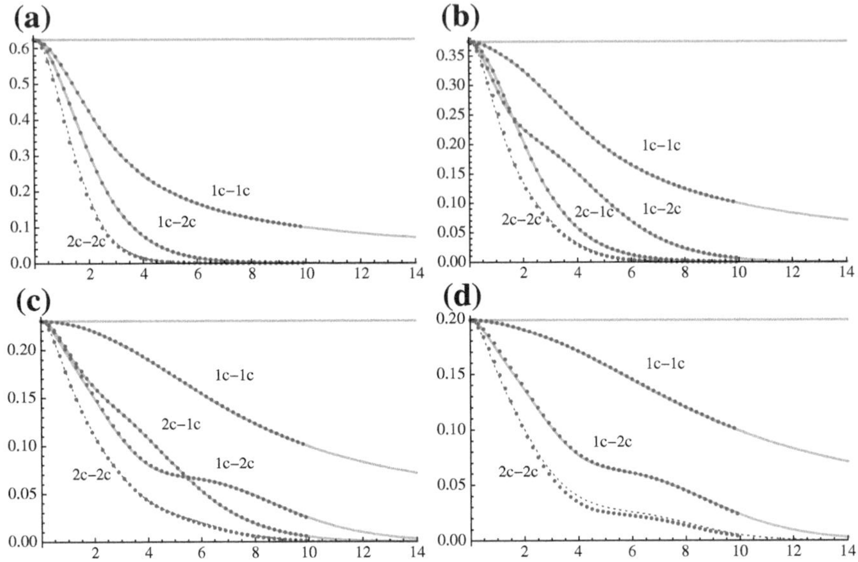

Figure 6. 3 Calculated electron repulsion integrals (points) for a diatomic system as a function of S. The results are compared with exact solutions for 1c-1c and 1c-2c integrals (thick lines), and an expansion in 144 Gaussians per density (thin dashed lines) for the 2c-2c integral. (a) $1s^2 1s^2$, (b) $1s^2 2s^2$, (c) $2s^2 3s^2$, and (d) $3s^2 3s^2$. For color version of this figure, the reader is referred to the online version of this chapter.

at around $S = 10$, we can replace the integrals by the multipole expansion. The L^2-errors peak in the intermediate range at around 10^{-7}; this is due to using the minimal $2k$-Sturmian basis set, and will decrease with a larger basis. Figure 6. 3 shows interelectron repulsion integrals for a diatomic system, as a function of scaled nuclear separation S, together with the exact solutions described in Section 2.6. The four-center integrals (labeled *2c-2c*) are compared to an approximate result using a large (and slow) Gaussian expansion.

3.2 Performance

The time needed to calculate interelectron repulsion integrals for a molecule can be broken down into three independent contributions:

Setup time per center-pair T_c: For every nuclear displacement, the $2k$-Sturmian overlap matrix $m_{\tau'\tau}(\mathbf{S})$ must be evaluated and diagonalized, and the matrix $J_{\mu'\mu}(\mathbf{S})$ evaluated. For a molecule with M atoms, this is done at most $M(M+1)/2$ times, but fewer if nuclear displacements are repeated, or if symmetry reductions can be exploited.

Setup time per orbital density T_2: For each orbital density ρ_{τ_1,τ_2}, we must compute the expansion coefficients $\mathbf{c}_{\tau_1,\tau_2}$ and the product $\mathbf{J}\mathbf{c}_{\tau_1\tau_2}$.

Computation time per interelectron repulsion integral T4: For producing the individual electron repulsion integrals, only the final dot product $\mathbf{c}_{\tau_1\tau_2} \cdot (\mathbf{J}\mathbf{c}_{\tau_3\tau_4})$ is required.

The following shows measurements performed on a 1.9 GHz Intel i7-3517U CPU, the author's laptop. Although the method lends itself well to vectorization, the benchmarks were done serially on a single CPU core.

$$\begin{aligned} T_c &= 4 \times 10^{-3} \text{ s}, \\ T_2 &= 13 \times 10^{-6} \text{ s}, \\ T_4 &= 40 \times 10^{-9} \text{ s}. \end{aligned} \tag{47}$$

Even though T_2 and T_c are many orders of magnitude slower than T_4, the following examples will illustrate that the contribution of T_4 to the total ERI computation time is dominant, because it has to be performed so many more times.

If we neglect all symmetry considerations, calculating all the electron repulsion integrals for an M-atom molecule with N atomic orbitals per nucleus requires

$$T = N_c T_c + N_2 T_2 + N_4 T_4, \tag{48}$$

where $N_c = M(M+1)/2, N_2 = (MN)^2$, and $N_4 = N_2(N_2+1)/2$. In practice, symmetry and other considerations allow us to reduce this significantly but it serves as an upper bound to the time spent calculating the electron repulsion integrals. The following back-of-the-envelope calculations illustrate how each term contributes to the total calculation time:

Example 1. Assume a trimer with $N = 14$ orbitals per nucleus. Then $N_c = 3,\ N_2 = 1764$, and $N_4 = 1.6 \times 10^6$. The total time needed to compute all integrals is

$$T = 0.01 \text{ s} + 0.03 \text{ s} + 0.06 \text{ s} = 0.1 \text{ s}. \tag{49}$$

For a trimer with $N = 30$ orbitals per nucleus, $N_2 = 8100,\ N_4 = 3.3 \times 10^7$, and

$$T = 0.01 \text{ s} + 0.13 \text{ s} + 1.3 \text{ s} = 1.4 \text{ s}. \tag{50}$$

Example 2. Given a molecule with $M = 60$ atoms and $N = 6$ atomic orbitals per nucleus, we get $N_c = 1770$, $N_2 = 129,600$, $N_4 = 8.4 \times 10^{10}$, and

$$T = 7.1 \text{ s} + 1.7 \text{ s} + 336 \text{ s} \approx 6 \text{ minutes}. \tag{51}$$

With 12 orbitals per nucleus, $N_2 = 518,400$, $N_4 = 1.3 \times 10^{11}$, and the total time is

$$T = 7.1 \text{ s} + 6.7 \text{ s} + 5380 \text{ s} \approx 1.5 \text{ hours}. \tag{52}$$

Notice that the setup times per nuclear displacement and per orbital density contribute negligibly, even for many-atom molecules.

4. DISCUSSION

4.1 Problem decomposition

The efficiency of the current method derives from our ability to decompose the problem into components for which the heavy parts can be precalculated and stored, and that can be efficiently recombined at runtime. Figure 6. 4 shows schematically this decomposition:

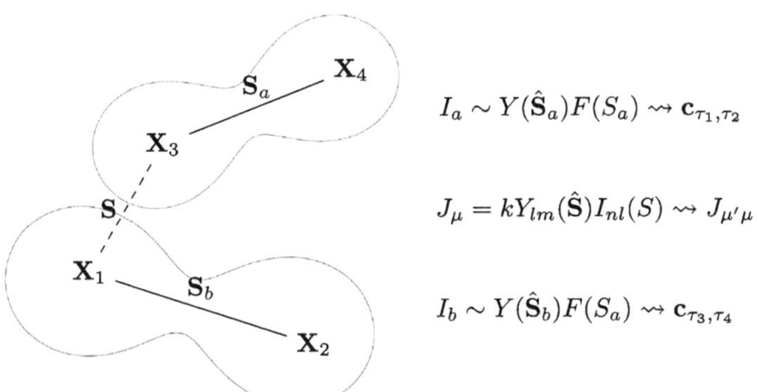

Figure 6. 4 Decomposition of ERI calculation. The full integral is a complicated function of nine spatial coordinates. The efficiency of the current work relies on relaying all the heavy work to offline precomputation. This is made possible by decomposing the full integral into smaller parts that factor into simple angular and complicated radial parts that can be precalculated and efficiently recombined at runtime.

- The fundamental integrals $I_{nl}(S)$ in Eq. 24 are precalculated and stored as functions of the scaled internuclear distance S. For each pair of nuclei, the matrix $J_{\mu'\mu}(\mathbf{S})$ is easily reconstructed, using the precalculated hyperangular coupling coefficients $\tilde{\mathfrak{c}}_{\mu'';\mu'\mu}$.
- For each two-center density $\rho_{\tau_1\tau_2}$, the quantity $I_a(\mathbf{S}_a)$ is expressed in terms of a few precalculated functions of the scaled distance S_a, combined with a spherical harmonic involving the angles, and precalculated angular coupling coefficients. The radial functions in S_a contain all the heavy computations. Hence, the expansion coefficients $c_{\tau_i\tau_j;\tau}$ can be obtained with little computational effort at runtime.

We are able to do this, in part, because of the simple relation of Coulomb Sturmians to hyperspherical harmonics. Once $J_{\mu'\mu}(\mathbf{S})$ and the density expansion coefficients are reconstructed, the full integrals can be computed as simple dot products.

4.2 Viability

The preliminary calculations indicate that the method improves substantially on current methods for evaluating electron repulsion integrals for ETOs, and that it makes molecular Coulomb Sturmians comparable in terms of speed with Gaussian orbitals. With the limited basis used, including 1s,..., 6g orbitals, the average time needed to compute a four-center integral is ≈40 ns on a 1.9 GHz processor, with an additional setup time of ≈4 ms per pair of nuclei in the molecule, and ≈13μs setup time per orbital density. A larger basis will increase both accuracy and runtime but only linearly. In addition, since the time determining computation is a simple dot product, it lends itself extremely well to vectorization and implementation on GPUs. However, in order to properly quantify the performance and accuracy, more rigorous testing is needed, as well as timing in realistic calculations.

The method is implemented in a set of `Mathematica` programs that performs all precalculation and automatically generates `C++` code, and a `C++` library `libsturm2k`, that can be used directly in user programs. Both of these will be made available during autumn 2013 on http://sturmian.kvante.org/coulomb/. The programs will be described in a separate publication.

APPENDIX A. ANGULAR AND HYPERANGULAR INTEGRATIONS

In a tree-dimensional space, the volume element is given by $dx_1 dx_2 dx_3$ in Cartesian coordinates or by $r^2 dr\, d\Omega$ in spherical polar coordinates. Thus we can write

$$dx_1 dx_2 dx_3 = r^2 dr\, d\Omega, \tag{A.1}$$

where $d\Omega$ is the element of solid angle. Similarly, in a d-dimensional space, we can write

$$dx_1 dx_2 \cdots dx_d = r^{d-1} dr\, d\Omega, \tag{A.2}$$

where r is the d-dimensional radius, and where $d\Omega$ is the element of generalized solid angle. We will now prove a general theorem for angular and hyperangular integration:

Theorem 1. *Let*

$$I(\mathbf{n}) \equiv \int d\Omega \left(\frac{x_1}{r}\right)^{n_1} \left(\frac{x_2}{r}\right)^{n_2} \cdots \left(\frac{x_d}{r}\right)^{n_d}, \tag{A.3}$$

where $x_1, x_2, \ldots, x_d$ are the Cartesian coordinates of a d-dimensional space, $d\Omega$ is the generalized solid angle, $\mathbf{r}$ is the hyperradius, defined by $r^2 \equiv x_1^2 + \cdots + x_d^2$, and where the n_j's are nonnegative integers. Then, letting $n \equiv n_1 + \cdots + n_d$,

$$I(\mathbf{n}) = \begin{cases} \dfrac{\pi^{d/2}}{2^{(n/2-1)}\Gamma\left(\frac{d+n}{2}\right)} \displaystyle\prod_{j=1}^{d} (n_j - 1)!! & \text{if all the } n_j'\text{s are even,} \\ 0 & \text{otherwise.} \end{cases} \tag{A.4}$$

Proof. Consider the integral

$$\int_0^\infty dr\, r^{d-1} e^{-r^2} \int d\Omega\, x_1^{n_1} x_2^{n_2} \cdots x_d^{n_d} = \prod_{j=1}^{d} \int_{-\infty}^{\infty} dx_j\, x_j^{n_j} e^{-x_j^2}. \tag{A.5}$$

If n_j is zero or a positive integer, then

$$\int_{-\infty}^{\infty} dx_j x_j^{n_j} e^{-x_j^2} = \begin{cases} \dfrac{(n_j - 1)!!\sqrt{\pi}}{2^{n_j/2}} & \text{if } n_j \text{ is even,} \\ 0 & \text{if } n_j \text{ is odd.} \end{cases} \tag{A.6}$$

Hence the right-hand side of Eq. (A.5) can be written as

$$\prod_{j=1}^{d}\int_{-\infty}^{\infty} dx_j\, x_j^{n_j} e^{-x_j^2} = \begin{cases} \dfrac{\pi^{d/2}}{2^{n/2}}\displaystyle\prod_{j=1}^{d}(n_j-1)!! & \text{if all the } n_j'\text{s are even,} \\ 0 & \text{otherwise.} \end{cases} \tag{A.7}$$

Using $\int_0^\infty \mathrm{d}r\, r^{d+n-1} e^{-r^2} = \frac{1}{2}\Gamma(\frac{d+n}{2})$, we can write the left hand side of Eq. (A.5) as

$$\int_0^\infty \mathrm{d}r\, r^{d+n-1} e^{-r^2} \int \mathrm{d}\Omega \left(\frac{x_1}{r}\right)^{n_1}\left(\frac{x_2}{r}\right)^{n_2}\cdots\left(\frac{x_d}{r}\right)^{n_d} = \frac{I(\mathbf{n})}{2}\Gamma\left(\frac{d+n}{2}\right). \tag{A.8}$$

Substituting (A.7) and (A.8) into (A.5), we obtain Eq. (A.4). Q.E.D. ■

Splitting into the even and odd cases, we can write this as (assuming $n_1, \ldots, n_d$ are even)

$$\begin{aligned} I(\mathbf{n}) &= (2\pi)^{d/2}\frac{(n_1-1)!!\cdots(n_d-1)!!}{(n+d-2)!!} && \text{when } d \text{ is even,} \\ I(\mathbf{n}) &= 2(2\pi)^{(d-1)/2}\frac{(n_1-1)!!\cdots(n_d-1)!!}{(n+d-2)!!} && \text{when } d \text{ is odd.} \end{aligned} \tag{A.9}$$

Notice that this allows us to easily perform d-dimensional angular integrals of multivariate polynomials, regardless of whether or not they are homogeneous, since if

$$P(\mathbf{x}) = \sum_{\mathbf{n}} c_{\mathbf{n}}\, x_1^{n_1}\cdots x_d^{n_d} \tag{A.10}$$

then

$$\int \mathrm{d}\Omega P(\mathbf{x}) = \sum_{\mathbf{n}} r^n c_{\mathbf{n}}\, I(\mathbf{n}). \tag{A.11}$$

The d-dimensional angular integration method that Eqs. (A.9) and (A.11) provide is furthermore extremely rapid: It requires only d multiplications and one division per nonzero polynomial coefficient. This means that angular and hyperangular integration of polynomials take only nanoseconds on current CPUs.

In the case when $n_1, \ldots, n_d$ are *not* required to be integers, but instead allowed to be real numbers greater than -1, the right- hand side of (A.5) can still be evaluated, yielding

$$\prod_{j=1}^{d}\int_{-\infty}^{\infty} x_j^{n_j} e^{-x_j^2} = \frac{1}{2}\prod_{j=1}^{d}(1+e^{i\pi n_j})\Gamma\left(\frac{n_j+1}{2}\right). \tag{A.12}$$

Thus, (A.4) is replaced by

$$I(\mathbf{n}) = \frac{1}{\Gamma\left(\frac{d+n}{2}\right)}\prod_{j=1}^{d}(1+e^{i\pi n_j})\Gamma\left(\frac{n_j+1}{2}\right). \tag{A.13}$$

This more general equation reduces to (A.4) when all the n_j's are nonnegative integers.

APPENDIX B. SOLVING UNDERSPECIFIED SYSTEMS OF EQUATIONS

It is an often occurring problem that one needs to solve an underspecified system of N linear equations, described by the matrix equation

$$\mathbf{Mc} = \mathbf{y} \tag{B.1}$$

i.e., where the rank L of $\mathbf{M}$ is smaller than N. In the present case, this occurs because the LCAO basis gradually becomes overcomplete as nuclear distance is decreased. This appendix describes how to deal both with the case of exact overcompleteness and with the case of near singularity, where ill-conditioned systems of equations introduce numerical error.

B.1 Removing exact linear dependence: factoring out the null-space

The null-space Ker($\mathbf{M}$) doesn't contribute to the solution of Eq. (B.1): If $\mathbf{Mc} = \mathbf{y}$, then for any vector $\mathbf{v} \in \text{Ker}(\mathbf{M})$,

$$\mathbf{M}(\mathbf{c}+\mathbf{v}) = \mathbf{Mc} + 0 = \mathbf{y}. \tag{B.2}$$

Hence, we can simply project away the null-space of $\mathbf{M}$ to obtain a linearly independent basis spanning the same space as the linearly dependent one. In the case where $\mathbf{M}$ is square and Hermitian, we can diagonalize it to obtain an orthonormal basis of eigenvectors $\mathbf{e}_j$. Writing the solution vector $\mathbf{c}$ and the right-hand side vector $\mathbf{y}$ in this coordinate system

$$\mathbf{c} = \sum_{j=1}^{N}\tilde{c}_j\mathbf{e}_j \text{ and } \mathbf{y} = \sum_{j=1}^{N}\tilde{y}_j\mathbf{e}_j \tag{B.3}$$

the linear equation takes the form

$$\mathbf{Mc} = \mathbf{y} \iff \sum_{j=1}^{N} \lambda_j \tilde{c}_j \mathbf{e}_j = \sum_{j=1}^{N} \tilde{y}_j \mathbf{e}_j \tag{B.4}$$

and the solution vector is simply the right-hand side divided by the nonzero eigenvalues $\tilde{c}_j = \frac{\tilde{y}_j}{\lambda_j}$ for $\lambda_j \neq 0$. The null-space Ker($\mathbf{M}$) is the eigenspace to the 0-eigenvalue, so to project it away we simply discard the eigenvectors corresponding to zero. Let $L = rank(\mathbf{M})$, i.e., the number (including multiplicity) of eigenvalues different from zero, and let $\mathbf{C}$ be the rectangular $L \times N$ matrix consisting of the L eigenvectors with $\lambda_j \neq 0$. Then

$$\begin{aligned}(\mathbf{CMC}^T)_{ij} &= \delta_{ij}\lambda_j, \\ \mathbf{Cy} &= \tilde{\mathbf{y}}.\end{aligned} \tag{B.5}$$

The solution vector in the original coordinate system is then

$$\mathbf{c} = \tilde{\mathbf{c}}\mathbf{C} \text{ where } \tilde{c}_j = \frac{\tilde{y}_j}{\lambda_j}. \tag{B.6}$$

For dense matrices, both inversion and diagonalization are $\mathcal{O}(N^3)$ operations, and hence removing overcompleteness doesn't add to the overall computational complexity of the method. In cases where $\mathbf{M}$ is not Hermitian or even square, we can use an almost identical method, simply replacing diagonalization by singular value decomposition in the above.

B.2 Numerical issues: near-linear dependence

Any singular systems of equations can be solved by projecting away the null-space in the manner described above. However, in many cases (including ours), we both have linear dependence and near-linear dependence. The near-linear dependence corresponds to very small eigenvalues of $\mathbf{M}$, resulting in dramatic numerical instability in the solutions, as can be inferred from Eq. (B.6).

A first solution is to throw away eigenvalues smaller than some chosen tolerance $\lambda_{\min}$. If we can assume $||\mathbf{c}||^2 \leq 1$, then we have a simple upper bound to the impact on the quality of the solution: assume $\mathbf{Mc} = \mathbf{y}$. What is the effect of discarding the eigenspaces with $|\lambda_j| < \lambda_{\min}$? Let L be the number of remaining eigenvectors, and set $\mathbf{c}' = \mathbf{c} - \sum_{|\lambda_j| < \lambda_{\min}} \tilde{c}_j \mathbf{e}_j$. Then

$$
\begin{aligned}
\|\mathbf{y}-\mathbf{y}'\|^2 &= \left\| \sum_{|\lambda_j|<\lambda_{\min}} \lambda_j \tilde{c}_j \mathbf{e}_j \right\|^2 = \sum_{|\lambda_j|<\lambda_{\min}} |\lambda_j \tilde{c}_j|^2 \\
&\leq \sum_{|\lambda_j|<\lambda_{\min}} |\lambda_j|^2 \\
&\leq (N-L)\lambda_{\min}.
\end{aligned} \tag{B.7}
$$

i.e., the introduced error in the solution is proportional to $\lambda_{\min}$ and the number of eigenvectors discarded. However, this simple method has two problems: first, when the largest and smallest used eigenvalues differ greatly in size, numerical instability is introduced even for only moderately small eigenvalues. Second, since the cutoff is discrete, this method introduces discontinuities in the results as eigenspaces go in and out of use. The solution is to assign weights to the eigenvectors that smoothly phase out their importance as they decrease in size. We weight the eigenvalues smaller than $\lambda_{\min}$ by a function that is equal to one at $\lambda_{\min}$, and phase them out exponentially as their sizes decrease:

$$
\lambda_j' = \lambda_j e^{\alpha \log\left(\frac{\lambda_{\min}}{|\lambda_j|}\right)}. \tag{B.8}
$$

This yields results that are both accurate, smooth, and numerically stable.

REFERENCES

1. Aquilanti, V.; Cavalli, S.; De Fazio, D.; Grossi, G. Hyperangular Momentum: Applications to Atomic and Molecular Science. In *New Methods in Quantum Theory*; Tsipis, C., Popov, V., Herschbach, D., Avery, J., Eds.; Klüwer: Dordrecht, 1996.
2. Aquilanti, V.; Cavalli, S.; Coletti, C. The *d*-Dimensional Hydrogen Atom; Hyperspherical Harmonics as Momentum Space Orbitals and Alternative Sturmian Basis Sets. *Chem. Phys.* **1997,** *214.*
3. Aquilanti, V.; Cavalli, S.; Coletti, C.; Di Domenico, D.; Grossi, G. Hyperspherical Harmonics as Sturmian Orbitals in Momentum Space: A Systematic Approach to the Few-body Coulomb Problem. *Int. Rev. Phys. Chem.* **2001,** *20,* 673.
4. Aquilanti, V.; Caligiana, A.; Cavalli, S.; Coletti, C. Hydrogenic Orbitals in Momentum Space and Hyperspherical Harmonics. Elliptic Sturmian Basis Sets. *Int. J. Quantum Chem.* **2003,** *92,* 212.
5. Boys, S. Electronic Wave Functions. I. A General Method of Calculation for the Stationary States of Any Molecular System. *Proc. Roy. Soc. London A* **1950,** *200,* 542–554.
6. Bray, I.; Fursa, D. Convergent Close-coupling Method: A "Complete Scattering Theory"? *Phys. Rev. Lett.* **1996,** *76,* 2674.
7. Calderini, D.; Cavalli, S.; Coletti, C.; Grossi, G.; Aquilanti, V. Hydrogenoid Orbitals Revisited: From Slater Orbitals to Coulomb Sturmians. *J. Chem. Sci.* **2012,** *124,* 187–192.

8. Fock, V. Zur Theorie des Wasserstoffatoms. *Z. Phys.* **1935,** *98,* 145.
9. Fock, V. Hydrogen Atoms and Non-euclidean Geometry. *Kgl. Norske Videnskab. Forh.* **1958,** *31,* 138.
10. Fursa, D.; Bray, I. Convergent Close-coupling Method for Positron Scattering from Noble Gasses. *New J. Phys.* **2012,** *14,* 035002.
11. Gill, P.; Gilbert, A. Resolutions of the Coulomb Operator: II. The Laguerre Operator. *Chem. Phys.* **2009,** *356,* 86–90.
12. Harris, F. Analytic Evaluation of Two-center STO Electron Repulsion Integrals via Ellipsoidal Expansion. *Int. J. Quantum Chem.* **2002,** *88,* 701–734.
13. Harris, F.; Michels, H. The Evaluation of Molecular Integrals for Slater-type Orbitals. *Adv. Chem. Phys.* **1967,** *13,* 205–266.
14. Hoggan, P. Four-center Slater-type Orbital Molecular Integrals Without Orbital Translations. *Int. J. Quantum Chem.* **2009,** *110* (1), 98–103. DOI:10.1002/qua.22213.
15. Hoggan, P. Slater-type Orbital Basis Sets: Reliable and Rapid Solution of the Schrödinger Equation for Accurate Molecular Properties. In *Solving the Schrödinger Equation: Has Everything Been Tried?* Popelier, P., Ed.; Imperial College Press (Chapter 7).
16. Hoggan, P.; Bouferguène, A. Why Specific ETOs are Advantageous for NMR and Molecular Interactions. In *Advances in the Theory of Quantum Systems in Chemistry and Physics*; Hoggan, P. E., Brändas, E. J., Eds.; Progress in Theoretical Chemistry and Physics; Springer, pp 83–102. ISBN 978-94-007-2075-6.
17. Mitnik, D.; Colavecchia, F.; Gasaneo, G.; Randazzo, J. Computational Methods for Generalized Sturmian Basis. *Comput. Phys. Commun.* **2011,** *182,* 1145–1155.
18. Pinchon, D.; Hoggan, P. E. New Index Functions for Storing Gaunt Coefficients. *Int. J. Quantum Chem.* **2007,** *107* (12), 2186–2196. DOI:10.1002/qua.21337. ISSN 1097-461X.
19. Safouhi, H. Efficient and Rapid Numerical Evaluation of the Two-electron Four-center, Coulomb Integrals Using Nonlinear Transformations and Useful Properties of Sine and Bessel Functions. *J. Comput. Phys.* **2002,** *176,* 1–19.
20. Savage, J.; Fursa, D.; Bray, I. Convergent Close-coupling Calculations of Positron-magnesium Scattering. *Phys. Rev. A* **2011,** *83,* 062709.
21. Shull, H.; Löwdin, P.-O. Superposition of Configurations and Natural Spin-orbitals. Applications to the He problem. *J. Chem. Phys.* **1959,** *30,* 617.
22. Steinborn, E. On the Evaluation of Exponential (Slater) Type Integrals. In *Methods in Computational Molecular Physics*; Dierksen, G., Wilson, S., Eds.; Reidel.
23. Talman, J. D. NumSBT: A Subroutine for Calculating Spherical Bessel Transforms Numerically. *Comput. Phys. Commun.* **2009,** *180,* 332–338. DOI:10.1016/j.cpc.2008.10.003.
24. Weatherford, C., Jones, H., Eds.; *International Conference on ETO Multicenter Integrals*; Reidel: Dordrecht, 1982.
25. Weniger, E.; Steinborn, E. The Fourier Transforms of Some Exponential-type Basis Functions and Their Relevance for Multicenter Problems. *J. Chem. Phys.* **1983,** *78,* 6121.

CHAPTER SEVEN

Three-Body Coulomb Problems with Generalized Sturmian Functions

G. Gasaneo[a], L.U. Ancarani[b], D.M. Mitnik[c], J.M. Randazzo[d], A.L. Frapiccini[a] and F.D. Colavecchia[d]

[a]Departamento de Física, Universidad Nacional del Sur and IFISUR-CONICET, Bahía Blanca, Argentina
[b]Théorie, Modélisation, Simulation, SRSMC, UMR CNRS 7565, Université de Lorraine, 57078 Metz, France
[c]Instituto de Astronomía y Física del Espacio (IAFE), CONICET and Departamento de Física, FCEyN, Universidad de Buenos Aires, Argentina
[d]División Colisiones Atómicas, Centro Atómico Bariloche and CONICET, Río Negro, Argentina

Contents

Advances in Quantum Chemistry, Volume 67
ISSN 0065-3276, http://dx.doi.org/10.1016/B978-0-12-411544-6.00007-8

Abstract

The study of structure and collision processes of three- and four-body problems has seen an extraordinary progress in the last decades. This progress has been in part associated to the incredible fast growth of the computer capabilities. However, the tools used to solve structure problems are different from those corresponding to the treatment of collision processes. In this review, we provide the theoretical framework and a selection of results for both structure as well as collision problems using only one technique that we have developed in the last few years, based on the use of Generalized Sturmian functions. We present results obtained in structure studies of isolated and confined two-electron atoms, and exotic and molecular systems. The same technique is applied to the study of various benchmark problems for the single ionization of hydrogen and the double ionization of helium by electron impact. In this way, we demonstrate that the Generalized Sturmian method can be successfully applied to the treatment of both types of problems.

1. INTRODUCTION

A wide variety of phenomena in atomic, molecular physics, or chemistry depend on the understanding of the three-body problem.[1] Many atomic and molecular processes in an extended range of energies present simultaneous excitation and ionization, double or multiple ionization of some of the electrons in the system. These processes cannot be treated with the standard *ab initio* methods designed to study structure, since emitted electrons are spread all over space. It is also difficult to deal with them using the methods designed specifically for atomic collision, since there, in general, the bound electrons are treated in approximate ways. Moreover, there are not many methods allowing the accurate and precise description of outgoing electrons and at the same time, providing on equal footing the correct initial bound state of a collision.

The study of atomic and molecular systems as presented in the literature can be separated into two branches, one devoted to the study of bound states and another focused on collision processes. For the bound state treatment of atomic and molecular systems, the use of spectral methods is a standard tool. The very well-established configuration interaction (CI) method rests on the use of the sum of basis functions to represent the wave function of the system.[1] The "configurations" are generally defined as the sum of products of basis functions associated with the description of the dynamics of two-particle interactions. For example, each of these functions is the solution of the electron–nucleus problem in the case of atoms, while the nucleus–nucleus interactions are considered for molecules. Approaches like VASP[2] or Gaussian[3] use basis functions very localized in configuration space to properly represent the dynamics of the electrons and the nuclei. These methods have proven to be very useful and robust, however, this does not necessarily result in efficiency. On the other hand, a well-known disadvantage of these methods is that they are unable to deal at all with the dynamics of a process involving the fragmentation of the system.

Quite recently, the idea of optimizing these existing methods for atomic and molecular structure calculations has taken a different line based on the replacement of Gaussian or Slater basis functions by Coulomb Sturmian functions (CSF),[4–6] with a focus on bound-type states only. A fantastic advantage has been observed when CSF were applied, e.g., to the structure calculation of many-electron atoms. It was found that all the basis functions scale with a length parameter. This property allows reduction of the calculations of all the matrix elements to basically two types of integrals that can be evaluated only once for all the electrons of the system.[5] From the collisional point of view, other difficulties arise.[7] First, the wave function of an ejected particle is completely delocalized. Second, outgoing particles should have the correct asymptotic behavior, which implies considering large values of the emitted particle coordinates and, therefore, the whole problem has to be solved over very large numerical domains.[8] Various techniques have been implemented to enforce those conditions or to circumvent them. Methods like the J-matrix[9–11] and the convergent-close-coupling (CCC),[12] among others, impose the asymptotic conditions at high values of one coordinate but they do not necessarily represent the correct behavior in the region where all the emitted particles are far from each other; such domain is the most important, for example, when studying double ionization processes. A different methodology is implemented in the exterior complex scaling (ECS) approach.[13] As often pointed out in the literature, this method avoids imposing the asymptotic condition because it sets an outgoing flux on the coordinates associated with the emitted (ionized) particles. Even if it is not yet totally clear how this happens,[14] it has been confirmed with many numerical results that the correct three-body outgoing behavior is conferred on the wave function when this methodology is applied. All these approaches work very efficiently on the treatment of collision processes involving just a very few particles, leading to accurate solutions to, for example, the ionization of hydrogen by electron impact.[15] The remarkable growth of computational capabilities has made possible the implementation of those recipes. However, computational limitations do not allow, for the moment, their extension for more than two interacting electrons. For example, full time-independent calculations of a simple four-body problem, such as the double ionization of Helium by electron impact, are not yet available. The main limitation is, of course, the size of the equation systems to be solved.

It is important to point out that the CCC approach, for example, does not consider all the electrons on the same footing (one of the emitted electrons is treated differently from the others) bringing in some problems in the approach.[16] On the other hand, the ECS treats the emitted electrons in the same way, rotating the coordinates to the complex plane for each electron and ending up in the continuum part of the spectra. The description of those electrons is done using spectral methods but with basis functions which, in most of the implementations, are purely numerical with no, or little, trace of the physics of the considered problem.

The synthesis of the aforementioned issues together with the aim of numerical basis optimization directed us to the Generalized Sturmian Functions (GSF) approach. Following the leading ideas of Avery[17–19] and Aquilanti,[20–23] on one side, and Rawitscher[24–28] and Macek,[29–34] on the other, starting nearly a decade ago[35] the present authors developed the approach thereafter.[36–43] The GSF approach makes use of a combination of the best features of the precedent methods. On one hand, it treats the dynamics of the system's electrons on equal footing as each of them is represented by GSF. On the other hand, the asymptotic behavior of each particle is taken into account by the basis allowing matching of the expansion region to the reaction region, where all the dynamics takes place. For electrons, for example, some may be in the continuum while others occupy bound states: within the GSF method, they will be represented by basis functions that have proper outgoing or bound-type behavior, respectively. It has been already numerically demonstrated that the use of GSF basis with outgoing components allows the enforcement of three-body type outgoing behavior in a similar way as the ECS method does. Generally speaking, a distinctive feature of the GSF method is that it employs basis functions possessing physical properties similar to those corresponding to the system under study. Moreover, the use of GSF greatly increases the convergence rate of various physical quantities when dealing with bound states and reduces the computational resources required for describing scattering processes. It is the purpose of this work to review some of the aspects of the Generalized Sturmian function method that we developed.

This paper is organized as follows: In Section 2 all the relevant information referred to the GSF for two-body problems is presented, as well as the theory corresponding to their application to bound and scattering problems involving only two-particles. Several relevant aspects of the method are introduced there, starting with the Sturm–Liouville equation for a general radial interaction, the boundary conditions, and the definition of the auxiliary and generating potentials of the GSF basis. A comparison with CSF is also presented. The remaining sections are devoted to the three-body problem. In Section 3, the use of GSF to study three body Coulomb problems for particles of arbitrary masses is presented, focusing on the calculation of atomic and molecular structure. Calculations for a wide variety of systems are discussed, ranging from the negative Hydrogen ion and the largely studied Helium atom to more exotic muonic systems. Ground and excited states including the double excited states of Helium are considered. Also, very accurate energy values are presented for a Helium atom confined inside a model cage representing a fullerene molecule, or in an impenetrable box. In Section 4, the GSF theory for three-body continuum states is described, both in spherical and in hyperspherical coordinates. The three-body asymptotic behavior for scattering problems is discussed, and the driven equations for two typical ionization processes are presented, illustrating the possible benefits of both systems of coordinates. In Section 5, several examples of simple collision systems are developed, centering the attention on the properties of the wave function, the convergence of the method, and the extraction of cross-sections from the asymptotic flux. Finally, in Section 6 the important features of the GSF method are summarized and some perspectives are outlined.

Atomic units are used through this paper, unless otherwise noted.

2. GENERALIZED STURMIAN FUNCTIONS

2.1 Definitions

The Generalized Sturmian functions to be studied and applied are all associated with the two-body problems described by the time-independent Schrödinger equation

$$\left[-\frac{1}{2\mu}\nabla^2 + \mathcal{U}(r) - E\right] \mathcal{S}(\mathbf{r}) = -\beta\, \mathcal{V}(r)\mathcal{S}(\mathbf{r}). \qquad (1)$$

We are assuming here that $\mathcal{U}(r)$ and $\mathcal{V}(r)$ are spherically symmetric potentials. Equation (1) has three main parameters in addition to the reduced mass

μ, they are: the energy E, the magnitude of $\mathcal{U}(r)$, and the magnitude β of the potential $\mathcal{V}(r)$. The idea behind the Generalized Sturmian functions is to consider the energy E and the magnitude of $\mathcal{U}(r)$ as externally fixed parameters and β the eigenvalue of the problem. Given the fact that we will be dealing with spherically symmetric potentials, we can propose the separation on radial and angular variables $\mathcal{S}(\mathbf{r}) = S_{nl}(r)\, Y_{lm}(\theta, \phi)/r$. With this definition, the radial part of Eq. (1) transforms into

$$\left[-\frac{1}{2\mu}\frac{\mathrm{d}^2}{\mathrm{d}r^2} + \frac{l(l+1)}{2\mu r^2} + \mathcal{U}(r) - E \right] S_{nl}(r) = -\beta_{nl}\, \mathcal{V}(r)\, S_{nl}(r). \qquad (2)$$

We name $\mathcal{U}(r)$ the *auxiliary* potential (e.g., Coulomb), and $\mathcal{V}(r)$ the *generating* potential defined in general to be of short-range i.e. vanishing in an outer region $r > R_S$. To transform the Schrödinger equation (2) into a Sturm–Liouville problem two boundary conditions are required. Throughout this paper, we seek solutions having a regular boundary condition at the origin

$$S_{nl}(r = 0) = 0. \qquad (3)$$

If the auxiliary potential $\mathcal{U}(r)$ is a Coulomb potential, then, the Kato cusp condition can also be imposed through an appropriate choice of the potentials $\mathcal{U}$ and $\mathcal{V}$, as discussed in. Ref. 42

In the outer region, where the second boundary condition is imposed, the generating potential vanishes. The radial Sturmian equation (2) reduces to

$$\left[-\frac{1}{2\mu}\frac{\mathrm{d}^2}{\mathrm{d}r^2} + \frac{l(l+1)}{2\mu r^2} + \mathcal{U}(r) - E \right] S_{nl}(r) = 0 \quad \text{for} \quad r > R_S, \qquad (4)$$

the solutions of which represent a particle of energy E moving under the influence of a potential $\mathcal{U}(r)$, and will have a unique asymptotic behavior, that is, independent of the eigenvalue β_{nl}.

Equation (2) together with the boundary conditions (3) and (4) define a Sturm–Liouville problem. Therefore, their solutions form a complete basis set with closure

$$\sum_n S_{nl}(r')\, \mathcal{V}(r)\, S_{nl}(r) = \delta(r - r'), \qquad (5)$$

and obey a potential-weighted orthogonality relation:

$$\int_0^\infty S_{n'l}(r)\,\mathcal{V}(r)\,S_{nl}(r)\,\mathrm{d}r = 0 \quad \text{for} \quad n' \neq n. \tag{6}$$

Two important issues related to this orthogonality relation are worth of notice. First, due to the fact that the generating potential $\mathcal{V}(r)$ is of short-range, the integration (6) can be performed numerically in a finite spatial region for any value of the energy E, without any additional requirement over the $S_{nl}(r)$ functions. The second point to note is that the integral is defined without taking the complex conjugate of the function $S_{n'l}(r)$. With this particular choice of the scalar product, the overlaps between the functions:

$$\mathcal{O}_{n'n} = \int_0^\infty S_{n'l}(r) S_{nl}(r)\,\mathrm{d}r \tag{7}$$

converge for any energy and potential (this was discussed in detail in Ref. 44). It is interesting to note that since the generating potential is of short-range, the boundary condition at large values of r can be imposed at any r value greater than R_S. This allows numerical construction of the functions in an efficient way. All the functions differ asymptotically from one another by a normalization factor.

Let us now illustrate that the $S_{nl}(r)$ functions have the same asymptotic behavior with different examples, beginning with the case of negative energies $E < 0$. For a short-range auxiliary potential $\mathcal{U}(r)$ (i.e., $\mathcal{U}(r)$ also vanishing in the outer region) the asymptotic solutions of (4) behave like

$$\lim_{r\to\infty} S_{nl} \propto e^{-\kappa r}, \tag{8}$$

where $\kappa = \sqrt{-2\mu E}$. Since E is a fixed parameter, all the basis functions have the same asymptotic behavior.

In the case of a Coulombic auxiliary potential $\mathcal{U}(r) = z_1 z_2/r$ with z_1 positive and z_2 negative, the solutions of (2) have an asymptotic behavior consisting of an exponentially decaying factor $\exp(-\kappa r)$ modified by a logarithmic factor

$$\lim_{r\to\infty} S_{nl} \propto e^{-\kappa r - \frac{z_1 z_2 \mu}{\kappa} \ln(2\kappa r)}, \tag{9}$$

which is, again, the same for all basis elements.

When dealing with positive energies the expressions given by Eqs. (8) and (9) transform into

$$\lim_{r\to\infty} S_{nl} \propto e^{ikr}, \tag{10}$$

and

$$\lim_{r\to\infty} S_{nl} \propto e^{ikr - i\frac{z_1 z_2 \mu}{k}\ln(2kr)}, \tag{11}$$

respectively. These are the asymptotic solutions of Eq. (4) corresponding, respectively, to a short-range and a Coulomb-type auxiliary potential. Similar expressions for incoming waves can be also used.

At this point we would like to emphasize the difference between our Generalized Sturmian functions with the Coulombic auxiliary potential, and the widely used Coulomb Sturmian functions. The CSF are solutions of

$$\left[-\frac{1}{2\mu}\frac{\mathrm{d}^2}{\mathrm{d}r^2} + \frac{l(l+1)}{2\mu r^2} - E\right] S^C_{nl}(r) = \beta_{nl}\,\frac{1}{r}\, S^C_{nl}(r), \tag{12}$$

i.e., the same Eq. (2) with $\mathcal{U}(r) = 0$ and $\mathcal{V}(r) = -1/r$. The discretized character of the solutions imposes $\beta_{nl} = n\kappa$. In this case, the asymptotic behavior of the CSF is

$$\lim_{r\to\infty} S^C_{nl} \propto e^{-\kappa r + n\ln(2\kappa r)} = (2\kappa r)^n\, e^{-\kappa r}, \tag{13}$$

which changes from one CSF element to another, due to the presence of the power n. For positive energies, the asymptotic condition reads

$$\lim_{r\to\infty} S^C_{nl} \propto e^{ikr + \frac{Z_n \mu}{k}\ln(2kr)}, \tag{14}$$

where $Z_n = \beta_n = in\sqrt{2\mu E}$. All the functions oscillate with the same wavenumber k (determined externally by the fixed energy) but each of them has a different logarithmic phase, depending on the eigenvalue.

2.2 Bound states

In this section, we briefly review the methodology for two-body bound states. For a given interaction potential $V(r)$, the aim is to solve the Schrödinger equation

$$\left[-\frac{1}{2\mu}\frac{d^2}{dr^2} + \frac{l(l+1)}{2\mu r^2} + V(r) - E \right] \Psi(r) = 0, \tag{15}$$

where, for bound states, the energy E is negative. The solution $\Psi_B(r)$ may be expanded as

$$\Psi_B(r) = \sum_n a_n S_{nl}(r). \tag{16}$$

Upon replacement in (15), and using (2) with energy E_s, we have

$$\sum_n a_n \left[V(r) - \mathcal{U}(r) - \beta_{nl}\mathcal{V}(r) - E + E_s \right] S_{nl}(r) = 0. \tag{17}$$

We may choose the auxiliary potential $\mathcal{U}(r)$ to be the interaction $V(r)$, so that the first two terms in Eq. (17) cancel each other. Projecting with basis functions $S_{n'l}(r)$ from the left, integrating over the coordinate, using the orthogonality relation (6) and the overlap matrix (7), we get

$$\sum_n \left[-\beta_{nl}\mathcal{V}_{n'n}\delta_{n'n} - (E - E_s)\mathcal{O}_{n'n} \right] a_n = 0, \tag{18}$$

which can be easily solved by standard matrix methods.[40]

2.2.1 Numerical implementation

Details of the numerical aspects involved in the generation of the Sturmian basis set are published in Ref. 43 and will not be repeated here. We will just mention that we use primarily a finite-differences scheme to solve Eq. (2). The boundary condition is imposed within this scheme, treating in a special way the last elements of the Hamiltonian matrix. The eigenvalues are initially calculated by using a complex-orthogonal transformation based on the theory explained by Luk and Qiao.[45] In previous work,[38,39] we performed a further relaxation of a random vector multiplied by the Hamiltonian matrix, until it converges to the desired eigenvalue. In that procedure, the corresponding eigenvectors were also obtained. We also developed another very fast iterative algorithm that allows us to obtain very accurate eigenvalues. For any eigenvalue, the corresponding eigenvector is computed by using a predictor–corrector method. The procedure includes a numerical (outwards) integration from the origin to a particular matching point and another integration (inwards) from the asymptotic region to the

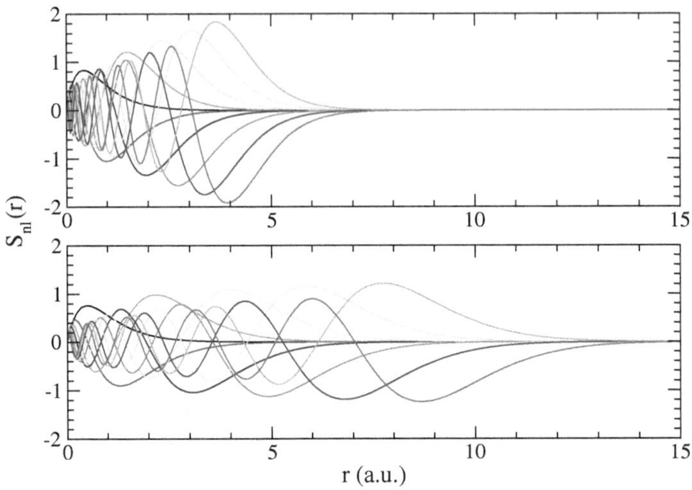

Figure 7.1 Up: First Generalized Sturmian functions S_{n0} for an auxiliary potential $\mathcal{U}(r) = 0$, a generating potential $\mathcal{V}(r) = e^{-0.375r}/r$, and an energy $E = -2.0$ a.u. Down: Coulomb Sturmian functions, for the same energy. For color version of this figure, the reader is referred to the online version of this chapter.

matching point. At this particular matching point the logarithmic derivatives are calculated for both solutions, and iterations are performed until the difference between both values becomes negligible.

For completeness, we should mention that we have also developed other implementations for GSF basis set generation. For example, in Ref. 41 we used a square integrable (L^2) basis set expansion with Laguerre-type basis functions[46] and B-spline functions.[47]

2.2.2 Example

As an illustration, consider the case in which $E = -2.0$ a.u. and $l = 0$. Figure 7.1 displays (in the upper curves) the first 10 GSF corresponding to the case of a Coulomb auxiliary potential

$$\mathcal{U}(r) = -\frac{Z}{r}, \tag{19}$$

where $Z = 0$, and a Yukawa generating potential

$$\mathcal{V}(r) = -\frac{e^{-a_s r}}{r}, \tag{20}$$

with $a_s = 0.375$. For comparison the Coulomb Sturmian functions for the same energy are plotted in the lower part of the figure. Clearly, the CSF are spread over a wide range, whereas the generating potential allows to concentrate the GSF basis set close to the origin while maintaining a realistic fall off behavior. It is usually in the inner region that the interacting part is important and needs a good numerical resolution. Thus, one may use the generating potential to tune the basis according to the particular physical problem under study.

2.3 Scattering states

For scattering solutions, one may set the full solution as[7]

$$\Psi(r) = \Psi_0(r) + \Psi_{sc}(r), \tag{21}$$

where $\Psi_0(r)$ is an asymptotic solution which solves a simplified, homogeneous, equation

$$\left[-\frac{1}{2\mu}\frac{\mathrm{d}^2}{\mathrm{d}r^2} + \frac{l(l+1)}{2\mu r^2} + V_0(r) - E\right]\Psi_0(r) = 0; \tag{22}$$

the scattering part $\Psi_{sc}(r)$ solves the inhomogeneous differential equation

$$\left[-\frac{1}{2\mu}\frac{\mathrm{d}^2}{\mathrm{d}r^2} + \frac{l(l+1)}{2\mu r^2} + V(r) - E\right]\Psi_{sc}(r) = -\left[V(r) - V_0(r)\right]\Psi_0(r). \tag{23}$$

Expanding $\Psi_{sc}(r)$ and $\left[V(r) - V_0(r)\right]\Psi_0(r)$ in Sturmian functions with externally fixed energy $E_s = E > 0$

$$\Psi_{sc}(r) = \sum_n a_n S_{nl}(r), \tag{24}$$

$$\left[V(r) - V_0(r)\right]\ \Psi_0(r) = \sum_n b_n \mathcal{V}(r) S_{nl}(r), \tag{25}$$

and using Eq. (2), the radial Schrödinger equation is converted into

$$\sum_n a_n \left[V(r) - \mathcal{U}(r) - \beta_{nl}\mathcal{V}(r)\right]\ S_{nl}(r) = -\sum_n b_n \mathcal{V}(r) S_{nl}(r). \tag{26}$$

Choosing again $\mathcal{U}(r) = V(r)$, only the generating potential remains on the LHS. Projecting to the left by $S_{n'l}(r)$, we end up again with a matrix problem

$$\beta_{n'} a_{n'} = b_{n'}. \tag{27}$$

The Sturmian basis functions transform the operator $[H - E]$ into a diagonal matrix whose elements are simply the Sturmian eigenvalues. This can be seen in an alternative form. Equation (23) can be rewritten as:

$$\Psi_{sc}^{\pm}(r) = G_l^{\pm}[V(r) - V_0(r)]\Psi_0(r) \tag{28}$$

in terms of Green's function $G_l^{\pm}$ which is responsible for providing the correct asymptotic behavior (outgoing (+) or incoming (−)) to $\Psi_{sc}^{\pm}(r)$. Now, Green's function satisfies the equation

$$\left[-\frac{1}{2\mu}\frac{d^2}{dr^2} + \frac{l(l+1)}{2\mu r^2} + V(r) - E\right] G_l^{\pm}\left(E, r, r'\right) = \delta\left(r - r'\right), \tag{29}$$

and can be expanded in terms of Sturmians functions as follows:

$$G_l^{\pm}\left(E, r, r'\right) = \sum_n g_{nl} S_{nl}^{\pm}(r') S_{nl}^{\pm}(r). \tag{30}$$

Replacing this expansion into (29), using Eq. (2), and taking $\mathcal{U}(r) = V(r)$, we find

$$-\sum_n g_{nl}\beta_{nl} S_{nl}^{\pm}(r') S_{nl}^{\pm}(r)\mathcal{V}(r) = \delta\left(r - r'\right). \tag{31}$$

Comparing this expression with the closure relation (5) we see that $g_{nl} = -1/\beta_{nl}$. This means that Green's function is diagonal in the Generalized Sturmian representation. Besides, the representation is optimized since the asymptotic region is associated with the range of the generating potential $\mathcal{V}(r)$; the asymptotic form of $G_l^{\pm}$ is directly given by the correct asymptotic behavior of the Sturmian functions. Indeed, since (2) can be written as

$$-\frac{1}{\beta_{nl}} S_{nl}^{\pm}(r) = G_l^{\pm}\mathcal{V}(r) S_{nl}^{\pm}(r), \tag{32}$$

the Sturmian functions are eigenfunctions of the operator $G_l^{\pm}\mathcal{V}(r)$ with eigenvalues $-1/\beta_{nl}$.[25–27]

It is clear that for scattering problems, the Schrödinger equation to be solved is considerably simplified when using GSF.

2.3.1 Numerical implementation

As pointed out above, a detailed description of the numerical procedures has been presented in Ref. 43. In order to calculate the Sturmian functions suitable for scattering calculations, we set the boundary conditions by choosing

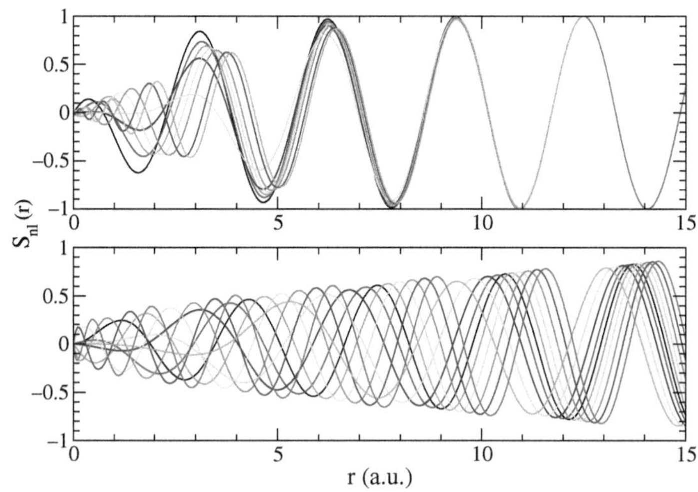

Figure 7.2 Up: First Generalized Sturmian functions S_{n0} for an auxiliary potential $\mathcal{U}(r) = 0$, a generating potential $\mathcal{V}(r) = e^{-0.375r}/r$, and an energy $E = 2.0$ a.u. Down: Coulomb Sturmian functions, for the same energy. For color version of this figure, the reader is referred to the online version of this chapter.

appropriately the auxiliary potential. For Coulombic outgoing/incoming boundary conditions, we calculate the regular $F(r)$ and irregular $G(r)$ radial Coulomb functions, and their derivatives, corresponding to a Sommerfeld parameter η. The asymptotic solution is then written as:

$$P(r \geq R_A) = \cos\delta\, F(r) + \sin\delta\, G(r). \tag{33}$$

where δ is the phase shift, calculated by assuming arbitrarily that at the point R_A the function satisfies $S_{nl}(R_A) = 1$, i.e.,

$$\begin{aligned} \cos\delta &= \frac{F(R_A) + iG(R_A)}{F(R_A)^2 + G(R_A)^2}, \\ \sin\delta &= \frac{G(R_A) - iF(R_A)}{F(R_A)^2 + G(R_A)^2}. \end{aligned} \tag{34}$$

In the same way as for the bound levels, the degree of mismatch of the solutions at the matching point allows estimation of the change $\Delta\beta_n$ needed for the next iteration step, until convergence is achieved in the slopes of the outward and inward functions.

As an example, consider the case in which $E = 2.0$ a.u. and $l = 0$, with the same auxiliary and generating potentials used in the previous bound state illustration. Figure 7.2 displays the first 10 GSF and, for comparison, the first 10 CSF corresponding to the same energy, and with outgoing boundary conditions. For GSF, the role played by the two potentials is now clear: the generating potential $\mathcal{V}(r)$ is used to compress the basis inside the range of the interaction, while the auxiliary potential $\mathcal{U}(r)$ dictates the asymptotic behavior which, for the GSF, is unique for all basis elements.

This is in contrast with CSF which possess different asymptotic behaviors. It has to be noted also that the CSF with outgoing flux behavior are divergent unless the energy is appropriately chosen to be complex.

2.3.2 Example

An application of the GSF method is given by the following example in which we define a distorted wave approach to the Coulomb wave equation, with a Sommerfeld parameter $\eta = z_1 z_2 \mu / k$. Consider the asymptotic function $\Psi_0(r)$ of Eq. (21) as given by

$$\begin{aligned} \Psi_0(r) &= \sin\left[kr - \left(\eta \ln(2kr) + \frac{\pi}{2} l\right) g(r)\right] \\ &\underset{r\to\infty}{\rightarrow} \frac{1}{2i}\left(-e^{-i\left[kr-\eta\ln(2kr)-\frac{\pi}{2}l\right]} + e^{i\left[kr-\eta\ln(2kr)-\frac{\pi}{2}l\right]}\right), \end{aligned} \tag{35}$$

where $g(r)$ can be any function growing faster than r at the origin and tending to one at large distances as, e.g., $g(r) = 1 - e^{-a_c r^2}$ (a_c is a positive real constant).The function $\Psi_0\,(r)$ solves, asymptotically, the Schrödinger equation:

$$\left[-\frac{1}{2\mu}\frac{\mathrm{d}^2}{\mathrm{d}r^2} + \frac{l(l+1)}{2\mu r^2} + \frac{z_1 z_2}{r} - E\right]\Psi_0\,(r) = \mathcal{O}\left(\frac{1}{r^2}\right). \tag{36}$$

By using $\Psi_0(r)$ instead of the free-particle function $rj_0(r)$—the spherical Bessel function—we are introducing a distorted-wave approach.[14] With these definitions the decomposition (21) is in agreement with standard scattering theory. Indeed, the driven Schrödinger equation to solve is

$$\begin{aligned} &\left[-\tfrac{1}{2\mu}\tfrac{\mathrm{d}^2}{\mathrm{d}r^2} + \tfrac{l(l+1)}{2\mu r^2} + \tfrac{z_1 z_2}{r} - E\right]\Psi_{sc}\,(r) \\ &= \left[E - \left(-\tfrac{1}{2\mu}\tfrac{\mathrm{d}^2}{\mathrm{d}r^2} + \tfrac{l(l+1)}{2\mu r^2}\right) - \tfrac{z_1 z_2}{r}\right]\Psi_0\,(r)\,. \end{aligned} \tag{37}$$

According to Eq. (36), the RHS of (37) goes as $1/r^2$ at large distances, so that one may require $\Psi_{sc}\,(r)$ to have outgoing behavior $A_l e^{i\left[kr-\eta\ln(2kr)-\frac{\pi}{2}l\right]}$, as scattering theory establishes. One may then extract the corresponding transition amplitude since, at large distances, we have

$$\begin{aligned} \Psi^+(r) \rightarrow &\frac{1}{2i}\left(-e^{-i\left[kr-\eta\ln(2kr)-\frac{\pi}{2}l\right]} + e^{i\left[kr-\eta\ln(2kr)-\frac{\pi}{2}l\right]}\right) \\ &+ A_l\, e^{i\left[kr-\eta\ln(2kr)-\frac{\pi}{2}l\right]}, \end{aligned} \tag{38}$$

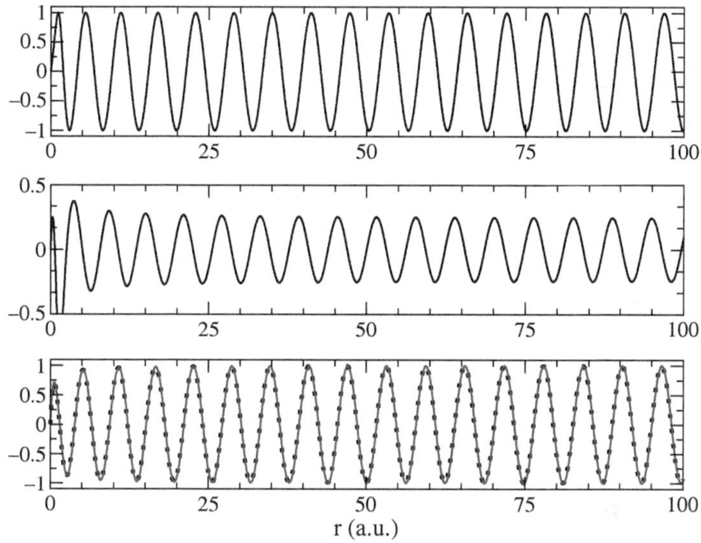

Figure 7.3 The functions $\Psi_0(r)$ (top panel), $\Psi_{sc}(r)$ (middle panel), and $\Psi(r)$ (bottom panel) of, respectively, Eqs. (35), (37), and (21) are plotted as a function of the radial coordinate r; we have taken $l = 0, k = 1, \mu = 1$, and $z_1 z_2 = -1$, and $a_c = 0.5$ in Eq. (35). In the bottom panel, the Coulomb wave function $F(r)$ is included (solid dots) for comparison. For color version of this figure, the reader is referred to the online version of this chapter.

where the transition matrix $A_l = e^{i\delta_l} \sin(\delta_l)$ will result from solving Eq. (37) for $\Psi_{sc}(r)$.

To illustrate the proposal, the distorted-wave driven Eq. (37) was solved numerically using GSF with outgoing asymptotic behavior.[14] In the top and middle panels of Figure 7.3, we plot the functions $\Psi_0(r)$ and $\Psi_{sc}(r)$. In the bottom panel the sum of these functions, $\Psi(r)$, as defined in Eq. (21) is shown and compared with the exact Coulomb wave function $F(r)$ (solid dots) with unit asymptotic norm. The facts that $\Psi_0(r)$ possesses the appropriated Coulomb distortion and that the sum (21) leads to the correct result, show that the driven Eq. (37) is well formulated and is in accordance with scattering theory.

3. THREE-BODY PROBLEMS: BOUND STATES

The configuration interaction method has been widely used to perform *ab initio* calculations of N-electron atomic and molecular systems.[5,17–19,48,49] The differences between various implementations reside in the type of radial basis set used to expand the solutions. The great advantage of the *CI* method is it simplicity and flexibility. It allows *ab initio* calculations of three or more particle systems in a relatively easy way. However, the convergence rate is strongly dependent on the quality of the chosen basis functions.

Among the many basis sets in spherical coordinates, a very efficient scheme for atomic systems is based on the use of Laguerre-type orbitals[48] which essentially define the Coulomb Sturmian functions.The parameters of the CSF basis for two-electron systems are numerically optimized to obtain the best values for the energies.[49] This Sturmian methodology was shown to be efficient in obtaining atomic bound states for two- and three-electron atoms, as well as for molecular systems.

Here we present an alternative *CI* methodology based on two-body GSF selecting parameters to construct optimized basis functions for bound state problems.

3.1 Systems with general masses

Let us consider three particles with masses (m_1, m_2, m_2) and charges (z_1, z_2, z_3), where m_3 is the heaviest particle. Separating the center of mass motion, and denoting by $\mathbf{r}_1$ and $\mathbf{r}_2$ the relative coordinates of particles m_1 and m_2 with respect to m_3, the Hamiltonian describing the relative motion of the two lighter particles is written as:

$$H = -\frac{1}{2\mu_{13}}\nabla^2_{\mathbf{r}_1} - \frac{1}{2\mu_{23}}\nabla^2_{\mathbf{r}_2} - \frac{1}{m_3}\nabla_{\mathbf{r}_1}\cdot\nabla_{\mathbf{r}_2} + \frac{z_1 z_3}{r_1} + \frac{z_2 z_3}{r_2} + \frac{z_1 z_2}{r_{12}}, \tag{39}$$

where $\mu_{ij} = (m_i m_j)/(m_i + m_j)$ is the reduced mass of particle i with respect to particle j, $r_{12} = |\mathbf{r}_1 - \mathbf{r}_2|$ and the operator $-\frac{1}{m_3}\nabla_{\mathbf{r}_1}\cdot\nabla_{\mathbf{r}_2}$ is the mass polarization term.The wave function $\Psi(\mathbf{r}_1, \mathbf{r}_2)$ satisfies the Schrödinger equation

$$H\Psi(\mathbf{r}_1, \mathbf{r}_2) = E\Psi(\mathbf{r}_1, \mathbf{r}_2), \tag{40}$$

where E is the energy of the three-particle system.

We may define the *CI* expansion of the wave function using two-body GSF for each coordinate $\mathbf{r}_1$ and $\mathbf{r}_2$. For a given total angular momentum L and projection M, it is written as:

$$\Psi(\mathbf{r}_1, \mathbf{r}_2) = \sum_{l_1 l_2}\sum_{n_1=1}^{N_1}\sum_{n_2=1}^{N_2} a^{l_1 l_2 LM}_{n_1 n_2}\,\mathcal{A}\frac{S_{n_1 l_1}(r_1)}{r_1}\frac{S_{n_2 l_2}(r_2)}{r_2}\mathcal{Y}^{LM}_{l_1 l_2}(\widehat{\mathbf{r}}_1, \widehat{\mathbf{r}}_2), \tag{41}$$

where $a^{l_1 l_2 LM}_{n_1 n_2}$ are the expansion coefficients and $\mathcal{A} = (1 + \epsilon P_{12})/\sqrt{2}$ is the symmetrization operator that accounts for the exchange of the particles.

The permutation operator P_{12} exchanges coordinates $r_1 \leftrightarrow r_2$, and $\epsilon = 1$ for singlet states or $\epsilon = -1$ for triplet states.

The angular part is expressed in terms of the coupled spherical harmonics[1]

$$\mathcal{Y}_{l_1 l_2}^{LM}(\widehat{\mathbf{r}}_1, \widehat{\mathbf{r}}_2) = \sum_{m_1 m_2} \langle l_1 m_2 l_2 m_2 | \, l_1 l_2 LM \rangle \, Y_{l_1 m_1}(\widehat{\mathbf{r}}_1) Y_{l_2 m_2}(\widehat{\mathbf{r}}_2), \tag{42}$$

where Y_{lm} are spherical harmonics and $\langle l_1 m_2 l_2 m_2 | \, l_1 l_2 LM \rangle$ are Clebsch–Gordan coefficients.The individual angular momenta l_1 and l_2 must satisfy the triangular rule and parity conservation $(-1)^L = (-1)^{l_1 + l_2}$.To avoid redundancies in expansion (41), the angular momenta are restricted to $l_1 \leq l_2$.Also, if $l_1 = l_2$, then $n_1 \leq n_2$. These restrictions ensure numerical stability in the calculation, avoiding repeated sets of equations given by symmetry.

Replacing expansion (41) into the Schrödinger Eq. (40) and projecting onto the Sturmian basis set, we obtain a generalized eigenvalue problem of the form

$$\mathbf{H}\,\mathbf{a}_\nu^{LM} = E_\nu \,\mathbf{S}\mathbf{a}_\nu^{LM}, \tag{43}$$

where for a given state ν, the vector $\mathbf{a}_\nu^{LM}$ contains the expansion coefficients $a_{n_1 n_2}^{l_1 l_2 LM}$ for fixed (L, M). The energy eigenvalues are $E = E_\nu$, while the Hamiltonian and overlapping matrices $\mathbf{H}$ and $\mathbf{S}$ have elements

$$[\mathbf{H}]_{n_1' n_2' n_1 n_2}^{LMl_1' l_2' l_1 l_2} = \left\langle \mathcal{Y}_{l_1' l_2'}^{LM} \frac{S_{n_1' l_1'}}{r_1} \frac{S_{n_2' l_2'}}{r_2} \middle| H \middle| \mathcal{A} \frac{S_{n_1 l_1}}{r_1} \frac{S_{n_2 l_2}}{r_2} \mathcal{Y}_{l_1 l_2}^{LM} \right\rangle, \tag{44}$$

$$[\mathbf{S}]_{n_1' n_2' n_1 n_2}^{LMl_1' l_2' l_1 l_2} = \left\langle \mathcal{Y}_{l_1' l_2'}^{LM} \frac{S_{n_1' l_1'}}{r_1} \frac{S_{n_2' l_2'}}{r_2} \middle| \mathcal{A} \frac{S_{n_1 l_1}}{r_1} \frac{S_{n_2 l_2}}{r_2} \mathcal{Y}_{l_1 l_2}^{LM} \right\rangle. \tag{45}$$

The matrices $\mathbf{H}$ and $\mathbf{S}$ can be computed as:

$$\begin{aligned}
[\mathbf{H}]_{n_1' n_2' n_1 n_2}^{LMl_1' l_2' l_1 l_2} = & \left[(E_1 + E_2)\, O_{n_1' n_1}^{l_1' l_1} \, O_{n_2' n_2}^{l_2' l_2} - \beta_{n_1 l_1} \, [\mathcal{V}_1]_{n_1' n_1}^{l_1' l_1} \, O_{n_2' n_2}^{l_2' l_2} \right. \\
& \left. -\beta_{n_2 l_2} \, [\mathcal{V}_2]_{n_2' n_2}^{l_2' l_2} \, O_{n_1' n_1}^{l_1' l_1} \right] \delta_{l_1' l_1} \delta_{l_2' l_2} \\
& + [\mathbf{T}_{12}]_{n_1' n_2' n_1 n_2}^{LMl_1' l_2' l_1 l_2} + \sum_{l=0}^{\infty} \frac{4\pi}{2l+1} R_{n_1' n_2' n_1 n_2}^{l_1' l_2' l_1 l_2 l} \sum_{m=-l}^{l} A_{l_1' l_2' l_1 l_2 l}^{LM} \\
& + \epsilon P(n_1 \to n_2, l_1 \to l_2),
\end{aligned} \tag{46}$$

$$[\mathbf{S}]_{n_1' n_2' n_1 n_2}^{LMl_1' l_2' l_1 l_2} = O_{n_1' n_1}^{l_1' l_1} \, O_{n_2' n_2}^{l_2' l_2} \, \delta_{l_1' l_1} \delta_{l_2' l_2} + \epsilon P(n_1 \to n_2, l_1 \to l_2). \tag{47}$$

For the sake of compactness, we do not provide here explicit expressions for the mass polarization matrix $\mathbf{T}_{12}$. Nine different terms are necessary to calculate it with no evaluation difficulties. For the electron–electron interaction term use is made of the standard spherical harmonics expansion

$$\frac{1}{r_{12}}=\sum_{l=0}^{\infty}\sum_{m=-l}^{l}\frac{4\pi}{2l+1}\frac{r_<^l}{r_>^{l+1}}Y_{lm}^*(\theta_1,\phi_1)Y_{lm}(\theta_2,\phi_2), \tag{48}$$

with $r_<=\min(r_1,r_2)$ and $r_>=\max(r_1,r_2)$. In the matrix expressions (46) and (47) we explicitly noted the action of the operator $P_{12}=P(n_1\to n_2,l_1\to l_2)$ onto the index of the basis set, and defined the following integrals:

$$O_{n'n}^{l'l}=\int_0^\infty \mathrm{d}r S_{n'l'}(r)S_{nl}(r), \tag{49a}$$

$$[\mathcal{V}_i]_{n'n}^{l'l}=\int_0^\infty \mathrm{d}r S_{n'l'}(r)\mathcal{V}_i(r)S_{nl}(r), \tag{49b}$$

$$R_{n_1'n_2'n_1n_2}^{l_1'l_2'l_1l_2l}=\int_0^\infty \mathrm{d}r_1\int_0^\infty \mathrm{d}r_2 S_{n_1'l_1'}(r_1)S_{n_2'l_2'}(r_2)\frac{r_<^l}{r_>^{l+1}}S_{n_1l_1}(r_1)S_{n_2l_2}(r_2), \tag{49c}$$

$$A_{l_1'l_2'l_1l_2l}^{LM}=\int \mathrm{d}\widehat{\mathbf{r}}_1\int \mathrm{d}\widehat{\mathbf{r}}_2\mathcal{Y}_{l_1'l_2'}^{LM*}(\widehat{\mathbf{r}}_1,\widehat{\mathbf{r}}_2)Y_{lm}^*(\widehat{\mathbf{r}}_1)Y_{lm}(\widehat{\mathbf{r}}_2)\mathcal{Y}_{l_1l_2}^{LM}(\widehat{\mathbf{r}}_1,\widehat{\mathbf{r}}_2). \tag{49d}$$

3.2 Two-electron atoms

We first look at the case where the two light particles are electrons ($m_1=m_2=1, z_1=z_2=-1$) bound to a heavy nucleus of infinite mass ($m_3\to\infty$), so the mass polarization term can be omitted

$$H_a=-\frac{1}{2\mu_{13}}\nabla_{\mathbf{r}_1}^2-\frac{1}{2\mu_{23}}\nabla_{\mathbf{r}_2}^2+\frac{z_1z_3}{r_1}+\frac{z_2z_3}{r_2}+\frac{z_1z_2}{r_{12}}. \tag{50}$$

We test our method computing the ground state of He ($z_3=2$) and H^- ($z_3=1$), and then extend our study to singly and doubly excited states of He.

The GSF for each electron can be defined for different sets of basis parameters ($\mathcal{U}(r),\mathcal{V}(r),E,N$). Our basis can represent either symmetric states where the electrons are equivalent (such as the ground state) or asymmetric excited states, where one electron has a lower energy than the other. Different basis parameters may be convenient to deal with continuum (for example, auto-ionizing) states. In all cases, we will use the notation E_i and N_i for $i = 1, 2$ to distinguish the electrons.

So far, we have considered the energy of the basis set as a real, negative parameter, so the GSF have an exponentially decaying asymptotic behavior. However, if the energy is chosen to be positive, outgoing wave asymptotic behavior can be imposed to the basis set, which is adequate for doubly excited, auto-ionizing states. The matrix obtained with this boundary condition is now complex symmetric and non-Hermitian, and the eigenvalues will be complex, too.

3.2.1 Partial-wave analysis of the ground state of He and H^-

In Table 7.1, we show the ground-state energy of He as a function of the angular configurations (l_1, l_2) considered for each electron. These results are, as far as we know, the best obtained with uncorrelated basis.[38,42] They were obtained with 20 GSF per angular momentum quantum number l_i, with a generating Yukawa potential with $a_s = 0.375$ and $E_i = -1.48385$ a.u. $(i = 1, 2)$. The calculation was performed up to $l_1 = l_2 = 12$, for which the maximum basis size was $N_{tot} = 2520$. We compare our results with those of Foumouo and collaborators[9] (obtained with 40 radial CSF functions per l_i for each electron) and with those of Bromley and Mitroy[49] (obtained with 20 radial functions per l_i for each electron). In these studies, a symmetric basis was used, composed of products of Laguerre-type orbitals with special choices of the scaling parameter λ. In the work of Foumouo, that parameter is fixed to a particular value, while in the work of Bromley it is varied

Table 7.1 Partial-wave analysis of the He ground state. l_i $(i = 1, 2)$ is the maximum angular configuration (l_1, l_2) considered. The second and third columns show calculations with 20 GSF and CSF per l_i, respectively, while the fourth column uses 40 CSF

	He ground-state energy		
l_i	GSF[38]	CSF[49]	CSF[9]
0	−2.879 028 654	−2.879 028 507 ($\lambda = 4.8$)	−2.879 027 97
1	−2.900 515 957	−2.900 515 873 ($\lambda = 7.8$)	−2.900 513 86
12	−2.903 710 272	−2.903 711 927 ($\lambda = 25.5$)	–
Exact[50]	**−2.903 724 377**		

independently for each l_i (almost 12 variational parameters for the best value obtained with 20 single-electron orbitals per l_i for $l_i = 12$). In both cases the use of Laguerre basis functions implies that the asymptotic behavior of the basis is not the one associated to the state that they are expanding, since the Coulombic logarithmic phase associated to CSF changes from one basis element to the other. This is not the case of GSF basis set where all the elements have the same asymptotic behavior, both exponential factor and power law. The asymptotic region does not need to be expanded with our basis, since it is already included exactly in all basis elements.

Note that optimizing only one parameter for the case $l_i = 0$, GSF give better results than those of Bromley and Mitroy for $l_i = 0$. Our result obtained with l_i up to 12 is compared with those of Foumouo and co-workers. The energy values presented in the table show that we have obtained more accurate results using half of the coefficients required by Foumouo. To obtain better values for l_i up to 12 we have to perform an optimization over a_s. We found that this optimization was not sufficient to reproduce Bromley's accuracy: we thus also varied the energy of the GSF basis. We found that by setting $a_s = 0.795$ and $E = 1.05$ a.u. for the 12 angular momenta, the resulting energy -2.903712009 a.u. is in better agreement with the exact value than any other calculations. Note also that this last calculation was performed adjusting both the asymptotic behavior of the basis set as well as the region where the He ground state is defined. The optimization of these two physical basis parameters is enough to avoid the complete optimization procedure (involving 13 parameters λ) implemented by Bromley and Mitroy.

Within the GSF method, the above calculation is relatively straightforward. However, the basis functions can be optimized even further to include more physics of the problem,[42] improving both their short distance and their asymptotic behavior. It is known that when one of the electrons is far from the nucleus, it is screened by the other, inner, electron. This can be associated to an effective charge "seen" by the outer electron. For that purpose we define the auxiliary potential as:

$$\mathcal{U}(r) = -\frac{Z_{as}}{r} + \frac{\left(-Z_{in} + Z_{as}\right) e^{-a_{tr} r}}{r}. \tag{51}$$

In many-electron wave function expansions, Z_{in} could be, for example, optimally chosen to be the electron–nucleus interaction weight, and Z_{as}

the asymptotic averaged charge of the electron in its outer region: i.e., $Z_{as} = Z_{in} - (N - 1)$, where N is the number of electrons.

Since the GSF equation includes both the generating and auxiliary potential, the Coulomb properties associated to $\mathcal{U}(r)$ should not be lost by the choice of $\mathcal{V}(r)$ either at $r \to 0$ or $r \to \infty$. An adequate selection is

$$\mathcal{V}(r) = -\frac{e^{-\lambda r}}{r}\left[r^{\delta}e^{-\gamma r^2} + (1 - e^{-\gamma r^2})\right], \tag{52}$$

with $\lambda > 0, \delta > 0$, and $\gamma > 0$. In this way, there are no further Coulomb singularities at $r \to 0$ nor logarithmic distortion at $r \to \infty$ apart from those associated to $\mathcal{U}(r)$. This fulfills the electron–nucleus Kato cusp conditions (see Ref. 42, for a more detailed description). With such basis functions we first performed a variational calculation for $l_i = 0$. This allowed us to obtain values for all the parameters of the potential $\mathcal{V}(r)$ and the parameter a_{tr} of the potential $\mathcal{U}(r)$. Then, we further improved the calculations with l_i up to 12 by adjusting E_i and λ. The values of the parameters are $E_i = -1.05$, $Z_{in} = -2$, $Z_{as} = -2$, $a_{tr} = 0$, $\lambda = 0.93$, $\delta = 0.3$, and $\gamma = 4$. With these values we obtained a ground-state energy of -2.903712820 a.u. for $l_i = 12$ and with 455 basis functions.

A similar calculation has been performed for H^-,[41] with 40 GSF for each electron per l_i, up to $l_i = 5$, obtaining a ground-state energy equal to $E_0 = -0.52772866$ a.u., in excellent agreement with the very precise variational result $E_0 = -0.52775101635$ a.u., given by Freund et al.[51]

3.2.2 Singly-and doubly excited states of He

We extend our study to calculate energies of asymmetrical (excited) states.[52] The improvement in accuracy and convergence reached in comparison with previous calculations is shown in Table 7.2 for the singlet and triplet states of He for $L = 3$. The size of the Sturmian basis for each electron is much smaller compared to that of the ground state, and few angular momenta configurations are needed to achieve convergence. The basis parameters are now different for each electron with $E_1 = -1.95$ a.u., $E_2 = -0.1$ a.u., $a_{s1} = 0.1$, $a_{s2} = 0.3$ and the one-electron basis sizes are $N_1 = 2$ and $N_2 = 12$. The same parameters were used for singlet and triplet cases.

In Table 7.3, we show the results for singly excited states of He with different values of angular momentum L and levels n. The energy values obtained were optimized first for the lowest n level. Although the basis parameters are those which give the best lower energy, they provide a good

Table 7.2 Convergence of the 4^1F and 4^3F He state energy as a function of the electron pair of angular momenta

l_1, l_2	Size	4^1F	4^3F
(0,3)	24	−2.031249	−2.031250
(1,2)	48	−2.031252256	−2.031252292
(2,3)	96	−2.03125509730	−2.03125511983
(3,4)	144	−2.03125512818	−2.03125514924
Ref. 57		**−2.03125514438175**	**−2.03125516840324**

Table 7.3 Energy of the first three excited states for singlet states of He for different L with a total of 168 basis functions

n	L	Present work	Ref. 57
3	2	−2.0556110426	−2.0556207328522456
4	3	−2.03125512987	−2.0312551443817490
5	4	−2.020000709670	−2.0200007108985847
6	5	−2.0138890317669	−2.0138890347542797

representation of the excited energy states as well. Of course this can be improved by changing the basis set values of the energy or the range of the Yukawa potential. However, we want to emphasize that the discretization provided by the finite spectral representation gives a good approximation for the lowest excited states without increasing the size or changing the basis set for each state under scrutiny.

Finally, we would like to point out that the precision of our values increases for higher total angular momenta. As mentioned before, the energies of the asymmetrical states are expected to give better accuracy than those of the ground states for the same number of basis elements. This is confirmed by our calculations.

Up to this point, we have shown that our method is able to deal with the ground and asymmetrical states for two-electron systems, for which exponential decay is the suitable asymptotic behavior. Our method can also be applied to calculate the doubly excited states of a two-electron atom. To this end, we choose a positive energy for the Sturmian basis, and the asymptotic behavior as outgoing wave, to meet the requirements of the doubly excited states where one of the electrons can escape and move far

Table 7.4 Energies for the Rydberg series of the singlet S states of He, using 30 GSF per electron

$(N, k)_n$	GSF		Ref. 58	
	$\Re(E)$	$\Im(E)$	$\Re(E)$	$\Im(E)$
$(2, 1)_2$	−0.777876955	−0.002060106	−0.777867636	−0.002270653
$(2, -1)_2$	−0.621817695	−0.000106535	−0.621927254	−0.000107818
$(3, 2)_3$	−0.351827523	−0.001406250	−0.353538536	−0.001504906

from the nucleus. The He eigenvalues E_i obtained with the diagonalization are then complex: the real part is the energies of each atomic state, while the imaginary one gives the width of the state lines. The choice of the basis parameters is made in a similar fashion as for the ground state, but the basis size must be increased for good accuracy.

In Table 7.4, we present the results for the real and imaginary part of the energy for some of the singlet S states of He. We use the $(N, k)_n$ nomenclature for the Rydberg series: the index N denotes the principal quantum number of the remaining ion once the outer electron is ionized. The index n denotes the principal quantum number of the outer electron while the index k determines the parabolic quantum number of the Stark-type state in which the inner electron resides.

The basis set parameters were chosen to give the best value of the first doubly excited state but note that good results for other states can be achieved. The size of the basis set can be increased not only to obtain more accurate results but also to find more eigenvalues that correspond to the Rydberg series.

Remarkable calculations have been performed by Piraux and collaborators using uncorrelated basis.[53–56] Using CSF, they obtained singly and doubly excited states comparable with the results provided by Drake with correlated basis functions. The advantage observed in their method is that the calculations required only one diagonalization and very few non-linear parameters. States with no natural parity were also studied with great success.[54,55]

3.3 Finite mass exotic and molecular systems

We shall now show that the *CI* approach with GSF is quite versatile, as it can be applied to any set of three particles, atomic or molecular. For bound states, we shall briefly consider several situations: two light particles of equal ($[\mu^-, \mu^-, {}^3\text{He}^{+2}]$) or unequal masses (muonic helium: $[e^-, \mu^-, \text{He}^{+2}]$),

Table 7.5 Ground-state energy of the $[\mu^-, \mu^-, {}^3\text{He}^{+2}]$ system. We used 20 radial functions per coordinate and for each partial-wave l_i

$[\mu^-, \mu^-, {}^3\text{He}^{+2}]$ Ground-state energy	
l_i	E_{l_i}
0	−576.874 110
4	−577.555 143
...	...
20	−577.593 497
Refs. 59,60	**−576.934 471**

three light particles (the Ps^- system), and two heavy particles with a light one (H_2^+).

3.3.1 Ground state of the $[\mu^-, \mu^-, {}^3\text{He}^{+2}]$ system

Consider a three charged particle atom with $z_1 = z_2 = -1$ and $z_3 = 2, m_3 = 5495.8852$ a.u. and $m_1 = m_2 = m_\mu$, where m_μ is the muon mass equal to 206.768262 a.u. Since the muons are heavy particles, they will remain much closer to the atomic nucleus than in the case of two electrons. In the calculations presented here we describe the muon–nucleus pairs with GSF which are defined on a radial domain of 0.050 a.u. and with basis energy $E = -300$ a.u. Table 7.5 shows our Sturmian expansion partial-wave results compared with the value given by Rodriguez et al.[59,60] obtained with a modest number of correlated basis functions.

3.3.2 Ground states of the $[\mu^-, e^-, {}^A\text{He}^{+2}]$ systems

Now we consider a He-like system where only one electron is replaced by a muon. This is a very asymmetrical system due to the difference between each particle's mass. Moreover, we consider two different nuclear masses of ${}^A\text{He}^{+2}$: $A = \infty$ and $A = 3$. As already mentioned, the muon–nuclei subsystem maintains a very short separation compared to the electron–nucleus pair. This produces a screening of the nuclear charge seen by the electron. We employed different GSF basis to describe the dynamics of the $\mu^- - He^{+2}$ and $e^- - He^{+2}$ pairs, defined over radial regions equal to 0.025 and 15 a.u. , respectively. The results were not so sensitive to the variation of the rest of the parameters appearing in the Sturmian equation. For these three–body

Table 7.6 Ground state energy of the $[\mu^-, e^-, {}^A\mathrm{He}^{+2}]$ systems. We used 20 radial functions per coordinate and for each partial-wave l_i

$[\mu^-, e^-, {}^A\mathrm{He}^{+2}]$ Ground-state energy		
l_i	$A = \infty$	$A = 3$
2	−414.036 397	−399.041 527
Refs. 59,60	**−414.036 395**	**−399.042 262**

systems the partial wave convergence is very fast; we present in Table 7.6 only the $l_i = 2$ calculations and compare with the results of Refs. 59,60.

3.3.3 Ground state of the Ps$^-$ system

We consider now the Ps$^-$ three-body system, which corresponds to $z_1 = z_2 = -1$ and $z_3 = 1$, and to reduced masses $\mu_{13} = \mu_{23} = 0.5$. Since now $m_3 = 1$ is finite, the mass polarization term will play an important role in the Hamiltonian (39), unlike in the previous two atomic systems. We have used the same generating and auxiliary potentials of Eqs. (51) and (52) with the following parameters: $E_s = -0.06371, Z_{in} = Z_{as} = -1, \alpha = 1, \lambda = 0.1$, $\delta = 0.3$, and $\gamma = 4$. As in the previous systems, this calculation is not fully variational. By

Table 7.7 Partial-wave convergence of the ground-state energy of Ps$^-$ system. We used 35 radial functions per coordinate and for each partial-wave l_i

Ps$^-$ Ground-state energy	
l_i	E_{l_i}
0	−0.257 240 143
1	−0.260 105 390
2	−0.261 496 276
...	...
12	−0.262 002 458
Exact[61]	**−0.262 005 070**

adjusting λ and the basis energy, including partial–waves up to $l_i = 6$ is enough to reach energy values below -0.262 a.u. Our best energy value (see Table 7.7), obtained considering up to $l_i = 12$, is in excellent agreement with the very accurate results of Drake et al.[61]

3.3.4 Ground state of H_2^+

Finally, we consider the molecular three-body system H_2^+. As we now have two heavy particles, there is no natural assignment for particle 3. By taking it to be one of the protons, we have $z_1 = z_3 = 1$ and $z_2 = -1$, and the reduced masses $\mu_{13} = m_p/2$ and $\mu_{23} \simeq 1$. Since $m_3 = m_p$ is finite, the product of gradient operators will play an important role in Eq. (39). Alternatively, exploiting the symmetry, we may take the electron as particle 3; in this case, $z_1 = z_2 = 1$ and $z_3 = -1$, and $\mu_{13} = \mu_{23} \simeq 1$. With this latter choice, the Schrödinger equation (39) does not differ too much from that of H^-, except by the mass polarization term, the center of coordinates being different. We emphasize that we do not consider the Born–Oppenheimer approximation in either case, as we treat the three particles in a fully quantum framework. Results are shown in Table 7.8.

3.4 Confined atoms

Scientists have paid much attention to the study of atoms and molecules under different compression regimes. This is due to the existence of diverse situations in physics and chemistry such as atoms trapped in cavities, in zeolite channels, or encapsulated in hollow cages of carbon-based

Table 7.8 Partial-wave convergence of the ground-state energy of H_2^+ molecule. We used 35 radial functions per coordinate and for each partial-wave l_i, except for $l_i = 0$ for which we used 44 functions per coordinate

H_2^+ Ground-state energy	
l_i	E_{li}
0	−0.513 721 047
2	−0.560 413 338
…	…
20	−0.596 299 557
Exact[62]	**−0.597 139 063**

nano-materials such as endohedral fullerenes. Models of confined atomic and molecular systems have also found applications in the analysis of the so-called artificial atoms or quantum dots due to their relevance in technological applications. The spherically enclosed atoms represent a model that has been applied in the analysis of several confined systems with different methodologies where compression is simulated through hard or soft walls.

As an example of the versatility and precision of our GSF method we present two calculations of an He atom confined by external potentials. In the first case, the He atom is confined inside a penetrable cage, like a fullerene molecule. In the second case, we study the He confined by an infinite potential well.

3.4.1 He confined in endohedral fullerene

Fullerene molecules are capable of enclosing atoms in their hollow interior, forming endohedrally confined atoms. Since these systems can lead to important applications, we devoted a paper[63] analyzing the dependence of the He energy levels on the strength of the confining potential. It is expected that the atomic properties of the confined atom (such as the wave functions, energy levels, the filling of electronic shells, polarizability, photo-absorption, and ionization, etc.) will be dramatically changed both quantitatively and qualitatively, from those characteristic of the free atoms, and change the physical processes where these species are involved (see, for example, Ref. 64). As the goal was to obtain a general qualitative understanding of these effects, instead of dealing with the real helium atom, the authors calculated the spectra of the Helium atom within the spherically symmetric model (Temkin–Poet model). A further simplification was made by modeling the endohedral environment by an attractive short-range spherical shell with a potential

$$\mathcal{U}(r) = \begin{cases} -\mathcal{U}_0 < 0 & r_c \leq r \leq r_c + \Delta \\ 0 & \text{otherwise} \end{cases} \tag{53}$$

where r_c is the inner radius and Δ its thickness. The values deduced by Xu et al.[65], $r_c = 5.75$ a.u. and $\Delta = 1.89$ a.u., which are specific for a C_{60} fullerene molecule were considered. The value of U_0, on the other hand, was varied from 0 to 10 a.u., in order to explore the general physics of the system, relevant to find other means of confining the atom (e.g., altering the number of carbon atoms in the fullerene cage). By varying the

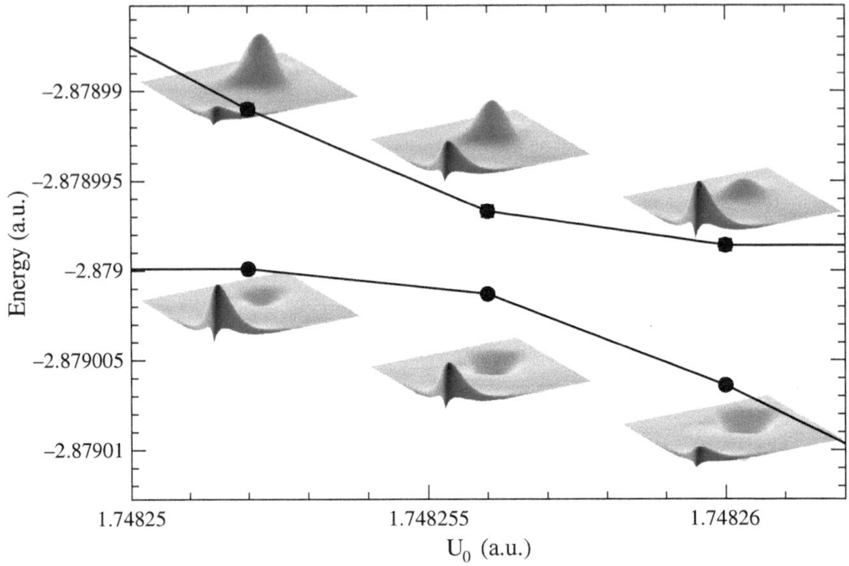

Figure 7.4 ψ_{1s^2} (He ground state) and ϕ_1^2 (two electrons in the fullerene cage) wave functions for different potential depths $\mathcal{U}_0$.

external confinement potential of an endohedrally confined hydrogen atom, Connerade et al.[66] studied the phenomenon of "mirror collapse". This effect occurs when an electron bound by the Coulombic potential falls into the outer potential well, but, at the same time, an excited level, having a bound orbital extended over the outer shell, collapses into the inner Coulombic attraction corresponding to the first level.

The investigation of a similar effect for endohedrally confined He was found to be extremely difficult, because the mirror collapse positions in terms of the confining potential are amazingly evasive. In order to catch the exact point at which the collapse occurs one needs a very fine tuning of the potential strength. As an example, we can see, in Figure 7.4, the way in which two different electron configurations interact. In this case, one wave function (ψ_{1s^2}) represents the two He electrons in the ground state (since it is an S-wave model, its energy is -2.879 a.u.). The other wave function corresponds to the two electrons confined inside the fullerene cage (ϕ^2). The energy is strongly dependent on the confinement potential and, at some particular value, a crossing between ψ_{1s^2} and ϕ^2 must occur, and the corresponding mirror collapse of the wave functions may be observed.

In order to capture the precise point at which the mirror collapse occurs, many calculations of the He spectra are needed, one for each confinement potential strength. It is necessary, therefore, to use a computational method capable of very fast calculations. Moreover, as Figure 7.4 shows, the collapse is extremely sensitive to the potential value. In this particular case, the effect occurs in a potential range between $U_0 = 1.74825$ and $U_0 = 1.74826$ a.u.. Thus, the computational method must also be very precise. Our GSF method is able to fulfill with both requirements.

3.4.2 He confined in an impenetrable spherical cage

Another example is the calculation of the atomic structure of an He atom inside an infinite potential well located at a given radius R. Many theoretical studies have been devoted to the effect of the confinement on atomic spectra. In particular, it has been observed that the degeneracy and relative ordering of the energy levels are both influenced significantly by the effect of confining potentials. Recently, considerable theoretical efforts have been made in performing more accurate computations on simple model systems, involving the hydrogen and helium atom, which could also serve as a benchmark for approximate methods. Within our GSF method, such calculations are straightforward. We only need to impose the confinement condition on the basis set, which automatically imposes the same condition on the total wave functions.

The spectrum of the He atom confined in a spherical cage of radius R is shown in Figure 7.5. In Table 7.9, the ground-state energies are shown as a function of the confinement radius R, and compared with

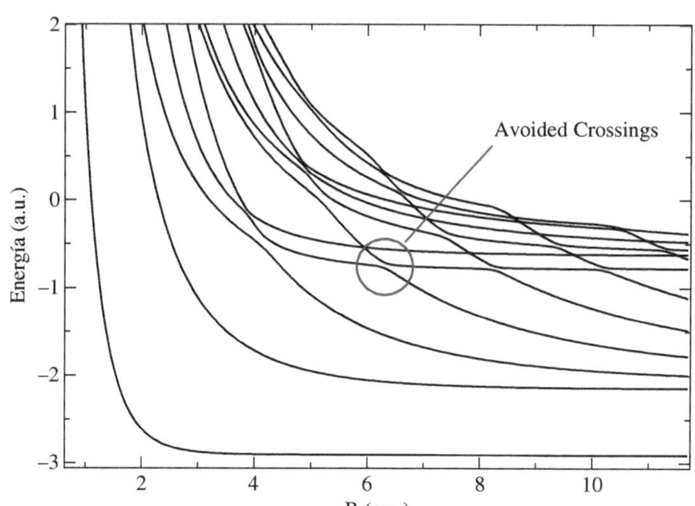

Figure 7.5 Confined He energies as a function of the cage radius *R*. For color version of this figure, the reader is referred to the online version of this chapter.

Table 7.9 Energies of the ground state of the confined He atom for different values of the confinement radius R

R	Present work	Ref. 68	Ref. 67	Ref. 71
0.6	13.318157	13.318340	13.3343	–
1.0	1.015794	1.015870	1.0183	1.0142
1.5	−1.906938	−1.906740	−1.9061	−1.9081
2.0	−2.604013	−2.603630	−2.5998	−2.6051
3.0	−2.872474	−2.871808	−2.8636	−2.8727
3.5	−2.893574	−2.892808	−2.8851	−2.8935
4.0	−2.900465	−2.899687	−2.8931	−2.9003
5.0	−2.903390	−2.902813	−2.8978	−2.9032
6.0	−2.903673	−2.903278	−2.8990	−2.9035
$R \to \infty$	−2.903712	−2.903513	−2.8999	−2.9037
Exact $R \to \infty$[50]	−2.903724			

other calculations. Those of Banerjee et al.[67] have been obtained using a basis with two variational parameters. The calculations of Aquino et al.[68] used generalized Hylleras functions, and Joslin and Goldman[71] performed Quantum Monte Carlo calculations, considered as the best available data values. Our calculations include up to the fifth partial wave and 20 radial functions per coordinate.

By linear interpolation one estimates the value of the critical cage radius at which the ground state crosses the $E = 0$ value. We found a value $R_c = 1.101125$ a.u., which is in very good agreement with the value $R_c = 1.1011$ a.u. obtained by Aquino.[68]

4. THREE-BODY PROBLEMS: SCATTERING STATES

4.1 Introduction

The calculation of scattering states (subscript *sc*) is much more difficult than that of bound states; while this is true also for the two-body case, it is even more complicated for the three-body case. As we shall see in the next subsection, the presence of different channels is one of the major difficulties. Another aspect that will immediately appear is that hyperspherical coordinates are more natural in the asymptotic domain where the three particles are well separated.

For the two-body case, *vide infra*, one of the advantages of using GSF is that they allow to successfully impose correct asymptotic scattering conditions. For the three-body case, the same strong statement cannot be made. The purpose of this section is to provide the theoretical Generalized Sturmian approach to solve three-body scattering problems. According to the channel one wants to study one of the two following expansions could be considered. The first one (essentially Eq. (41)) uses spherical coordinates, and is based on an adequate linear combination of outgoing two-body GSF, one in each coordinate $\mathbf{r}_1$ and $\mathbf{r}_2$, and may be written as:

$$\Psi_{sc}^{+}\left(\mathbf{r}_1,\mathbf{r}_2\right) = \sum_{LM} \sum_{l_1 l_2} \sum_{n_1} \sum_{n_2} a_{n_1 n_2}^{LMl_1 l_2} \mathcal{A} \frac{S_{n_1 l_1}^{+}(r_1)}{r_1} \frac{S_{n_2 l_2}^{+}(r_2)}{r_2} \mathcal{Y}_{l_1 l_2}^{LM}(\widehat{\mathbf{r}}_1,\widehat{\mathbf{r}}_2). \tag{54}$$

Alternatively, one may use hyperspherical coordinates (see Section 4.5) $\rho = \sqrt{r_1^2 + r_2^2}$ and $\omega_5 = \{\alpha, \widehat{\mathbf{r}}_1, \widehat{\mathbf{r}}_2\}$, with $\alpha = \tan^{-1}(r_2/r_1)$, and write

$$\Psi_{sc}^{+}\left(\mathbf{r}_1,\mathbf{r}_2\right) = \sum_{m\nu} \tilde{a}_{m\nu} \frac{S_{m\nu}^{+}(\rho)}{\rho^{5/2}} \Omega_\nu(\omega_5). \tag{55}$$

Above, outgoing behavior (+) was chosen for the scattering states, $a_{n_1 n_2}^{LMl_1 l_2}$ and $\tilde{a}_{m\nu}$ are linear coefficients, and the label ν collects all angular quantum numbers and summation indexes.

Other approaches for atoms and molecules based on Sturmian functions can be found in the literature. In the work of Piraux and collaborators, CSF are used to study different ionization processes produced by photon absorption and electron impact on atoms.[9,10,11] CSF have been also used by Papp and collaborators[69,70] to deal with three-body problem involving Coulomb plus short-range interactions within a Faddeev approach. Regarding issues of interest in quantum chemistry as applied to the electronic structure of atoms and molecules, Coulomb Sturmians pertaining to coordinate systems alternative to the spherical ones have also been developed and tested numerically for three-body problems.[20,23] A different line has been implemented by Ovchinnikov.[31] However, all these approaches are very different to the one reviewed here.

4.2 Asymptotic behaviors

To illustrate the different three-body asymptotic channels, consider the collision of an electron with hydrogen. Depending on the energy of the projectile, one or more processes may occur, in particular elastic scattering, scattering with simultaneous excitation of the residual target, or ionization with two electrons escaping in the field of a proton. The three-body wave function solving the electron–hydrogen Schrödinger equation must contain simultaneously—i.e., it couples—all these possible processes. When solving the differential equation two major difficulties appear: (i) numerically expensive large grids are required since scattering states are spread and the probability of finding the particles at any point of the configuration space is different from zero; and (ii) it is far from obvious to impose as boundary conditions the well-defined asymptotic behaviors in different spacial domains.

Consider first the Ω_0 region, the one corresponding to all three particles far from each other. As all interparticle distances are large, no contributions coming from bound states are expected in the asymptotic three-body scattering wave function. Various approximated solutions are known for this region.[72–74,76–78] An hyperspherical wave is known to be also a correct solution[8]:

$$\Psi^{+}_{as,c}\left(\mathbf{r}_1,\mathbf{r}_2\right) \underset{\rho\to\infty}{\to} \frac{(2\pi i)^{1/2}}{(2\pi)^3} K^{3/2} T_{\tilde{\mathbf{k}}_1,\tilde{\mathbf{k}}_2} \frac{e^{iK\rho - i\lambda_0 \ln(2K\rho) - i\sigma_0}}{\rho^{5/2}}, \tag{56}$$

where λ_0 is a Coulomb parameter, σ_0 is a phase and $T_{\tilde{\mathbf{k}}_1,\tilde{\mathbf{k}}_2} = T\left(\frac{K}{\rho}\mathbf{r}_1, \frac{K}{\rho}\mathbf{r}_2\right)$ is the ionization transition amplitude. The coordinate-dependent momenta $\tilde{\mathbf{k}}_j$ ($j = 2, 3$) were defined originally by Alt and Mukhamedzhanov,[74] while $K = \sqrt{k_1^2 + k_2^2}$ is the hyper-momentum of the particles.

Next, take the Ω_i regions (with $i = 1, 2, 3$) which correspond to two of the particles close to each other and the third is far away from the pair.[75] We may have the situation in which one of the electrons is close to the nucleus, forming a bound state, and the other is far away. In this case, the wave function should have the following asymptotic form:

$$\Psi^{+}_{as,1}(\mathbf{r}_1,\mathbf{r}_2) \underset{r_1\to\infty}{\to} \Psi_i(\mathbf{r}_1,\mathbf{r}_2) - \frac{1}{2\pi}\sum_n F_n(k_n\hat{\mathbf{r}}_1,\mathbf{k}_i)\frac{e^{ik_n r_1 - i\eta_1 \ln(2k_n r_1)}}{r_1}\phi_n(\mathbf{r}_2), \tag{57}$$

where $\Psi_i(\mathbf{r}_1,\mathbf{r}_2)$ represents the initial state which is defined before the collision. Here we are assuming that electron 1 is the projectile and η_1 is the corresponding Sommerfeld parameter. The second term in this expression represents the excitation of the target and the dispersion of the projectile; $F_n\left(k_n\hat{\mathbf{r}}_1,\mathbf{k}_i\right)$ represents the excitation amplitude. The symmetric case is the situation where the other electron is far away and the initial one ends up bound at the target

$$\Psi^{+}_{as,2}(\mathbf{r}_1,\mathbf{r}_2) \underset{r_2\to\infty}{\to} -\frac{1}{2\pi}\sum\!\!\!\!\!\!\int_m G_m(k_m\hat{\mathbf{r}}_2,\mathbf{k}_i)\,\frac{e^{ik_m r_2 - i\eta_2\ln(2k_m r_2)}}{r_2}\phi_m(\mathbf{r}_1), \tag{58}$$

where $G_m\left(k_m\hat{\mathbf{r}}_2,\mathbf{k}_i\right)$ represents the capture of the incoming electron and η_2 the Sommerfeld parameter for electron 2. The summations in Eq. (57) and (58) run over bound and continuum states. The third Ω_i region corresponds to the situation where both electrons are close to each other and far away from the nucleus.

When the collision process is studied and the Schrödinger equation is solved numerically, all the channels, represented by the asymptotic wave functions $\Psi^{+}_{as,1}(\mathbf{r}_1,\mathbf{r}_2)$, $\Psi^{+}_{as,2}(\mathbf{r}_1,\mathbf{r}_2)$, and $\Psi^{+}_{as,c}(\mathbf{r}_1,\mathbf{r}_2)$, are coupled and are incorporated simultaneously into the solution. In other words, the full solution should have the following general form at large distances:

$$\Psi^{+}\left(\mathbf{r}_1,\mathbf{r}_2\right) \underset{\rho\to\infty}{\to} \Psi^{+}_{as,1}\left(\mathbf{r}_1,\mathbf{r}_2\right) + \Psi^{+}_{as,2}\left(\mathbf{r}_1,\mathbf{r}_2\right) + \Psi^{+}_{as,c}\left(\mathbf{r}_1,\mathbf{r}_2\right). \tag{59}$$

Numerically, it is extremely difficult to impose these conditions to the wave function. As mentioned in the introduction, however, various methods have succeeded but not without difficulties. In this section, we shall see how GSF can be used to tackle the problem. Applications for several processes will be given in Section 5.

4.3 Driven equation for three-body scattering problems

A standard procedure to solve three-body scattering problems consists in transforming the homogeneous three-body Schrödinger equation into an inhomogeneous one (driven equation). As in Section 2.3, Eq. (21), the three-body the solution is separated as the sum of an initial channel wave function $\Psi_i(\mathbf{r}_1,\mathbf{r}_2)$ and the scattering wave function $\Psi^{+}_{sc}(\mathbf{r}_1,\mathbf{r}_2)$ which contains all the information about the collision process:

$$\Psi^{+}(\mathbf{r}_1,\mathbf{r}_2) = \Psi_i(\mathbf{r}_1,\mathbf{r}_2) + \Psi^{+}_{sc}(\mathbf{r}_1,\mathbf{r}_2). \tag{60}$$

When this proposal is substituted into the Schrödinger equation (40) the following driven equation results for $\Psi^{+}_{sc}(\mathbf{r}_1,\mathbf{r}_2)$

$$[H-E]\,\Psi^{+}_{sc}(\mathbf{r}_1,\mathbf{r}_2) = -W(\mathbf{r}_1,\mathbf{r}_2)\,\Psi_i(\mathbf{r}_1,\mathbf{r}_2), \tag{61}$$

where $W(\mathbf{r}_1, \mathbf{r}_2)$ represents the interactions not solved by $\Psi_i(\mathbf{r}_1, \mathbf{r}_2)$.

It is interesting to notice that the transition amplitude $T_{\tilde{\mathbf{k}}_1,\tilde{\mathbf{k}}_2}$ appearing in Eq. (56) can be extracted directly from the solution of (61) which, formally, may be written as:

$$\Psi_{sc}^{+}(\mathbf{r}_1, \mathbf{r}_2) = \int d\mathbf{r}_1' d\mathbf{r}_2' G^{+}\left(\mathbf{r}_1, \mathbf{r}_2, \mathbf{r}_1', \mathbf{r}_2'\right) W\left(\mathbf{r}_1', \mathbf{r}_2'\right) \Psi_i\left(\mathbf{r}_1', \mathbf{r}_2'\right), \quad (62)$$

where $G^{+}\left(\mathbf{r}_1, \mathbf{r}_2, \mathbf{r}_1', \mathbf{r}_2'\right)$ is the three-body Coulomb Green's function. In the Ω_0 region its asymptotic limit reads[8]

$$G^{+}\left(\mathbf{r}_1, \mathbf{r}_2, \mathbf{r}_1', \mathbf{r}_2'\right) \to \frac{(2\pi i)^{1/2}}{(2\pi)^3} K^{\frac{3}{2}} \frac{e^{i[K\rho - \lambda_0 \ln(2K\rho) - \sigma_0]}}{\rho^{\frac{5}{2}}} \Psi_{\tilde{\mathbf{k}}_1,\tilde{\mathbf{k}}_2}^{-}\left(\mathbf{r}_1', \mathbf{r}_2'\right), \quad (63)$$

where the function $\Psi_{\tilde{\mathbf{k}}_1,\tilde{\mathbf{k}}_2}^{-}\left(\mathbf{r}_1', \mathbf{r}_2'\right)$ is the exact solution of the three-body problem with incoming wave asymptotic behavior.

From these two relations, one finds that the scattering wave function $\Psi_{\mathbf{k}_1,\mathbf{k}_2}^{-}\left(\mathbf{r}_1', \mathbf{r}_2'\right)$ will have the asymptotic behaviour (56) in which the transition amplitude $T_{\tilde{\mathbf{k}}_1,\tilde{\mathbf{k}}_2}$ is given by

$$T_{\tilde{\mathbf{k}}_1,\tilde{\mathbf{k}}_2} = \langle \Psi_{\tilde{\mathbf{k}}_1,\tilde{\mathbf{k}}_2}^{-}(\mathbf{r}_1, \mathbf{r}_2) \mid W(\mathbf{r}_1, \mathbf{r}_2) \mid \Psi_i(\mathbf{r}_1, \mathbf{r}_2)\rangle. \quad (64)$$

This is the standard definition for the transition amplitudes required by scattering theory. Thus, once the driven Eq. (61) is solved with appropriate asymptotic conditions, one can extract the ionization transition amplitude from the evaluation—at large distances—of the scattering wave function itself. However, since all possible channels are coupled, it is possible to extract also the other transition amplitudes appearing in Eqs. (57) and (58).

4.3.1 Ionization of hydrogen by electron impact

In the case of an electron-hydrogen collision, the Hamiltonian H is given by Eq. (50) where $z_1 = z_2 = -1$ and $z_3 = Z = 1$. The initial state may be chosen as a plane wave for the incident electron $e^{i\mathbf{k}_i.\mathbf{r}_1}$ multiplied by an hydrogenic ground state e^{-Zr_2}; actually, depending whether one looks at singlet ($S = 0$) or triplet ($S = 1$) states, one may take the following combination

$$\Psi_i(\mathbf{r}_1, \mathbf{r}_2) = \mathcal{A}\frac{1}{\sqrt{\pi}}e^{-r_1}e^{i\mathbf{k}_i.\mathbf{r}_2},$$

where $k_i = \sqrt{2(E - (-Z^2/2))}$. The RHS of the driven equation (61) is then

$$-\left[\left(\frac{1}{r_{12}} - \frac{1}{r_2}\right)\Psi_i(\mathbf{r}_1, \mathbf{r}_2) + (-1)^S(1 \leftrightarrow 2)\right]. \tag{65}$$

Other proposals including correlation can be defined for the initial state. For example, a C3-like approach where the projectile–target interaction is explicitly included presents the advantage of leading to a short-range $W(\mathbf{r}_1, \mathbf{r}_2)$ function.[79] However, it has the disadvantage of coupling all the coordinates and making more difficult the task of solving the driven equation.

4.3.2 Double ionization of helium by high-energy electron impact

Consider now the double ionization of helium atoms by high-energy electron impact. As shown in Ref. 80 and briefly sketched below, this four-body scattering problem can be reduced to a three-body problem, and thus we end up with a driven equation of the kind (61).

The non-relativistic four-body Hamiltonian for three electrons and an infinite mass helium nucleus of charge $Z = 2$ is given by

$$H_{4b} = -\frac{1}{2}\nabla_0^2 - \frac{1}{2}\nabla_1^2 - \frac{1}{2}\nabla_2^2 - \frac{Z}{r_0} - \frac{Z}{r_1} - \frac{Z}{r_2} + \frac{1}{r_{01}} + \frac{1}{r_{02}} + \frac{1}{r_{12}}, \tag{66}$$

where particle 0 labels the electron projectile (incident with momentum $\mathbf{k}_i$, and scattered with momentum $\mathbf{k}_f$), while particles 1 and 2 are the target electrons. We first define a projectile Hamiltonian

$$H_p(Z_p) = \left(-\frac{1}{2}\nabla_0^2 - \frac{Z_p}{r_0}\right), \tag{67}$$

which includes only Coulomb projectile–nucleus interaction with a model charge Z_p. We then decompose the four-body Hamiltonian as follows:

$$H_{4b} = H_0\left(Z_p\right) + \bar{W}\left(Z_p\right), \tag{68}$$

where

$$H_0\left(Z_p\right) = H_p(Z_p) + H_a, \tag{69a}$$

$$\bar{W}\left(Z_p\right) = -\frac{Z - Z_p}{r_0} + \frac{1}{r_{01}} + \frac{1}{r_{02}}. \tag{69b}$$

Above H_a is the three-body helium Hamiltonian (50); let $\Phi^{(0)}\left(\mathbf{r}_1, \mathbf{r}_2\right)$ represent its ground state. As well as all kinetic operators, the Hamiltonian $H_0\left(Z_p\right)$ includes all interactions of the subsystem (1,2) through H_a, and a projectile–nucleus interaction $-Z_p/r_0$ through $H_p(Z_p)$. The two subsystems are coupled through the interaction $\bar{W}(Z_p)$ which may be considered as a perturbation.

To study electron-impact double ionization processes we need to find a scattering solution, with outgoing-type behavior, of the four-body Schödinger equation

$$\left[H_0\left(Z_p\right) + \bar{W}\left(Z_p\right) - E\right]\, \Psi^+\left(\mathbf{r}_0, \mathbf{r}_1, \mathbf{r}_2\right) = 0. \tag{70}$$

Using a perturbation expansion in terms of the magnitude of $\bar{W}\left(Z_p\right), \Psi^+ = \sum_n \lambda^n \Psi^{(n)+}$, we obtain a system of differential equations for $\Psi^{(n)+}$. The function

$$\Psi^{(0)+}\left(\mathbf{r}_0, \mathbf{r}_1, \mathbf{r}_2\right) = \frac{1}{(2\pi)^{3/2}} e^{i\mathbf{k}_i \cdot \mathbf{r}_0} \Phi^{(0)}\left(\mathbf{r}_1, \mathbf{r}_2\right), \tag{71}$$

solution of the Hamiltonian H_0 (which is separable in the two subsystems (1,2) and 0), represents the initial state of the system (zeroth-order equation). For presentation purposes, the projectile–nucleus interaction is neglected and a plane wave is taken ($Z_p = 0$). Alternatively (see Ref. 80, for details), it can be properly represented through a Coulomb wave function with charge $Z_p = Z$[81] which includes the projectile–nucleus interaction in both initial and final channels.

Next, we explore the first-order equation in which the interaction $\bar{W}\left(Z_p\right)$ is included only once. Taking a plane wave for the scattered electron, the two ejected electrons solution satisfies the driven equation[80]

$$[E_a - H_a]\ \Phi_{sc}^{(1)+}(\mathbf{r}_1,\mathbf{r}_2) = W_{fi}(\mathbf{r}_1,\mathbf{r}_2)\,\Phi^{(0)}(\mathbf{r}_1,\mathbf{r}_2), \tag{72}$$

where $E_a = E - k_f^2/2$ denotes the energy of two electrons in interaction with the nucleus in the final state, and the driven term includes

$$W_{fi}(\mathbf{r}_1,\mathbf{r}_2) = \langle \mathbf{k}_f | \bar{W} | \mathbf{k}_i \rangle = \frac{1}{(2\pi)^3}\frac{4\pi}{q^2}\left(-Z + e^{i\mathbf{q}\cdot\mathbf{r}_1} + e^{i\mathbf{q}\cdot\mathbf{r}_2}\right) \tag{73}$$

with $\mathbf{q} = \mathbf{k_i} - \mathbf{k_f}$ the momentum transfer. In this way, the four-body problem is reduced to a pure three-body one as described by (61) where the dynamics of the two ejected electrons in the presence of the heavy nucleus is described by Eq. (72). The four-body scattering problem is well formulated and its first-order solution possesses all the information contained in the first Born approximation. Following similar steps used for the three-body case, we can extract a transition amplitude (64) that in this case reads

$$T_{\tilde{\mathbf{k}}_1,\tilde{\mathbf{k}}_2} = \frac{1}{(2\pi)^3}\frac{4\pi}{q^2}\langle \Psi^-_{\tilde{\mathbf{k}}_1,\tilde{\mathbf{k}}_2}(\mathbf{r}_1,\mathbf{r}_2)\,|-Z + e^{i\mathbf{q}\cdot\mathbf{r}_1} + e^{i\mathbf{q}\cdot\mathbf{r}_2}|\Phi^{(0)}(\mathbf{r}_1,\mathbf{r}_2)\rangle \tag{74}$$

and one recovers the standard first Born approximation, see, e.g., Ref. 82,83. In the calculations presented in the literature, approximated wave functions, or numerical ones, have been used for $\Psi^-_{\tilde{\mathbf{k}}_1,\tilde{\mathbf{k}}_2}(\mathbf{r}_1,\mathbf{r}_2)$. In the transition matrix element (74) the exact solution of the three-body problem $\Psi^-_{\tilde{\mathbf{k}}_1,\tilde{\mathbf{k}}_2}(\mathbf{r}_1,\mathbf{r}_2)$ should be used; the corresponding three-body Schrödinger equation has been given, for the first time, in Ref. 80.

Whether for single ionization of hydrogen, or double ionization of helium by fast incident electrons, the scattering problem is transformed into a three-body driven equation (Eq. (61) or (72)) with outgoing boundary conditions. These equations can be solved with GSF either in spherical or hyperspherical coordinates.

4.4 Solving the driven equation with GSF (spherical coordinates)

In order to solve the driven equation (61), we may use a *CI* expansion (54). Projecting over the basis elements a matrix problem results,

$$[\mathbf{S}(E - E_1 - E_2) - \mathbf{H}]\ \mathbf{a} = \mathbf{b}, \tag{75}$$

where the Hamiltonian $\mathbf{H}$ and overlap $\mathbf{S}$ matrices have elements $[\mathbf{H}]^{l_1' l_2' l_1 l_2}_{n_1' n_2' n_1 n_2}$ and $[\mathbf{S}]^{l_1' l_2' l_1 l_2}_{n_1' n_2' n_1 n_2}$ similar to those described for the three-body bound states, except that the GSF have outgoing behavior.

The elements of the vector $\mathbf{b}$ are the expansion coefficients of the driven term

$$W(\mathbf{r}_1, \mathbf{r}_2)\, \Psi_i(\mathbf{r}_1, \mathbf{r}_2) = \sum_{LMl_1 l_2} \sum_{n_1} \sum_{n_2} b^{LMl_1 l_2}_{n_1 n_2} \mathcal{V}_{g_1}(r_1) \mathcal{V}_{g_2}(r_2)$$
$$\mathcal{A}\left(\frac{S^+_{n_1 l_1}(r_1)}{r_1} \frac{S^+_{n_2 l_2}(r_2)}{r_2} \mathcal{Y}^{LM}_{l_1 l_2}(\widehat{\mathbf{r}}_1, \widehat{\mathbf{r}}_2) \right). \tag{76}$$

Note that, using the generating potentials $\mathcal{V}_{g_1}(r_1)$ and $\mathcal{V}_{g_2}(r_2)$, the orthogonality property of the GSF makes the matrices diagonal and leads to a largely simplified calculation. One of the advantages of using the GSF is that they remove from the equation the kinetic energy of each electron, and the electron–nucleus potentials by taking them as auxiliary potentials in Eq. (2). In this way, only the generating potentials and the electron–electron interaction have to be evaluated to obtain Hamiltonian $\mathbf{H}$.

4.5 Solving the driven equation with GSF (hyperspherical coordinates)

We shall now present an hyperspherical approach. To do so, we first need to define the coordinates, set the scattering problem, and then propose hyperspherical GSF to solve it.

4.5.1 Hyperspherical coordinates

The three-body Hamiltonian with particles of masses m_1, m_2, and m_3 can be written in terms of mass-scaled Jacobi coordinates $\mathbf{x}$ and $\mathbf{X}$.[21,22,84,85] From these vectors, which represent any of the three existing pairs in three-body problems, hyperspherical coordinates can be defined: a hyperradius ρ, defined as $\rho^2 = x^2 + X^2$, and five hyperangular coordinates (denoted collectively by ω_5) that include the hyperangle $\tan\alpha = X/x$ and the polar angles θ_x, ϕ_x and θ_X, ϕ_X defining the orientations $\widehat{\mathbf{x}}$ and $\widehat{\mathbf{X}}$ of the Jacobi vectors in the center-of-mass reference frame. The previous definitions allow one to write $x = \rho\cos\alpha$ and $X = \rho\sin\alpha$; below $x = r_1$ and $X = r_2$.

The three-body Hamiltonian is now written as $H = T + V(\rho, \omega_5)$ where the kinetic energy operator takes the form

$$T = -\frac{1}{2\mu}\left[\frac{1}{\rho^5}\frac{\partial}{\partial\rho}\left(\rho^5\frac{\partial}{\partial\rho}\right) - \frac{\Lambda^2}{\rho^2}\right],$$

where $\mu = \sqrt{m_1 m_2 m_3/(m_1 + m_2 + m_3)}$ is the three-body reduced mass, and Λ^2 is the grand orbital angular momentum operator.[84]
The interaction potentials, say between two electrons (positions $\mathbf{r}_1, \mathbf{r}_2$) and a nucleus of charge Z, may be written as[86,30]

$$V(\rho, \omega_5) = -\frac{Z}{r_1} - \frac{Z}{r_2} + \frac{1}{r_{12}} = \frac{C(\omega_5)}{\rho}, \tag{77}$$

where

$$C(\omega_5) = -\frac{Z}{\cos\alpha} - \frac{Z}{\sin\alpha} + \sum_{l=0}^{\infty}\frac{4\pi}{2l+1}$$
$$\sum_{m=-l}^{l}(-1)^m Y_{l-m}(\hat{\mathbf{r}}_2)\, Y_{lm}(\hat{\mathbf{r}}_1)\left\{\begin{array}{c}\sec\alpha\tan^l\alpha \\ \csc\alpha\cot^l\alpha\end{array}\right\}, \tag{78}$$

the top choice (between brackets) being for $0 \le \alpha \le \frac{1}{4}\pi$ and the bottom one for $\frac{1}{4}\pi \le \alpha \le \frac{1}{2}\pi$.

4.5.2 Scattering driven equation

The Schrödinger equation to be considered is $[H - E]\,\Psi(\rho, \omega_5) = 0$. As proposed in spherical coordinates (see Eq. (60)), the wave function $\Psi(\rho, \omega_5)$ for a collision process may be separated in two parts $\Psi(\rho, \omega_5) = \Psi_i(\rho, \omega_5) + \Psi_{sc}(\rho, \omega_5)$. Again, $\Psi_i(\rho, \omega_5)$ is a known initial state, eigensolution of an approximate Hamiltonian $H_0 = H - W$. Recall that since $H = T + V(\rho, \omega_5)$ is the full Hamiltonian, W is the neglected, unsolved interaction not included in the initial prepared state. $\Psi_{sc}(\rho, \omega_5)$, on the other hand, is a wave function solving all the interactions $V(\rho, \omega_5)$. According to this separation, the function $\Psi_{sc}(\rho, \omega_5)$ satisfies a driven Schrödinger equation

$$[T + V(\rho, \omega_5) - E]\Psi_{sc}(\rho, \omega_5) = -W(\rho, \omega_5)\Psi_i(\rho, \omega_5) = \varphi(\rho, \omega_5), \tag{79}$$

where $\varphi(\rho, \omega_5)$ denotes the driving term. Equation (79) must be solved imposing outgoing behavior to $\Psi_{sc}(\rho, \omega_5)$ for large values of ρ ($\rho \to \infty$).

Let K be the hyperspherical momentum related to the total energy through $E = K^2/2\mu$. When all particles are far from each other, the Peterkop-type asymptotic behavior

$$\Psi_{sc}^{+}(\rho, \omega_5) \propto \frac{e^{iK\rho - i\frac{C(\omega_5)}{K}\ln(2K\rho)}}{\rho^{5/2}} \tag{80}$$

indicates that an hyperspherical approach could be more adequate than the use of interparticle (spherical) coordinates. It also shows that, in the asymptotic region, the Coulomb interactions couple the angles with the hyperradius in a particular form, through a Coulomb logarithmic phase.

Two important issues when solving this collision problem are (i) the range of the interaction $W(\rho, \omega_5)$ and (ii) the use of an appropriate matrix definition for the operator $(H - E)$ or for Green's function $(H - E)^{-1}$. The basis set to be used has to take into account both issues: it has to be complete in the region where the interaction $W(\rho, \omega_5)$ is not negligible, and outside that region has to possess the correct asymptotic behavior corresponding to all three Coulomb interactions.

4.5.3 Hyperspherical Generalized Sturmian functions

There are various ways of defining a Sturmian strategy in hyperspherical coordinates. Two of them were discussed in Refs. 84,86. Here we briefly review the approach presented in Ref. 84.

First, as in Ref. 30, we define a set of angular Sturmian functions $\Omega_\nu(\omega_5)$ depending on the angular coordinates ω_5. They are solutions of the following angular Sturmian eigenvalue equation

$$\left[\Lambda^2 + 2\mu\rho_\nu C(\omega_5)\right] \Omega_\nu(\omega_5) = \nu(\nu+4)\Omega_\nu(\omega_5), \tag{81}$$

where ρ_ν or, alternatively, ν can be considered as the eigenvalues. If ν is used as eigenvalue, then ρ_ν is considered as an externally fixed parameter ρ_{eff} (there are various ways of defining ρ_{eff}, some of them have been discussed in Ref. 84). In this case, the eigenfunctions $\Omega_\nu(\omega_5)$ satisfy the following orthogonality and closure relations

$$\int d\omega_5 \Omega_{\nu'}(\omega_5)\, \Omega_\nu(\omega_5) = \delta_{\nu'\nu}, \tag{82a}$$

$$\sum_{\nu} \Omega_{\nu}\left(\omega_{5}'\right) \Omega_{\nu}\left(\omega_{5}\right) = \delta\left(\omega_{5} - \omega_{5}'\right), \tag{82b}$$

where $d\omega_5$ represents the five-dimensional volume element and $\delta\left(\omega_5 - \omega_5'\right)$ is the symbolic product of the Dirac delta corresponding to all five angular coordinates.

In the specific case of the S-wave models to be considered in the applications (Section 5), the eigenfunctions are the Jacobi polynomials[21,22,84,86]

$$\Omega_n(\alpha) = \frac{4(n+1)}{\sqrt{\pi}}\, {}_2F_1\left(-n, n+2, \frac{3}{2}; \sin^2\alpha\right), \tag{83}$$

where ${}_2F_1$ represents the Gauss hypergeometric function.[87] These angular functions satisfy the eigenvalue equation

$$\Lambda^2 \Omega_n(\alpha) = \nu_n(\nu_n + 4)\Omega_n(\alpha), \tag{84}$$

with $\nu_n = 2n\,(n = 0, 1, \ldots)$, form a complete set and satisfy the orthonormality relation

$$\int_0^{\pi/2} \Omega_{n'}(\alpha)\Omega_n(\alpha) \sin^2\alpha \cos^2\alpha \; d\alpha = \delta_{n'n}. \tag{85}$$

For the hyperradial coordinate we can introduce a set of radial functions $\bar{S}_{m\nu}(\rho)$ satisfying the Sturmian equation:

$$\left[-\frac{1}{2\mu}\frac{1}{\rho^5}\frac{\partial}{\partial\rho}\left(\rho^5\frac{\partial}{\partial\rho}\right) + \frac{\nu(\nu+4)}{2\mu\rho^2} + \mathcal{U}(\rho) - E\right] \bar{S}_{m\nu}(\rho) = -\beta_m \mathcal{V}_g(\rho)\, \bar{S}_{m\nu}(\rho). \tag{86}$$

The potential $\mathcal{U}(\rho)$ can be of short- or long-range; $\mathcal{V}_g(\rho)$ is a generating potential. Introducing the reduced function $\bar{S}_{m\nu}(\rho) = S_{m\nu}(\rho)/\rho^{\frac{5}{2}}$ into (86) leads to

$$\left[-\frac{1}{2\mu}\frac{\partial^2}{\partial\rho^2} + \frac{\nu(\nu+4) + \frac{15}{4}}{2\mu\rho^2} + \mathcal{U}(\rho) - E\right] S_{m\nu}(\rho) = -\beta_m \mathcal{V}_g(\rho)\, S_{m\nu}(\rho), \tag{87}$$

which has the same form as Eq. (2) presented in Section 2.1 (the difference appearing only in the definition of the $1/\rho^2$ term) and similar boundary conditions can then be used. Note that the parameter ν can be a fixed parameter or the eigenvalue of the angular Sturmian functions. The hyper-radial eigenfunctions $S_{m\nu}(\rho)$ form an orthogonal and complete set such that

$$\int d\rho S_{m'\nu}(\rho)\mathcal{V}_g(\rho)\, S_{m\nu}(\rho) = \delta_{m'm}, \tag{88a}$$

$$\sum_m S_{m\nu}\left(\rho'\right) \mathcal{V}_g(\rho)\, S_{m\nu}(\rho) = \delta\left(\rho - \rho'\right) \tag{88b}$$

and the algebra maps easily to the one related to the spherical radial Sturmian functions.

4.5.4 Hyperspherical Generalized Sturmian functions applied to scattering problems

To solve the driven Eq. (79), we use expansion (55) for the wave function $\Psi_{sc}\left(\rho, \omega_5\right)$. We also propose the following expansion for the driven term

$$W\left(\rho, \omega_5\right) \Psi_i\left(\rho, \omega_5\right) = \sum_{m\nu} c_{m\nu} \mathcal{V}_g(\rho) \frac{S_{m\nu}(\rho)}{\rho^{\frac{5}{2}}} \Omega_\nu\left(\omega_5\right), \tag{89}$$

whose range dictates that of the generating potential $\mathcal{V}_g(\rho)$ to be chosen. Replacing in (79) we obtain

$$\begin{aligned}\sum_{m\nu} \tilde{a}_{m\nu} \left[\left(\mathcal{U}(\rho) + \beta_m \mathcal{V}_g(\rho)\right) + \rho_{eff} \frac{C(\omega_5)}{\rho^2} + \frac{C(\omega_5)}{\rho}\right] S_{m\nu}(\rho)\, \Omega_\nu\left(\omega_5\right) \\ = W\left(\rho, \omega_5\right) \Psi_i\left(\rho, \omega_5\right),\end{aligned} \tag{90}$$

One convenient election for the potential $\mathcal{U}(\rho)$ could be $\mathcal{U}(\rho) = -Z/\rho$. Projecting Eq. (90) over the basis functions, and using the orthogonality properties (82a) and (88a), we find

$$\begin{aligned}\sum_{m\nu}\left[\left(\left[\frac{-Z}{\rho}\right]_{mm',\nu} + \beta_m \delta_{mm'}\right)\delta_{\nu\nu'} + \left(\rho_{eff}\left[\frac{1}{\rho^2}\right]_{mm',\nu} + \left[\frac{1}{\rho}\right]_{mm',\nu}\right)\left[C(\omega_5)\right]_{\nu\nu'}\right] \\ \tilde{a}_{m\nu} = c_{m'\nu'},\end{aligned} \tag{91}$$

where the matrix elements $\left[C(\omega_5)\right]_{\nu\nu'}$ are

$$\left[C(\omega_5)\right]_{\nu\nu'} = \int \mathrm{d}\omega_5\, \Omega_{\nu'}\left(\omega_5\right) C(\omega_5)\Omega_{\nu}\left(\omega_5\right), \tag{92}$$

and the matrix elements $[A]_{mm',y}$ are defined by one-dimensional integrals

$$[A]_{mm',\nu} = \int d\rho S_{m\nu}(\rho) A\left(\rho\right) S_{m'\nu}(\rho). \tag{93}$$

Solving the matrix problem (91) provides the expansion coefficients $\tilde{a}_{m\nu}$. In the next section, we see that building the solution in this way provides the scattering functions with the correct asymptotic behavior (80).

5. THREE-BODY SCATTERING STATES: APPLICATIONS

5.1 Introduction

Model calculations, used as benchmarks, can be found throughout the collision literature. S-wave models, though restricted to zero angular momentum states, serve as a test bed as they contain most of the features and difficulties associated to the full physical problem but, require less computational resources. They are useful, in general, as they allow us to put on a strong footing different numerical methods which do not necessarily yield converging results when applied to complicated scattering processes. For the three-body problem, for example, before solving the full electron–Hydrogen ionization problem, S-wave model (often referred to as Temkin–Poet (TP) model[88,89]) calculationss have played a very important role in the development of theoretical methods.

In this section we consider three-body scattering S-wave models solved using GSF. We start with a recently proposed break-up model problem, then consider the electron–Hydrogen TP model $(e, 2e)$ processes, and finally a model calculation for high impact energy $(e, 3e)$ processes.

In the TP model, the electron–electron repulsion $1/r_{12}$ is spherically averaged and thus replaced by $1/r_>$ where $r_> = \max(r_1, r_2)$.The three-body Coulomb potential (two electrons and a nucleus of charge Z) becomes

$$V\left(r_1, r_2\right) = -\frac{Z}{r_1} - \frac{Z}{r_2} + \frac{1}{r_>}. \tag{94}$$

In hyperspherical coordinates, the "charge" $C(\omega_5)$ defined through (77) is replaced by $\tilde{C}(\alpha)$ with
Thus only an α dependence is retained in the potential: $V\left(\rho, \omega_5\right) = \tilde{C}(\alpha)/\rho$.

$$\tilde{C}(\alpha) = -\frac{Z}{\cos\alpha} - \frac{Z}{\sin\alpha} + \begin{cases} \frac{1}{\cos\alpha} & 0 \leq \alpha \leq \frac{1}{4}\pi \\ \frac{1}{\sin\alpha} & \frac{1}{4}\pi \leq \alpha \leq \frac{1}{2}\pi. \end{cases} \quad (95)$$

The Schrödinger equation corresponding to the TP model provides a simplified version of the physical problem.

5.2 Three-body *S*-wave model problem

Recently,[90] a three-body S-wave model has been proposed. Though apparently rather simple, this model contains the essential difficulties of the real problem (Coulomb potentials and non-separability). As the analytical solution of the model can be written out, it provides a very interesting and unique tool to validate numerical methods.

Differently from the TP model, the authors of Ref. 90 suggested replacing $\mathcal{C}(\omega_5)$ by a constant charge $\mathcal{C}$, i.e., the following model Coulomb potential

$$V(\rho) = \frac{\mathcal{C}}{\rho}, \quad (96)$$

which can be either attractive ($\mathcal{C} < 0$) or repulsive ($\mathcal{C} > 0$). Note that the potential is not one of the three Coulomb interactions that appear in the physical case $V(\rho, \omega_5)$. Although seemingly simple in hyperspherical coordinates, it is not separable in spherical coordinates r_1 and r_2; assuming, for example, that $r_2 < r_1$, their coupling is very particular

$$\frac{\mathcal{C}}{\sqrt{r_1^2 + r_2^2}} = \frac{\mathcal{C}}{r_1}\left[1 - \frac{1}{2}\left(\frac{r_2}{r_1}\right)^2 + \cdots\right] = \frac{\mathcal{C}}{\rho\cos\alpha}\left[1 - \frac{1}{2}\tan^2\alpha + \cdots\right], \quad (97)$$

where the second equality illustrates the expansion in terms of the hyperangle α. The lowest order (first term) of expansion (97) corresponds to retaining the first term of the real potential (77). The model potential (96) therefore tests the r_1 and r_2 dependence in a way which differs substantially from the TP model. On top of that, it offers the possibility to make a detailed and interesting asymptotic investigation of the scattering wave function. Indeed, for a given "angular" set (fixed ω_5), the angular dependent charge $C(\omega_5)$ takes a constant value and thus the physical three-body potential reduces to the model. The knowledge of the analytic solution allows one to investigate, in an original manner, for what hyperradius the corresponding asymptotic regime is actually reached. By varying the angles, one may explore different asymptotic domains, and related convergence issues.

In the proposed model, the initial state is taken to be a symmetrized bound-free product of a standing spherical wave in the relative coordinate r_1 between the incoming particle and the center of the target and a bound-like state in the target coordinate r_2; the interaction neglected in the initial channel is given, for example, by a Yukawa potential $e^{-a\rho}/\rho$. More specifically, we take the following source (driving term)

$$\varphi\left(\rho,\alpha\right)=\rho^{t}e^{-a\rho}\frac{1}{2}\left[\frac{\sin r_1}{r_1}\frac{\sinh r_2}{r_2}+\frac{\sin r_2}{r_2}\frac{\sinh r_1}{r_1}\right], \tag{98}$$

with a parameter a such that $\Re(a)>1$ and a real parameter $t\geq -1$. This source has a known single series expansion in Jacobi polynomials $\Omega_n(\alpha)$, the coefficients being functions of ρ.

The Coulomb potential is simple in hyperspherical but not in spherical coordinates. The driven term is not separable in either set of coordinates, not even asymptotically. Thus the model equation

$$\left[-\frac{1}{2\mu}\frac{1}{\rho^5}\frac{\partial}{\partial\rho}\left(\rho^5\frac{\partial}{\partial\rho}\right)+\frac{\Lambda^2}{2\mu\rho^2}+\frac{\mathcal{C}}{\rho}-E\right]\Psi\left(\rho,\alpha\right)=\varphi\left(\rho,\alpha\right), \tag{99}$$

together with the source (98) provides a physically meaningful Coulomb scattering problem which presents typical three-body problem difficulties including non-separability.

To numerically solve this three-body problem hyperspherical Sturmian functions, built as a product of coupled functions (55), were employed.[90,91] For the angular part, Jacobi polynomials $\Omega_n(\alpha)$ were used and can be generated either through their analytic definition (83) or by solving numerically the hyperangular eigenvalue equation (84) by discretizing the functions on a uniform angular lattice.[91] Within a finite-difference scheme and using a second-order approximation, a discretized version of this equation is obtained and solved efficiently using standard matrix diagonalization

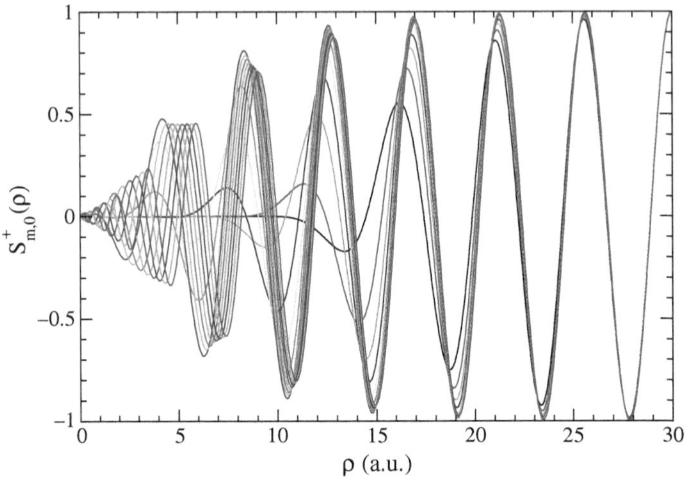

Figure 7.6 The first 15 Generalized Hyperspherical Sturmian basis set $S^+_{mn}(\rho)$, for $n = 0$ and $E = 1$ a.u. For color version of this figure, the reader is referred to the online version of this chapter.

routines, such as those from Lapack package.[92] Coupled to these angular functions, for a given n, we take as hyperradial basis functions, the Sturmian functions $S_{mn}(\rho)$ solving numerically Eq. (87) where $\nu = \nu_n = 2n$ and E is externally fixed as the energy of the system. The auxiliary potential $\mathcal{U}(\rho)$ is taken to be equal to the interaction potential $\mathcal{C}/\rho$, while the generating potential $\mathcal{V}_g(\rho)$ is set as a Yukawa potential $-\exp(-a_s\rho)/\rho$. With this choice, asymptotically, Eq. (87) reduces to a Coulomb homogeneous equation providing all basis functions (and thus the hypersherical GSF) a unique—and appropriate—asymptotic behavior

$$\lim_{\rho\to\infty} S^+_{mn}(\rho) \propto e^{iK\rho - i\eta\ln(2K\rho)}, \tag{100}$$

where K is the hyperspherical momentum and $\eta = \mathcal{C}\mu/K$ the Sommerfeld parameter.

In Figure 7.6, the functions $S^+_{mn}(\rho)$, corresponding to the first 15 hyperradial quantum numbers m, and for the hyperangular quantum number $n = 0$, obtained with $a_s = 0.2$, are plotted for a model attractive charge $\mathcal{C} = -1$, assuming a reduced mass $\mu = 1$, and for an energy $E = 1$ a.u. One easily appreciates how every function in the set achieves the asymptotic behavior (100) smoothly and the set is dense for low hyperradial values. Therefore, any well-behaved function that vanishes at $\rho < 25$ a.u., can be perfectly expanded by this basis. For higher n values, the basis is dense for higher ρ values, allowing the expansion at a more extended range.

Let us expand the numerical solution of the scattering problem (99) with such a basis set

$$\Psi^{NUM}(\rho,\alpha) = \frac{1}{\rho^{\frac{5}{2}}} \sum_m \sum_n a_{mn} \; S^+_{mn}(\rho) \; \Omega_n(\alpha). \tag{101}$$

Since both $\Psi^{NUM}\,(\rho,\alpha)$ and $S^+_{mn}\,(\rho)$ have the same asymptotic behavior, the above expansion is restricted to the internal region where particle interaction occurs. Using the eigenvalue Eq. (84) for the angular part, Eq. (87) for the radial part, projecting over $S^+_{pq}(\rho)\Omega_q(\alpha)/\rho^{\frac{5}{2}}$ and using the orthonormality relation (85), the unknown coefficients a are given by the following matrix equation

$$\sum_m \beta_{p,m} \, [\mathbf{V}]_{p,qm} \, a_{pm} = I_{pq}, \tag{102}$$

where the matrix elements are defined as

$$[\mathbf{V}]_{p,qm} = \int_0^\infty \; S^+_{pq}(\rho) \; \mathcal{V}_g\,(\rho) \; S^+_{mq}(\rho) \; d\rho\,. \tag{103}$$

The RHS vector I_{pq} elements are defined by

$$I_{pq} = \int_0^{\pi/2} \Omega_p(\alpha) \sin^2\alpha \cos^2\alpha d\alpha \int_0^\infty \; S^+_{pq}\,(\rho)\, \varphi\,(\rho,\alpha)\, \rho^{\frac{5}{2}} \, d\rho, \tag{104}$$

which, for the source (98) considered, reduce to the product of one-dimensional integrals. The driven equation (99) is transformed into an algebraic problem which can be easily solved using standard matrix techniques.[43]

The general solution of the complete scattering problem is known analytically.[90] It is built as a linear combination of products of hyperangular $\Omega_n(\alpha)$ functions times hyperradial functions[90,93] built to have the desired outgoing asymptotic behavior

$$\Psi^+(\rho,\alpha) \to f(\alpha) \frac{e^{iK\rho - i\eta \ln(2K\rho)}}{\rho^{\frac{5}{2}}}, \tag{105}$$

expected for a Coulomb scattering problem.[94,95] This limit provides an analytical expression also for the transition amplitude $f(\alpha)$. The latter can be easily extracted also from $\Psi^{NUM}(\rho,\alpha)$ using the outgoing asymptotic behavior (100), since

$$\Psi^{NUM}(\rho,\alpha) \to \sum_n \left(\sum_m a_{n,m}\right) \Omega_n(\alpha) \frac{e^{iK\rho - i\eta \ln(2K\rho)}}{\rho^{\frac{5}{2}}}, \quad (106)$$

yielding, by comparison with (105),

$$f(\alpha) = \sum_n \left(\sum_m a_{n,m}\right) \Omega_n(\alpha). \quad (107)$$

In Ref. 90,91, several kinematic situations were studied and overall excellent agreement was found between the numerical hyperspherical expansion $\Psi^{NUM}(\rho,\alpha)$ and the analytical solution. An example is provided in Figure 7.7 for a fixed value $\alpha = \pi/4$. We show the comparison in two regions, close to the origin (left) and at very large hyperradii (right).

All hyperradial basis elements not only diagonalize the kinetic energy and the interaction, but also possess the same appropriate asymptotic behavior; thus, they only need to expand the solution in the interaction region. These properties strongly accelerate the expansion convergence rate for the scattering wave function, and allow for a straightforward extraction of the transition

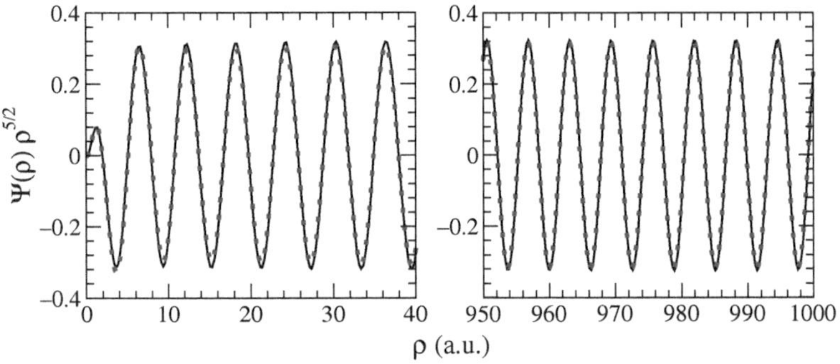

Figure 7.7 Real part of the numerical (full line) scattering solution $\Psi^{NUM}(\rho,\alpha)\rho^{5/2}$ given by Eq. (101) as a function of ρ along the cut $\alpha = \pi/4$, for $K = 1$ a.u. (i.e., the energy $E = 0.5$ a.u.), the parameters of the source are taken to be $a = 2$ and $t = 0$, and the interaction charge as $C = -1$. The solid dots correspond to the real part of the analytical solution $\Psi^{+}(\rho,\alpha)\rho^{5/2}$. For color version of this figure, the reader is referred to the online version of this chapter.

amplitude. Excellent agreement with the analytical results is found with only very few expansion terms. The numerical results shown in Figure 7.7 were obtained with only 8 hyperangular n- and 15 hyperradial m-terms.

The model problem allowed the authors[90,91] to explore how the scattering wave function is modified in behavior for different hyperradial domains, and how far one should go to extract the transition amplitude from the wave function itself. It was found that the required hyperradial distances are very large, especially for low energies. With the GSF hyperspherical method, one can reach the truly outgoing asymptotic region, where no other numerical method (besides the propagations performed by Malegat et al.[96]) can handle the calculations.

5.3 *S*-wave model of (*e*, 2*e*) processes on hydrogen

In this section, we review our results for the ionization of hydrogen by electron impact. To avoid all the difficulties introduced by the angular dependence of the wave functions and trying to clarify the crucial numerical and physical issues, we consider here the TP model:

$$\left[-\frac{1}{2r_1^2}\frac{\partial}{\partial r_1}\left(r_1^2\frac{\partial}{\partial r_1}\right)-\frac{1}{2r_2^2}\frac{\partial}{\partial r_2}\left(r_2^2\frac{\partial}{\partial r_2}\right)-\frac{Z}{r_1}-\frac{Z}{r_2}+\frac{1}{r_>}-E_a\right]$$
$$\Psi_{sc}^{+}\left(r_1,r_2\right)=-\left[\left(\frac{1}{r_>}-\frac{1}{r_2}\right)\Psi_i\left(r_1,r_2\right)+(-1)^S(1\leftrightarrow 2)\right]. \tag{108}$$

This is the standard test for any numerical method to be implemented for three-body ionization calculations, as it presents the difficulties associated to non-separability and to the long-range of the Coulomb potentials. As described in Section 4.4, to solve Eq. (108), one may use a *CI* expansion (54) with a linear combination of properly symmetrized products of two-body outgoing Sturmian functions $S^{+}_{n_1l_1}(r_1)S^{+}_{n_2l_2}(r_2)$. In Ref. 97, for each coordinate, a short-range generating potential, i.e., $\mathcal{V}(r_i)\to 0$ for $r_i > r_c$, was taken. On the other hand, choosing the auxiliary potential $\mathcal{U}(r_i)$, with a Coulomb tail, $-Z_i/r_i$ for $r_i > r_c$, provides an asymptotic behavior of the basis functions associated with an outgoing wave of energy E_i, distorted by the charge Z_i. The best way of defining the asymptotic behavior in each coordinate r_i corresponds to $E_i = E$ and $Z_i = Z - 1$. It actually corresponds to an appropriate description for the Ω_α region: one particle is at finite distance with close-to-zero or negative energy and sees the full

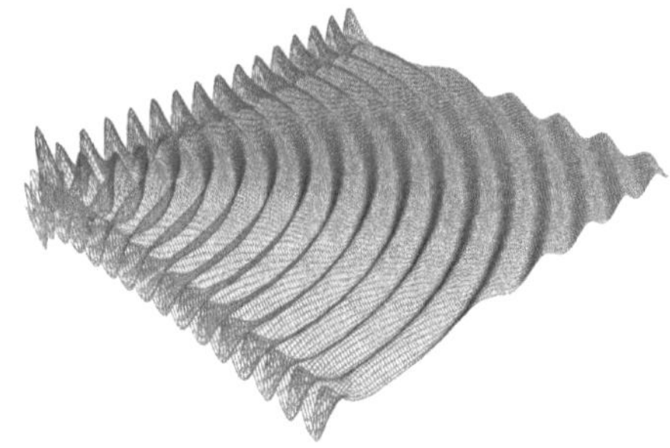

Figure 7.8 Schematic view of the real part of the scattering wave function evaluated with the Sturmian expansion. For color version of this figure, the reader is referred to the online version of this chapter.

nuclear charge Z while the other is at infinite distances carrying most of the system's energy but sees a screened charge Z_i. Since this screened value does not correspond to the atomic value at short distances, in order to have a more consistent picture, we define the potential $\mathcal{U}_i(r_i)$ by parts, i.e., $\mathcal{U}_i(r_i) = -Z/r_i$ for the inner region ($r < r_c$), and $\mathcal{U}_i(r_i) = -(Z-1)/r_i$ for the outer region ($r > r_c$) (no major differences were observed between the results obtained with smooth or sharp charge transitions). The choice of the outer charge has a considerable effect in yielding a *CI* expansion convergence toward the correct asymptotic behavior and, at the same time, a smooth inner solution. On the other hand, the inner charge is not so important since in the inner region the expansion has to deal with the potentials not removed by the basis elements; by choosing $Z_i = Z$ in the inner region, though, the Sturmian functions diagonalize not only the kinetic energy but also the electron–nucleus potentials Z/r_i.

The typical form of a two-electron double continuum wave function obtained after solving Eq. (108) is shown in Figure 7.8 as a function of r_1 and r_2. As expected from Eq. (56) an hyperspherical wave front can be clearly identified in a wide domain which includes the ionization region where $r_1 = r_2$ are both large. As we already explained in Section 4.1, the scattering wave function contains all channels at the same time and they are all coupled. On the borders of the figure, close to the axis $r_i = 0$, a different type of structure can be noticed and corresponds to the presence of excitation channels.

Once the wave function is obtained, various techniques can be applied to extract the cross-sections.

Before describing the SDCS results, we would like to add some comments on the convergence properties of the different approaches presented in the literature. As we have already mentioned, most of the computational methods are based on the transformation of the Schrödinger equation into a linear system of equations. Its size is related to the number of basis elements employed. Computational limitation arises of course due to the limited size of the computer clusters. To reach convergence in a given spatial domain, a minimum number n_{min} of basis elements is required. This number depends on the kind of problem and efficiency of the basis employed. In Ref. 97 a detailed comparative study of the convergence rate of the GSF method in comparison with finite element method (FEM) was performed. The former was shown to perform better (faster and more stable). In order to compare the efficiency of the Sturmian expansion with respect to other techniques, we compared their density of basis elements d, a quantity defined as the number of total basis elements divided by the two-dimensional expanded area. First we can mention the work of Bartlett[98] who used the so-called propagating ECS (PECS) (designed to increase the numerical efficiency of the ECS), with a discrete variable representation with different grid regions. Results in a square domain of $R_0 = 220$ a.u. divided in $\simeq 625$ intervals were presented, yielding a density $d_{PECS} \simeq 626^2/220^2 \simeq 8.1$. Time-dependent calculation was performed by Pindzola and Robicheaux[100] using grid points characterized by $\Delta r = 0.2$ a.u. and domains of size $R_0 = 100$ a.u. to $R_0 = 500$ a.u. in steps of 100 a.u.. The best result for the TP model obtained with that method

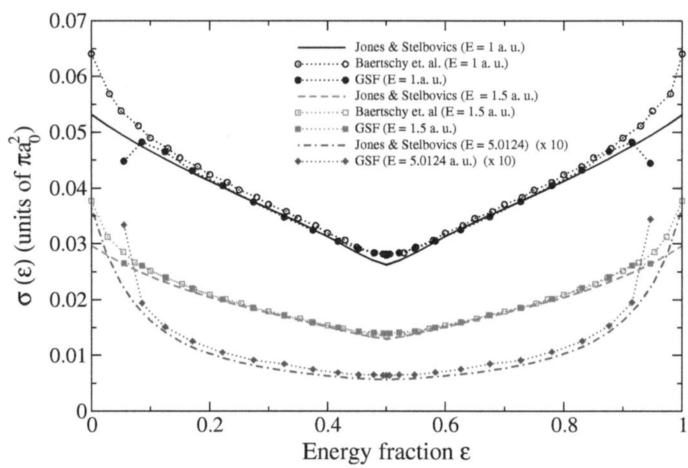

Figure 7.9 SDCS for different impact energies calculated with the GSF. The curves are constructed by evaluating the wave function flux at different hyperradius and then extrapolating the results to infinity. Also shown are the benchmark results of Jones and Stelbovics[100] and Baertschy and co-workers.[101] For color version of this figure, the reader is referred to the online version of this chapter.

required a box of $R_0 = 500$ a.u., resulting into a density $d_{TD} = 25$. With the GSF expansion, converged SDCS were extracted from the wave function evaluated in a spatial domain with $r_c = 130$ a.u., and calculated with $N_s = 175$ basis functions per coordinate.[97] The resulting density $d_{GSF} = 0.9112$ illustrates the GSF method high efficiency.

As shown in Section 4.3, ionization amplitudes can be extracted from the scattering wave function (see Eq. (56)). The corresponding SDCS, calculated with the GSF method, are shown in Figure 7.9 for three different impact energies. They are evaluated at different and finite values of hyperradius ρ and then extrapolated to $\rho \to \infty$.[97] For comparison, benchmark calculations of Jones and Stelbovics[100,101] are also presented. All GSF results were performed with 150 Sturmian functions per coordinate and in a domain of 130 a.u..

Near the region $\alpha = 0$ and $\alpha = \pi/2$, the influence of the excitation channel appears, leading to differences in the SDCSs as the latter are obtained from the wave function itself. We are currently developing a technique to separate the individual channels contributions of the scattering wave function to be able to define appropriately such SDCS.

5.4 *S*-wave model of (*e*, 3*e*) processes on helium

Kinematically complete $(e, 3e)$ experiments, in which the three outgoing particles are detected in coincidence, provide the most detailed information of electron impact double ionization of atoms.[82] Absolute fivefold differential cross-sections for helium have been measured by the Orsay group[12,83] in kinematic conditions such that the first Born approximation should be suitable. In spite of this, no theoretical study has yet managed to describe satisfactorily all the data. What is more confusing and difficult to explain, is that several *ab initio* methods provide different answers both in cross-section shapes and magnitudes (see a review in Ref. 102). From a theoretical point of view, the description of an $(e, 3e)$ process on helium requires the solution of a pure four-body Coulomb problem. However, as discussed in Section 4, reduction to a three-body problem can be performed in the case of high-energy projectiles as those used in the Orsay experiments.

In view of this unsatisfactory situation, the authors of Ref. 80 looked for a simplified problem for which agreement between theoretical methods could possibly be found. They considered an *S*-wave $(e, 3e)$ model with energy and geometry conditions used by the experimental Orsay group. The corresponding three-body model differs from that investigated in

Refs. 103,104, suitable for low-energy incident electrons, where the full four-body problem was considered.

Instead of considering the solution for the full first-order equation (72), they considered the following S-wave model[80]

$$\left[-\frac{1}{2r_1^2}\frac{\partial}{\partial r_1}\left(r_1^2\frac{\partial}{\partial r_1}\right)-\frac{1}{2r_2^2}\frac{\partial}{\partial r_2}\left(r_2^2\frac{\partial}{\partial r_2}\right)-\frac{Z}{r_1}-\frac{Z}{r_2}+\frac{1}{r_>}-E_a\right]$$
$$\Phi_{sc}^{(1)+}\left(r_1,r_2\right)=\mathcal{F}\left(r_1,r_2\right), \tag{109}$$

where

$$\mathcal{F}\left(r_1,r_2\right)=-\frac{1}{(2\pi)^3}\frac{4\pi}{q^2}\left[-Z+j_0(qr_1)+j_0(qr_2)\right]\,\Phi^{(0)}\left(r_1,r_2\right), \tag{110}$$

where $j_0(x)$ represents the spherical Bessel function of zeroth order, $\mathbf{q}=\mathbf{k_i}-\mathbf{k_f}$ is the momentum transfer, and $\Phi^{(0)}\left(r_1,r_2\right)$ is the ground-state solution of the S-wave helium equation Eq. (109) with the RHS set to zero. Moreover, since the idea was to provide benchmark values, all unnecessary ingredients were simplified and a simple ground state was taken: the product of screened exponentials $\Phi^{(0)}\left(r_1,r_2\right)=(Z_e^3/\pi)e^{-Z_e(r_1+r_2)}$ with $Z_e=Z-5/16$.

The model equation (109) was numerically investigated with both Sturmian approaches described in Section 4. For the spherical coordinates approach, the numerical technique is essentially the one used above for the electron–Hydrogen problem and does not need to be repeated. For the hyperspherical approach, expansion (55) was used with the hyperradial Sturmian functions $S_{mn}^{+}(\rho)$ satisfying equation (87). The generating potential was taken to be of short-range (vanishing faster than ρ^{-1} as $\rho\to\infty$), and the auxiliary potential as a Coulomb potential with charge Z, thus imposing to the hyperradial basis functions the desired asymptotic outgoing boundary condition (100) (with $\eta=\mu Z/K=\mu Z/\sqrt{2\mu E}$). Upon replacing either spherical (54) or hyperspherical (55) expansions into the scattering equation (109), and projecting onto the basis elements, one obtains a linear system

$$[\mathbf{H}-(E-\tilde{E})\mathbf{O}]\boldsymbol{a}=\boldsymbol{F}, \tag{111}$$

where $\mathbf{H}$ and $\mathbf{O}$ are the matrix representation of the Hamiltonian and overlap, and $\boldsymbol{F}$ represents the RHS projected onto the basis set; $\boldsymbol{a}$ is the vector of coefficients that builds the solution. For the spherical expansion $\tilde{E}=E_1+E_2$; as the best choice of the Sturmian parameters is

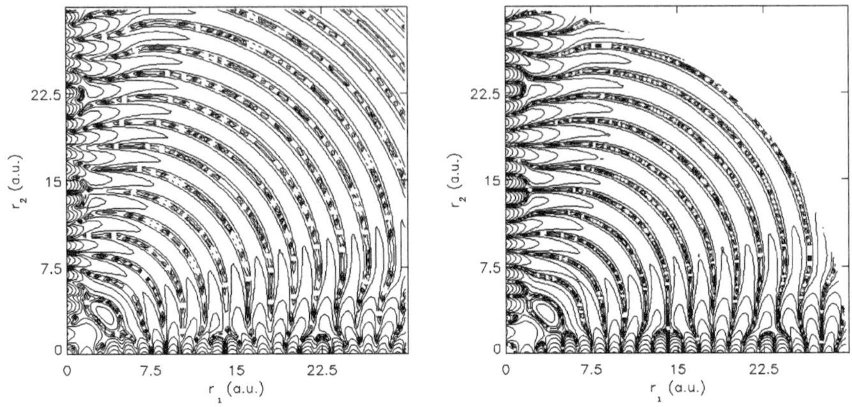

Figure 7.10 Left: Real part of the scattering wave function $\Phi_{sc}^{(1)+}(r_1, r_2)\times\rho^{5/2}$ as a function of the ejected electrons' radial coordinates r_1 and r_2, for $E_a = 0.791$ a.u. and $q = 0.24$ a.u. Right: real part of $\Phi_{sc}^{(1)+}(\rho, \alpha)\times\rho^{5/2}$.

$E_1 = E_2 = E$, the overlap matrix elements have to be calculated. In the case of the hyperspherical expansion, $\tilde{E} = E$, so that no overlap matrix elements are required.

The (singlet, $S = 0$) solution $\Phi_{sc}^{(1)+}$ of Eq. (109) has been calculated with the two Sturmian expansions for several kinematical situations.[80] One of them is for a momentum transfer $q = 0.24$ a.u. which corresponds to the initial and final projectile energies of, respectively, $E_i = 5599$ eV and $E_f = 5500$ eV, and a deflection of $0.45°$, used in the $(e, 3e)$ Orsay experiment.[83] These values, together with the exact ground-state energy of the bound initial state, define the energy of the final three-body subsystem (1,2) equal to $\simeq 20$ eV. For an equal energy sharing situation, this corresponds to 10 eV per electron, as in the experiments.[83]

The real part of the scattering solution, actually $\Phi_{sc}^{(1)+} \times \rho^{5/2}$, is shown as a function of r_1 and r_2 in the contour plots presented in Figure 7.10. The factor $\rho^{5/2}$ was chosen in order to keep the amplitude of the ionization (the hyperspherical outgoing wave) uniform as $\rho \to \infty$; it has to be noted that this factor increases the amplitude of single ionization channels (the peaks close the axis $r_1 = 0$ or $r_2 = 0$, i.e., the Ω_α regions) by the factor $\rho^{1/2}$. The result of the spherical expansion is shown in the left panel.

In the domain $r_1, r_2 > 5$ a.u., in which the driven term vanishes, the equation has the corresponding homogeneous equation solution. The basis functions for r_i values larger than r_c are simply products of outgoing

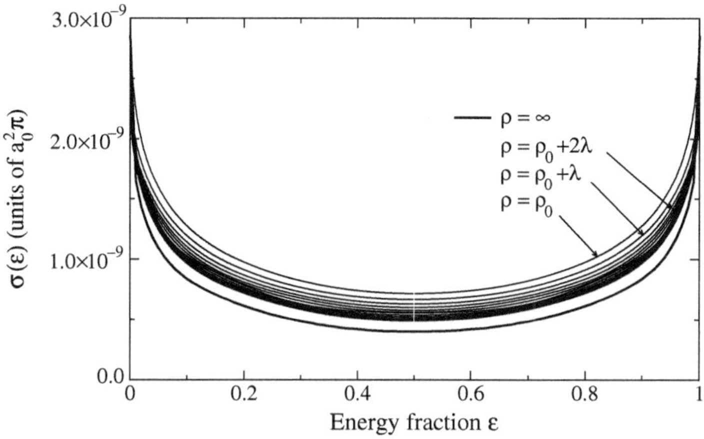

Figure 7.11 Single differential cross-section for the *S*-wave model $(e, 3e)$ process for $q = 0.24$ a.u. The equal energy fraction value corresponds to ionized electron energies equal to 15 *eV*. SDCS evaluated with the flux formula for different values of $\rho_n = \rho_0 + n\lambda$ ($\lambda = \frac{2\pi}{\sqrt{2E}}$, $\rho_0 = 2\lambda$) are shown. The bottom solid line represents the $\rho \to \infty$ extrapolated result.

waves in spherical coordinates. However, as the figure shows, in the inner region they manage to generate the appropriate solution with a hyperspherical outgoing front, which is also observed in the electron–Hydrogen TP model. For comparison, the result obtained with the hyperspherical expansion is shown in the right panel; in this case the hyperspherical wave front is naturally generated by the basis. Thus, two completely independent codes and methods are leading to the same solution and the exact solution is obtained by enforcing outgoing type flux conditions on the basis set.

In Ref. 80, the main aim was to provide $e^- - He$ double ionization benchmark data within the above *S*-wave model. The transition amplitude for the double ionization process can be extracted from evaluating $|\Phi_{sc}^{(1)+}(\rho, \alpha)\, \rho^{5/2}|^2$ at large values of ρ. This technique is equivalent to taking the *S*-wave component of the transition amplitude defined by the integral (74); besides, it provides a verification of the scattering wave function's accuracy. From the transition amplitude one obtains a SDCS, noted $\sigma(q, \alpha)$, which should be independent of the hyperradial coordinate; effectively, one evaluates numerically σ_ρ at different values of ρ and then extrapolates the result to infinite distances with a form $\sigma \simeq \sigma_\rho + O[\rho^{-1}]$.[95] For a given q value, $\sigma(q, \alpha)$ is a singly differential cross-section which—through α (with $k_1 = K \cos\alpha$ and $k_2 = K \sin\alpha$)—describes how the energy is shared between the two ejected electrons. Such SDCS for *S*-wave double ionization of the

S-wave helium model have been presented as benchmark values in Ref. 80. A further example is provided by Figure 7.11, for two ejected electrons sharing 30 eV (the projectile's energy is 5600 eV and the momentum transfer $q = 0.24$ a.u. as in the Orsay experiment). This SDCS was obtained with the spherical GSF and confirmed by the hyperspherical Sturmian approach. No comparison with other calculations could be presented since they are the first calculations of the process at the considered projectile energy. The study aimed to stimulate other numerical methods: if agreement can be found for the present model, one would then attribute the existing differences for the real $(e, 3e)$ process[102] to $L \neq 0$ and/or convergence issues.

6. SUMMARY AND PERSPECTIVES

In this paper we presented a review of some applications of the Generalized Sturmian Method. We want to stress once again that remarkable work has been done by other researchers but mostly using Coulomb Sturmian functions, and not generalized ones. The work of Macek and Ovchinnikov,[29–34] and Rawistcher[24–27] on Generalized Sturmian functions had a profound influence on our research. The work of Piraux,[9] Avery,[17,18] Aquilanti,[21,22,105] Goscinski,[106] Manakov,[107] Shakeshaft,[108] Rotenberg,[109] Szmytkowski,[110,111] and Maquet,[112,113] among others, have also been of great importance for the development of most of our investigations.

In Section 2 we gave a short description of the Generalized Sturmian Function theory and how this is applied in two-body problems. We gave also a numerical example where the two-body Coulomb problem is reformulated within the short-range scattering theory. The application of the method to three-body bound states is discussed in Section 3. We started by reviewing the results for the He and H^- systems. We showed that the convergence of the energies can be substantially improved when the GSF are used instead of CSF. The best energies obtained with uncorrelated basis have been produced using GSF.[42] We proved that optimizing the asymptotic property of the basis functions, the appropriate behavior can be generated in the three-body wave functions. This was shown for He-like atoms, for the H_2^+ molecule, as well as some exotic systems, but even more clearly for He doubly excited states. In this case imposing outgoing-type behavior to the GSF we obtained not only the real part of the energies but also their corresponding lifetimes. The method is robust enough to allow the inclusion of general auxiliary potentials in the Sturmian equation. This facilitates, for example, treatment of confined systems. We exemplified its robustness performing energy calculations of He atoms confined in a model cage representing a C_{60} fullerene, and in an impenetrable cage. All these calculations can be performed with high accuracy; the latter can be increased as much as necessary to recognize the avoided crossing structure appearing in both systems.

The description of scattering problems is based on the solution of a driven equation. The wave function of the problem is separated into a prepared initial state and a scattering state. All the physics about the collision is condensed into the scattering part of the wave function, which satisfies the driven equation. In Section 2 we showed how to apply the GSF method to two-body problems. As an example of the application we defined a distorted wave approach to enable inclusion of Coulomb potentials into the standard scattering theory. In Section 4, we first derived the driven equations for the single ionization of hydrogen atom by electron impact and for the double ionization of He by high-energy electron impact. Both problems lead to the same type of driven equation. Two Sturmian approaches were discussed as tools to solve it. One is based on the product of GSF in $\mathbf{r}_1, \mathbf{r}_2$ coordinates. In an alternative approach, the radial coordinates r_1, r_2 are replaced by ρ and α, the hyperspherical radius and angle. In both cases the scattering wave function is forced to have purely outgoing behavior at large values of ρ. This condition is built in the spherical approach, while it is natural in the case of the hyperspherical approach, for the cases under scrutiny.

In Section 5, we presented the application of the GSF method to the solution of three benchmark model problems. All of them correspond to S-wave models. The first one possesses analytical solution in hyperspherical coordinates. We used our hyperspherical GSF recipe to solve the problem and found perfect agreement for all the cases considered. Secondly, we evaluated the electron-impact ionization of Hydrogen within the S-wave model. In this case we compared our results with those provided by other methods. As a third example, we considered an S-wave model for the double ionization of He by high-energy electron impact. As the model was introduced only very recently, no other calculations are available for comparison. However, we checked our results by computing the solution of the problem using the spherical and the hyperspherical approaches: agreement between both methods is remarkable. In all the cases it can be verified that the expected hyperspherical wave front is effectively built by the GSF method. The corresponding cross-sections can be extracted directly from the asymptotic limit of the wave function.

Results for the full $(e, 2e)$ and $(e, 3e)$ processes as well as for double photoionization of atoms are presently being obtained and soon will be published elsewhere.

In the present review, we presented GSF in spherical and hyperspherical coordinates. However, as mentioned in Ref. 35 parabolic coordinates could be very useful to deal with many interesting problems where Coulomb potentials appear. Indeed, very important studies have been performed using these coordinates. The group of Aquilanti and collaborators has done remarkable progress on the treatment of bound states using parabolic CSF.[114,115] Ojha,[116] Burgdörfer,[117] Piraux,[118,119] and Zaytsev,[120–124] among others, have considered using CSF in parabolic coordinates for the treatment of two- and three-body problems involving in many cases ionization. The study and implementation of GSF and quasi-Sturmian functions in parabolic[123] and spherical coordinates[125] is the object of the actual investigations which are being performed in collaboration with Zaytsev. These investigations extend and formalize the studies initiated in the 1990s by Garibotti's[78,126,75,77] and Miraglia's[76,127] groups.

The application of GSF to the study many-electron atoms and molecular systems is the purpose of our current research.

ACKNOWLEDGMENTS

The authors kindly acknowledge the enlightening discussions with Prof. Rawitscher and his careful reading of this paper. The authors also thank the careful revisions performed by Profs. Piraux, Aquilanti, and Avery. The authors very much acknowledge the invaluable support given by different agencies and universities along these years: the Agencia Nacional de Promoción Científica y Tecnológica (ANPCyT) PICT 0934/08 and PICT 04/20548); the Consejo Nacional de Investigaciones Científicas y Técnicas (CONICET), PIP 200901/552 and PIP 5595. The authors also acknowledge support from the Universidad de Buenos Aires (UBACyT), Universidad Nacional de Cuyo (06/C279, 06/C229, 06/C348 grants), Universidad Nacional del Sur (PGI), and Université de Lorraine, and the French-Argentinian program ECOS-Sud A10E01.

REFERENCES

1. Drake, G.W. F. *Springer Handbook of Atomic, Molecular, and Optical Physics*; Springer: New York, 2005.

2. Kresse, G.; Hafner, J. Ab Initio Molecular Dynamics for Liquid Metals. *Phys. Rev. B* **1993**, *47*, 558.Kresse, G.; Hafner, J. Ab Initio Molecular-Dynamics Simulation of the Liquid-Metalamorphous-Semiconductor Transition in Germanium. *Phys. Rev. B* **1994**, *49*, 14251.
3. Frisch, M. J. et al. *Gaussian 09, Revision A.1*, Gaussian Inc., Wallingford, CT, 2009.
4. Avery, J. *Hyperspherical Harmonics and Generalized Sturmians*; Kluwer Academic Publishers: Dordrecht, The Netherlands, 2000.
5. Avery, J.; Avery, J. *Generalized Sturmians and Atomic Spectra*; World Scientific: Singapore, 2006.
6. Avery, J.; Avery, J. Coulomb Sturmians as a Basis for Molecular Calculations. *Mol. Phys.* **2012**, *110*, 10.
7. Newton, R. G. *Scattering Theory of Waves and Particles*; Dover Publications Inc.: New York, 2002.
8. Kadyrov, A. S.; Mukhamedzhanov, A. M.; Stelbovics, A. T.; Bray, I.; Pirlepesov, F. Asymptotic Behavior of the Coulomb Three-Body Scattered Wave. *Phys. Rev. A* **2003**, *68*, 022703.
9. Foumouo, E.; Kamta, G. L.; Edah, G.; Piraux, B. Theory of Multiphoton Single and Double Ionization of Two-Electron Atomic Systems Driven by Short-Wavelength Electric Fields: An Ab Initio Treatment. *Phys. Rev. A* **2006**, *74*, 063409.
10. Knyr, V. A.; Nasyrov, V. V.; Popov, Yu. V.; Ya Stotland, L. Application of the J-Matrix Method to Problems of Single Ionization of Atoms by Fast Electrons. *J. Exp. Theor. Phys.* **1996**, *82*, 190.
11. Silenou Mengoue, M.; Kwato Njock, M. G.; Piraux, B.; Popov, Yu. V.; Zaytsev, S. A. Electron-Impact Double Ionization of He by Applying the Jacobi Matrix Approach to the Faddeev-Merkuriev Equations. *Phys. Rev. A* **2011**, *83*, 052708.
12. Kheifets, A.; Bray, I.; Lahmam-Bennani, A.; Duguet, A.; Taouil, I. A Comparative Experimental and Theoretical Investigation of the Electron-Impact Double Ionization of He in the keV Regime. *J. Phys. B* **1999**, *32*, 5047.
13. McCurdy, C. W.; Horner, D. A.; Rescigno, T. N.; Martín, F. Theoretical Treatment of Double Photoionization of Helium Using a B-Spline Implementation of Exterior Complex Scaling. *Phys. Rev. A* **2004**, *69*, 032707.
14. Gasaneo, G.; Ancarani, L. U.; Mitnik, D. M. On the Applicability of the Exterior Complex Scaling Method for Scattering Problems Including Coulombic Potentials. *Eur. Phys. J. D* **2012**, *66*, 91.
15. Rescigno, T. N.; Baertschy, M.; Isaacs, W. A.; McCurdy, C. W. Collisional Breakup in a Quantum System of Three Charged Particles. *Science* **1999**, *286*, 2474.
16. Stelbovics, A. T. Calculation of Ionization within the Close-Coupling Formalism. *Phys. Rev. Lett.* **1999**, *83*, 1570.
17. Avery, J.; Avery, J. Generalized Sturmian Solutions for Many-Particle Schrödinger Equations. *J. Phys. Chem. A* **2004**, *108*, 8848–8851.
18. Avery, J.; Avery, J. The Generalized Sturmian Method for Calculating Spectra of Atoms and Ions. *J. Math. Chem.* **2003**, *33*, 145.
19. Avery, J.; Shim, R. Molecular Sturmians. Part 1. *Int. J. Quantum Chem.* **2001**, *83*, 1–10.
20. Aquilanti, V.; Cavalli, S.; Coletti, C.; Grossi, G. Alternative Sturmian Bases and Momentum Space Orbitals: An Application to the Hydrogen Molecular Ion. *Chem. Phys.* **1996**, *209*, 405.
21. Aquilanti, V.; Cavalli, S.; De Fazio, D. J. Hyperquantization Algorithm. I: Theory for Triatomic Systems. *Chem. Phys.* **1998**, *109*, 3792.
22. Aquilanti, V.; Cavalli, S.; De Fazio, D.; Volpi, A.; Aguilar, A.; Gimenez, X.; Lucas, J. M. Hyperquantization Algorithm. II: Implementation for the *F*1 H_2 Reaction Dynamics Including Open-Shell and Spin-Orbit Interactions. *J. Chem. Phys.* **1998**, *109*, 721.

23. Aquilanti, V.; Cavalli, S.; Coletti, C.; Di Domenico, D.; Grossi, G. Hyperspherical Harmonics as Sturmian Orbitals in Momentum Space: A Systematic Approach to the Few-Body Coulomb Problem. *Int. Rev. Phys. Chem.* **2001**, *20*, 673.
24. Rawitscher, G. Positive Energy Weinberg States for the Solution of Scattering Problems. *Phys. Rev. C* **1982**, *25*, 2196.
25. Rawitscher, G. Separable Representation of the Two-Body Reid Soft Core T Operator in Terms of Positive Energy Weinberg States. *Phys. Rev. C* **1989**, *39*, 440.
26. Rawitscher, G. In *Topics in Theoretical Physics*, Aguilera-Navarro, V. C., Galetti, D., Pimentel, B. M., Tomio, L. Instituto de Fisica Teórica, **1995**, pp 255.
27. Rawitscher, G. Iterative Solution of Integral Equations on a Basis of Positive Energy Sturmian Functions. *Phys. Rev. E* **2012**, *85*, 026701.
28. Rawitscher, G. Iterative Evaluation of the Effect of Long-Range Potentials on the Solution of the SchrÖdinger equation. *Phys. Rev. A* **2013**, *87*, 032708.
29. Yu, S. Electron-Energy and Angular-Distribution Theory for Low-Energy Ion-Atom Collisions. *Phys. Rev. Lett.* **1995**, *75*, 2474.
30. Macek, J. H.; Yu Ovchinnikov, S. Hyperspherical Theory of Three-Particle Fragmentation and Wannier Threshold Law. *Phys. Rev. A* **1996**, *54*, 544.
31. Yu Ovchinnikov, S.; Macek, J. H. Positive Energy Sturmian States for Two-Coulomb-Center Problems. *Phys. Rev. A* **1997**, *55*, 3605.
32. Gasaneo, G.; Macek, J. H. Hyperspherical Adiabatic Eigenvalues for Zero-Range Potentials. *J. Phys. B* **2002**, *35*, 2239.
33. Macek, J. H.; Yu Ovchinnikov, S.; Gasaneo, G. Solution for Boson-Diboson Elastic Scattering at Zero Energy in the Shape-Independent Model. *Phys. Rev. A* **2005**, *72*, 032709.
34. Macek, J. H.; Yu Ovchinnikov, S.; Gasaneo, G. Exact Solution for Three Particles Interacting via Zero-Range Potentials. *Phys. Rev. A* **2006**, *73*, 032704.
35. Gasaneo, G.; Colavecchia, F. D. Two-Body Coulomb Wavefunctions as Kernel for Alternative Integral Transformations. *J. Phys. A* **2003**, *36*, 8443.
36. Frapiccini, A. L.; Gasaneo, G.; Colavecchia, F. D.; Mitnik, D. Sturmian Functions in a L^2 Basis: Critical Nuclear Charge for n-Electron Atoms. *J. Electron. Spectrosc. Relat. Phenom.* **2007**, *161*, 199–203.
37. Frapiccini, A. L.; Gonzalez, V. Y.; Randazzo, J. M.; Colavecchia, F. D.; Gasaneo, G. L^2 Discretization of Sturmian Wave Functions for Coulomb-like Potentials. *Int. J. Quant. Chem.* **2007**, *107*, 832–844.
38. Randazzo, J. M.; Frapiccini, A. L.; Colavecchia, F. D.; Gasaneo G. *Phys. Rev. A* **2009**, *79*, 022507; Randazzo, J. M. Métodos ab initio para el problema de tres cuerpos con interacciones Coulombianas, Ph.D. thesis, Instituto Balseiro, Universidad Nacional de Cuyo, Argentina, 2009.
39. Randazzo, J. M.; Frapiccini, A. L.; Colavecchia, F. D.; Gasaneo, G. Discrete Sets of Many-Body Sturmians. *Int. J. Quantum Chem.* **2009**, *109*, 125.
40. Randazzo, J. M.; Gasaneo, G.; Colavecchia, F. D. A General Method to Obtain Sturmian Functions for Scattering and Bound State Problems. *Int. J. Quantum Chem.* **2010**, *110*, 963–974.
41. Frapiccini, A. L.; Randazzo, J. M.; Gasaneo, G.; Colavecchia, F. D. *J. Phys. B* **2010**, *43*, 101001; Frapiccini, A. L. Descripción de la dinámica de electrones en interacción con átomos, Ph.D. thesis, Instituto Balseiro, Universidad Nacional de Cuyo, Argentina, 2010.
42. Randazzo, J. M.; Ancarani, L. U.; Gasaneo, G.; Frapiccini, A. L.; Colavecchia, F. D. Generating Optimal Sturmian Basis Functions for Atomic Problems. *Phys. Rev. A* **2010**, *81*, 042520.
43. Mitnik, D. M.; Colavecchia, F. D.; Gasaneo, G.; Randazzo, J. M. Computational Methods for Generalized Sturmians Basis. *Comput. Phys. Commun.* **2011**, *182*, 1145.
44. Ambrosio, M. J.; Del Punta, J. A.; Rodriguez, K. V.; Gasaneo, G.; Ancarani, L. U. Mathematical Properties of Generalized Sturmian Functions. *J. Phys. A* **2012**, *45*, 015201.

45. Luk, F. T.; Qiao, S. A Fast Eigenvalue Algorithm for Hankel Matrices. *Linear Algebra Appl.* **2000**, *316*, 171.
46. Broad, J. T. Calculation of Two-Photon Processes in Hydrogen with an L^2 Basis. *Phys. Rev. A* **1985**, *31*, 1494.
47. de Boor, C. *A Practical Guide to Splines*; Springer: New York, 1978.
48. Shull, H.; Löwdin, P. O. Superposition of Configurations and Natural Spin Orbitals. Applications to the He Problem. *J. Chem. Phys.* **1959**, *30*, 617.
49. Bromley, M. W. J.; Mitroy, J. Convergence of the Partial Wave Expansion of the He Ground State. *Int. J. Quantum Chem.* **2007**, *107*, 1150.
50. Goldman, S. P. Uncoupling Correlated Calculations in Atomic Physics: Very High Accuracy and Ease. *Phys. Rev. A.* **1998**, *57*, R677.
51. Freund, D. E.; Huxtable, B. D.; Morgan, J. D. Variational Calculations on the Helium Isoelectronic Sequence. *Phys. Rev. A* **1984**, *29*, 980.
52. Frapiccini, A. L.; Randazzo, J. M.; Gasaneo, G.; Colavecchia, F. D. Sturmian Expansion of Two-Electron Atomic Systems: Singly and Doubly Excited States. *Phys. Rev. A* **2010**, *82* 042503
53. Eiglsperger, J.; Piraux, B.; Madronero, J. Spectral Representation of the Three-Body Coulomb Problem: Perspectives for Highly Doubly Excited States of Helium. *Phys. Rev. A* **2009**, *80*, 022511.
54. Eiglsperger, J.; Piraux, B.; Madronero, J. Spectral Representation of the Three-Body Coulomb Problem. I: Non-Autoionizing Doubly-Excited States of High Angular Momentum in Helium. *Phys. Rev. A* **2010**, *81*, 042527.
55. Eiglsperger, J.; Piraux, B.; Madronero, J. Spectral Representation of the Three-Body Coulomb Problem. II: Autoionizing Doubly-Excited States of Unnatural Parity in Helium. *Phys. Rev. A* **2010**, 042528.
56. Eiglsperger, J.; Schnwetter, M.; Piraux, B.; Madroero, J. Spectral Data for Doubly Excited States of Helium with Non-Zero Total Angular Momentum. *At. Data Nuclear Data Tables* **2012**, *98*, 120–148.
57. Drake, G. W. F. Variational Eigenvalues for the Rydberg States of Helium: Comparison with Experiment and with Asymptotic Expansions. *Phys. Rev. Lett.* **1990**, *65*, 2769; Drake, G. W. F.; Yan, Z. C. Energies and Relativistic Corrections for the Rydberg States of Helium: Variational Results and Asymptotic Analysis. *Phys. Rev. A* **1992**, *46*, 2378.
58. Bürgers, A.; Wintgen, D.; Rost, J. M. Highly Doubly Excited S States of the Helium Atom. *J. Phys. B* **1995**, *28*, 3163.
59. Rodriguez, K. V. Bases Correlacionadas Aplicadas al Estudio de Sistemas de Tres Cuerpos Generales, Ph.D. Thesis, Departamento de Física, Universidad Nacional del Sur, Argentina, 2010.
60. Ancarani, L. U.; Rodriguez, K. V.; Gasaneo, G. Correlated $n^{1,3}$ S States for Coulomb Three-Body Systems. *Int. J. Quantum Chem.* **2011**, *111*, 4255.
61. Drake, G. W. F.; Grigorescu, M. Binding Energy of the Positronium Negative Ion: Relativistic and QED Energy Shifts. *J. Phys. B* **2005**, *38*, 3377.
62. Korobov, V. I. Coulomb Three-Body Bound-State Problem: Variational Calculations of Nonrelativistic Energies. *Phys. Rev. A* **2000**, *61*, 064503.
63. Mitnik, D. M.; Randazzo, J.; Gasaneo, G. Endohedrally Confined Helium: Study of Mirror Collapses. *Phys, Rev. A* **2008**, *78*, 062501.
64. Colavecchia, F. D.; Gasaneo, G.; Mitnik, D. Double Photoionization of Endohedrally Confined Atoms. *J. At. Mol. Opt. Phys.* **2011**, *2011*, 817034.
65. Xu, Y. B.; Tan, M. Q.; Becker, U. Oscillations in the Photoionization Cross Section of C_60. *Phys. Rev. Lett.* **1996**, *76*, 3538.

66. Connerade, J. P.; Dolmatov, V. K.; Lakshmi, P. A.; Manson, S. T. Electron Structure of Endohedrally Confined Atoms: Atomic Hydrogen in an Attractive Shell. *J. Phys. B* **1999**, *32*, L239.
67. Banerjee, A.; Kamal, C.; Chowdhury, A. Calculation of Ground- and Excited-State Energies of Confined Helium Atom. *Phys. Lett. A* **2006**, *350*, 121.
68. Aquino, N.; Flores-Riveros, A.; Rivas-Silva, J. F. The Compressed Helium Atom Variationally Treated Via a Correlated Hylleraas Wave Function. *Phys. Lett. A* **2003**, *307*, 326.
69. Papp, P.; Plessas, W. Coulomb-Sturmian Separable Expansion Approach: Three-Body Faddeev Calculations for Coulomb-like Interactions. *Phys. Rev. C* **1996**, *54*, 50.
70. Papp, Z.; Darai, J.; Hu, C.-Y.; Hlousek, Z. T.; Końya, Z. T.; Yakovlev, S. L. Resonant-State Solution of the Faddeev-Merkuriev Integral Equations for Three-Body Systems with Coulomb Potentials. *Phys. Rev. A* **2002**, *65*, 032725.
71. Joslin, C.; Goldman, S. Quantum Monte Carlo Studies of Two-Electron Atoms Constrained in Spherical Boxes. *J. Phys. B* **1992**, *25*, 1965.
72. Garibotti, C. R.; Miraglia, J. E. Ionization and Electron Capture to the Continuum in the H^+-Hydrogen-Atom Collision. *Phys. Rev. A* **1980**, *21*, 572.
73. Brauner, M.; Briggs, J. S.; Klar, H. Triply-Differential Cross Sections for Ionisation of Hydrogen Atoms by Electrons and Positrons. *J. Phys. B* **1989**, *22*, 2265.
74. Alt, E. O.; Mukhamedzhanov, A. M. Asymptotic Solution of the Schrödinger Equation for Three Charged Particles. *Phys. Rev. A* **1993**, *47*, 2004.
75. Colavecchia, F. D.; Gasaneo, G.; Garibotti, C. R. Separable Wave Equation for Three Coulomb Interacting Particles. *Phys. Rev. A* **1998**, *57*, 1018.
76. Macri, P.; Miraglia, J. E.; Garibotti, C. R.; Colavecchia, F. D.; Gasaneo, G. Approximate Analytical Solution for Two Electrons in the Continuum. *Phys. Rev. A* **1997**, *55*, 3518.
77. Colavecchia, F. D.; Gasaneo, G.; Garibotti, C. R. Electron-Ion Correlation Effects in Ion-Atom Single Ionization. *J. Phys. B* **2000**, *33*, L467.
78. Gasaneo, G.; Colavecchia, F. D.; Garibotti, C. R.; Miraglia, J. E.; Macri, P. Correlated Continuum Wave Functions for Three Particles with Coulomb Interactions. *Phys. Rev. A* **1997**, *55*, 2809.
79. Gasaneo, G.; Ancarani, L. U. Use of Double-Bound Three-Body Coulomb Distorted-Wave-like Basis Sets for Two-Electron Wave Functions. *Phys. Rev. A* **2008**, 77, 012705.
80. Gasaneo, G.; Mitnik, D. M.; Randazzo, J. M.; Ancarani, L. U.; Colavecchia, F. D. S-Model Calculations for High-Energy-Electron-Impact Double Ionization of Helium. *Phys. Rev. A* **2013**, *87*, 042707.
81. Joachain, C. J. *Quantum Collision Theory*, North-Holland Publishing Company, 1983.
82. Berakdar, J.; Lahmam-Bennani, A.; Dal Cappello, C. The Electron-Impact Double Ionization of Atoms: An Insight into the Four-Body Coulomb Scattering Dynamics. *Phys. Rep.* **2003**, *374*, 91.
83. Lahmam-Bennani, A., et al. Origin of Dips and Peaks in the Absolute Fully Resolved Cross Sections for the Electron-Impact Double Ionization of He. *Phys. Rev. A* **1999**, *59*, 3548.
84. Gasaneo, G.; Mitnik, D. M.; Frapiccini, A. L.; Colavecchia, F. D.; Randazzo, J. M. Theory of Hyperspherical Sturmians for Three-Body Reactions. *J. Phys. Chem. A* **2009**, *113*, 14573.
85. Colavecchia, F. D.; Mrugala, F.; Parker, G. A.; Pack, R. T. Accurate Quantum Calculations on Three-Body Collisions in Recombination and Collision-Induced Dissociation. II: The Smooth Variable Discretization Enhanced Renormalized Numerov Propagator. *J. Chem. Phys.* **2003**, *118*, 10387.
86. Gasaneo, G.; Ancarani, L. U. A Spectral Approach Based on Generalized Sturmian Functions for Two- and Three-Body Scattering Problems. *J. Phys. A* **2012**, *45*, 045304.

87. Abramowitz, M.; Stegun, I. A. *Handbook of Mathematical Functions*; Dover: New York, 1972.
88. Temkin, A. Nonadiabatic Theory of Electron-Hydrogen Scattering. *Phys. Rev.* **1962**, *126*, 130.
89. Poet, R. The Exact Solution for a Simplified Model of Electron Scattering by Hydrogen Atoms. *J. Phys. B* **1978**, *11*, 3081.
90. Ancarani, L. U.; Gasaneo, G.; Mitnik, D. M. An Analytically Solvable Three-Body Break-Up Model Problem in Hyperspherical Coordinates. *Eur. Phys. J. D* **2012**, *66*, 270/1-11.
91. Mitnik, D. M.; Gasaneo, G.; Ancarani, L. U. Use of Generalized Hyperspherical Sturmian Functions for a Three-Body Break-Up Model Problem. *J. Phys. B* **2013**, *46*, 015202.
92. Dongarra, J., et al. *Lapack User's Guide*; 3rd ed.; SIAM: Philadelphia, USA, 1999.
93. Ancarani, L. U.; Gasaneo, G. Two-Body Coulomb Problems with Sources for the J-Matrix Method. *J. At. Mol. Sci.* **2011**, *2*, 203.
94. Rudge, M. R. H. Theory of the Ionization of Atoms by Electron Impact. *Rev. Mod. Phys.* **1968**, *40*, 564.
95. Peterkop, R. K. *Theory of Ionization of Atoms by Electron Impact*; Colorado Associated University Press: Boulder, CO, 1977.
96. Malegat, L.; Selles, P.; Kazansky, A. K. Absolute Differential Cross Sections for Photo Double Ionization of Helium from the Ab Initio Hyperspherical R-Matrix Method with Semiclassical Outgoing Waves. *Phys. Rev. Lett.* **2000**, *85*, 4450.
97. Randazzo, J. M.; Buezas, F.; Frapiccini, A. L.; Colavecchia, F. D.; Gasaneo, G. Solving Three-Body-Breakup Problems with Outgoing-Flux Asymptotic Conditions. *Phys. Rev. A* **2011**, *84*, 052715.
98. Bartlett, P. L. A Complete Numerical Approach to Electronhydrogen Collisions. *J. Phys. B* **2006**, *39*, R379.
99. Pindzola, M. S.; Robicheaux, F. Differential Cross Sections in the Ejected Energy for an L = 0 Model of the Electron-Impact Ionization of Hydrogen. *Phys. Rev. A* **1997**, *55*, 4617.
100. Jones, S.; Stelbovics, A. T. Efficient Solution of Three-Body Quantum Collision Problems: Application to the Temkin-Poet Model. *Phys. Rev. A* **2002**, *66*, 032717.
101. Baertschy, M.; Rescigno, T. N.; Isaacs, W. A.; McCurdy, C. W. Benchmark Single-Differential Ionization Cross Section Results for the s-Wave Model of Electron-Hydrogen Scattering. *Phys. Rev. A* **1999**, *60*, R13.
102. Ancarani, L. U.; Dal Cappello, C.; Gasaneo, G. Double Ionization of Two-Electron Systems. *J. Phys.: Conf. Ser.* **2010**, *212*, 012025.
103. Bartlett, P. L.; Stelbovics, A. T. Electron-Helium S-Wave Model Benchmark Calculations. I: Single Ionization and Single Excitation. *Phys. Rev. A* **2010**, *81*, 022715.
104. Bartlett, P. L.; Stelbovics, A. T. Electron-Helium S-Wave Model Benchmark Calculations. II: Double Ionization, Single Ionization with Excitation, and Double Excitation. *Phys. Rev. A* **2010**, *81*, 022716.
105. Aquilanti, V.; Cavalli, S.; Coletti, C. Hyperspherical Symmetry of Hydrogenic Orbitals and Recoupling Coefficients among Alternative Bases. *Phys. Rev. Lett.* **1998**, *80*, 3209.
106. Goscinski, O. *Preliminary Research Report No. 217*, Quantum Chemistry Group, Uppsala University, 1968.
107. Krylovetsky, A. A.; Manakov, N. L.; Marmo, S. I. Generalized Sturm Expansions of the Coulomb Green Function and Two-Photon Gordon Formulas. *J. Exp. Theor. Phys.* **2001**, *92*, 37.
108. Shakeshaft, R. Sturmian Expansion of Green Function and its Application to Multiphoton Ionization of Hydrogen. *Phys. Rev. A* **1986**, *34*, 244.

109. Rotenberg, M. Theory and Application of Sturmian Functions. *Adv. At. Mol. Phys.* **1970**, *6*, 233.
110. Szmytkowski, R. The Continuum SchrödingerCoulomb and DiracCoulomb Sturmian Functions. *J. Phys. A* **1998**, *31*, 4963.
111. Szmytkowski, R. Remarks on Completeness of Many-Electron Sturmians. *J. Phys. A* **2000**, *33*, 4553.
112. Maquet, A. Use of the Coulomb Green's Function in Atomic Calculations. *Phys. Rev. A* **1977**, *15*, 1088.
113. Maquet, A.; Véniard, V.; Marian, T. A. The Coulomb Green Function and Multiphoton Calculations. *J. Phys. B* **1998**, *31*, 3743.
114. Aquilanti, V.; Cavalli, S.; Coletti, C. The d-Dimensional Hydrogen Atom: Hyperspherical Harmonics as Momentum Space Orbital and Alternative Sturmian Basis Sets. *Chem. Phys.* **1997**, *214*, 1.
115. Aquilanti, V.; Calagiana, A.; Cavalli, S. Hydrogenic Elliptic Orbitals, Coulomb Sturmian Sets, and Recoupling Coefficients Among Alternative Bases. *J. Quantum Chem.* **2003**, *92*, 99.
116. Ojha, P. C. The Jacobi-Matrix Method in Parabolic Coordinates: Expansion of Coulomb Functions in Parabolic Sturmians. *J. Math. Phys.* **1987**, *28*, 392.
117. Yoshida, S.; Reinhold, C. O.; Burgdorfer, J. Accurate Ionization Thresholds of Atoms Subject to Half-Cycle Pulses. *Phys. Rev. A* **1998**, *58*, 2229.
118. Bugacov, A.; Piraux, B.; Pont, M.; Shakeshaft, R. Asymmetry in Ionization of Oriented Rydberg States of Hydrogen by a Half-Cycle Pulse. *Phys. Rev. A* **1995**, *51*, 4877.
119. Bugacov, A.; Piraux, B.; Pont, M.; Shakeshaft, R. Ionization of Rydberg Hydrogen by a Half-Cycle Pulse. *Phys. Rev. A* **1995**, *51*, 1490.
120. Zaytsev, S. A. One- and two-Dimensional Coulomb Green Function Matrices in Parabolic Sturmians Basis. *J. Phys. A* **2008**, *41*, 265204.
121. Zaytsev, S. A. The Parabolic Sturmian-Function Basis Representation of the Six-Dimensional Coulomb Green Function. *J. Phys. A* **2009**, *42*, 015202.
122. Zaytsev, S. A. Representation of the Three-Body Coulomb Green's Function in Parabolic Coordinates: Paths of Integration. *J. Phys. A* **2010**, *43*, 385208.
123. Zaytsev, S. A.; Gasaneo, G. Solving a Three-Body Continuum Coulomb Problem with Quasi-Sturmian Functions. *J. At. Mol. Scs.* **2013**, *4*, 302–320.
124. Zaytsev, S. A.; Popov, Yu V.; Piraux, B. Parabolic Sturmians Approach to the Three-Body Continuum Coulomb Problem. *Phys. At. Nuclei* **2013**, *76*, 365.
125. Del Punta, J. A.; Ambrosio, M. J.; Gasaneo, G.; Zaytsev, S. A.; Ancarani, L. U. Non-Homogeneous Solutions of a Coulomb Schrödinger equation as basis set for scattering problems. *J. Phys. A*, **2013**.
126. Gasaneo, G.; Colavecchia, F. D.; Garibotti, C. R.; Miraglia, J. E.; Macri, P. Multivariable Hypergeometric Solutions for Three Charged Particles. *J. Phys. B* **1997**, *30*, L265.
127. Miraglia, J. E.; Bustamante, M. G.; Macri, P. A. Approximate Wave Functions for Two Electrons in the Continuum of a Coulomb Charge. *Phys. Rev. A* **1999**, *60*, 4532.

CHAPTER EIGHT

Further Improvements on $\psi^{(\alpha^*)}$—ETOs with Hyperbolic Cosine Functions and Their Effectiveness in Atomic Calculations

S. Aksoy[a], **S. Firat**[a] **and M. Ertürk**[a]
[a]Department of Physics, Faculty of Arts and Sciences, Canakkale Onsekiz Mart University, Canakkale, 17100, Turkey

Contents

Abstract

In the last few years, exponential type orbitals became very important in electronic structure calculations of atoms and molecules. In this work, improvements on effectiveness of the $\psi^{(\alpha^*)}$-exponential type orbitals ($\psi^{(\alpha^*)}$-ETOs) ($-\infty < \alpha^* < 3$) containing different hyperbolic cosine functions are presented for the ground states of neutral atoms and their ions. The Hartree–Fock–Roothaan energies within the minimal basis set framework for some atoms up to $Z=18$ and their ions are listed and compared with the results obtained with other exponential type orbitals such as conventional double-zeta Slater, noninteger-n Slater with different hyperbolic cosine basis sets and numerical Hartree–Fock values. The accuracy of $\psi^{(\alpha^*)}$-ETOs is greatly improved for all atomic systems studied. The optimal noninteger values of α^* are determined for each atomic system examined in this work.

1. INTRODUCTION

It is well known that the construction of highly accurate wave functions is one of the most complicated problems in the Hartree–Fock (HF) scheme for atoms, molecules, and solids. In this solution of the HF

Advances in Quantum Chemistry, Volume 67
ISSN 0065-3276, http://dx.doi.org/10.1016/B978-0-12-411544-6.00008-X

equations, the basis set expansion or Hartree–Fock–Roothaan (HFR) approach[1,2] is used. In HFR method, the wave functions are expanded in a complete basis set using suitable basis functions. Therefore, chosen basis functions are important for electronic structure calculations. We note that the accuracy of results of calculation and computational cost in the HFR method directly depend on the quality and the number of reliable basis functions adopted for the calculations.

Among the basis functions most frequently used in quantum chemical calculations are Gaussian type orbitals (GTOs) and their advantages are well known. In addition to GTOs, exponential type orbitals (ETOs) are also used especially for atomic calculations. In the case of molecules, the GTOs are dominantly used as alternative to ETOs because they are computationally more convenient and tractable than ETOs in the evaluation of multicenter integrals for molecules. However, ETOs have more correct form, namely, the cusp at the nuclear center and the exponential decay at a long distance from origin than GTOs.[3,4] For these reasons, ETOs must be preferred as basis functions comparing to GTOs.

Slater type orbitals (STOs) are one of the well-known and most widely used ETOs in quantum chemistry. However, the STOs are not orthogonal with respect to the principal quantum numbers and they are also monomial and have no radial nodes for $n - l - 1 > 0$. Note that the nodal structure of radial functions is crucially important for determination of molecular properties. For this reason, the results of calculation could not satisfy the high quality of the fundamental physical properties of atomic and molecular systems when the STOs are used as basis functions. In order to fulfill the cusp condition and the correct behavior of the wave function, different ETOs such as Coulomb–Sturmian, Lambda, B function (BTOs) and $\psi^{(\alpha^*)}$-ETOs have been proposed by different authors.[5–11] Nowadays, there is a growing interest in the applications of ETOs in electronic structure calculation of atoms and molecules and related programing codes[12–29] (see references therein). Therefore, the development in the accuracy of ETOs is of paramount importance for future work on ETOs and its application to electronic structure calculations.

A significant improvement of the accuracy was achieved by introducing hyperbolic cosine type functions as indicated in Refs. 30–34. Use of the well-known hyperbolic cosine $Cosh(\beta r)$ (HC) and modified hyperbolic cosine $Cosh(\beta r + \gamma)$ (MHC) functions to better describe to radial functions of STOs has been criticized because their efficiency greatly improves minimal basis sets calculations. Recently, in this context, the noninteger n-STOs with HC function (NISTO-HC)[30] and noninteger n-STOs with

MHC (MNISTO-HC)[31] function as radial basis functions have been proposed. With this analogy, noninteger n-generalized exponential type orbitals with HC[33] and MHC functions[34] (NGETO-HC and MNGETO-HC, respectively) were successfully constructed to apply to atomic calculations.

More recently, we have reported[35] that the significantly improved HFR results for some neutral atoms were obtained using noninteger α^* values in $\psi^{(\alpha^*)}$-ETOs. It is also found that the $\psi^{(\alpha^*)}$-ETOs with noninteger α^* are more accurate than well-known Coulomb–Sturmian, Lambda, and BTOs functions. Therefore, the $\psi^{(\alpha^*)}$-ETOs are prominent than other ETOs.

In this work, the aim of our study is to construct the efficient and improved $\psi^{(\alpha^*)}$-ETOs basis functions by using two different hyperbolic cosine functions. These new basis functions (abbreviated as $\psi^{(\alpha^*)}$ -ETOHC and $\psi^{(\alpha^*)}$ -METOHC hereafter) have been used in atomic self-consistent field calculations. The performance and accuracy of these basis sets are compared with those of NISTO-HC, MNISTO-HC, and numerical Hartree–Fock (NHF) results. It is shown that the basis sets constructed in this work are as accurate as those obtained from NISTO-HC and MNISTO-HC in which the principal quantum numbers are nonintegral. Thus, by the use of hyperbolic cosine type functions, the effectiveness of $\psi^{(\alpha^*)}$-ETOs showed in Ref. 35 is greatly improved. Hartree atomic units are used throughout this work.

2. GENERAL DEFINITIONS AND PROPERTIES

In the case of noninteger indices of α^*, the $\psi^{(\alpha^*)}$ -ETOHC and $\psi^{(\alpha^*)}$ -METOHC in standard convention are defined as ${}^{H}\psi^{(\alpha^*)}$ and ${}^{M}\psi^{(\alpha^*)}$, respectively:

$$ {}^{H}\psi^{(\alpha^*)}_{nlm}(\zeta,\beta;\vec{r}) = {}^{H}R^{(\alpha^*)}_{n}(\zeta,\beta;r)S_{lm}(\theta,\varphi), \tag{1a}$$

$$ {}^{M}\psi^{(\alpha^*)}_{nlm}(\zeta,\beta,\gamma;\vec{r}) = {}^{M}R^{(\alpha^*)}_{n}(\zeta,\beta,\gamma;r)S_{lm}(\theta,\varphi), \tag{2a}$$

where n, l, and m are the quantum numbers and they match those of the occupied atomic orbitals. $S_{lm}(\theta,\phi)$ are well-known complex or real spherical harmonics. The radial functions ${}^{H}R^{(\alpha^*)}_{n}$ and ${}^{M}R^{(\alpha^*)}_{n}$ are defined as

$$ {}^{H}R^{(\alpha^*)}_{n}(\zeta,\beta;r) = \frac{(2\zeta)^{3/2}}{\Gamma(p_l^*+1)}\left[\frac{\Gamma(q_n^*+1)}{(2n)^{\alpha^*}(n-l-1)!}\right]^{1/2} e^{-\frac{x}{2}}x^l\,{}_1F_1(-[n-l-1];\, p_l^*+1;\, x)Cosh(\beta r), \tag{1b}$$

$$
{}^{M}R_{n}^{(\alpha^{*})}(\zeta,\beta,\gamma;r) = \frac{(2\zeta)^{3/2}}{\Gamma(p_l^{*}+1)}\left[\frac{\Gamma(q_n^{*}+1)}{(2n)^{\alpha^{*}}(n-l-1)!}\right]^{1/2}
e^{-\frac{x}{2}}x^{l}{}_{1}F_{1}(-[n-l-1];\ p_l^{*}+1;\ x)Cosh(\beta r+\gamma). \tag{2b}
$$

Here, $x=2\zeta r, p_l^{*}=2l+2-\alpha^{*}, q_n^{*}=n+l+1-\alpha^{*}$, and $-\infty<\alpha^{*}<3$. ${}_1F_1$ and $\Gamma(x)$ are the confluent hypergeometric function and gamma function, respectively. The ${}^{H}\psi^{(\alpha^{*})}$ and ${}^{M}\psi^{(\alpha^{*})}$ functions contain variational parameters α^{*},ζ,β, and γ; the parameter γ can be either positive, zero, or negative and $\zeta > \beta\geq 0$. It should be noted that the ${}^{M}\psi^{(\alpha^{*})}$ reduces to ${}^{H}\psi^{(\alpha^{*})}$ for a special case of $\gamma = 0$.

In order to simplify the calculations of matrix elements, it is useful to express the $\psi^{(\alpha^{*})}$-ETOs[11] as finite linear combinations of STOs:

$$
\psi_{nlm}^{(\alpha^{*})}(\zeta,\vec{r}) = \sum_{n'=l+1}^{n} \omega_{nn'}^{(\alpha^{*})l}\chi_{n'lm}(\zeta,\vec{r}), \tag{3}
$$

where the coefficient $\omega_{nn'}^{(\alpha^{*})l}$ is given by

$$
\omega_{nn'}^{(\alpha^{*})l} = \frac{1}{\Gamma(2l+3-\alpha^{*})}\left[\frac{\Gamma(n+l+2-\alpha^{*})(2n')!}{(2n)^{\alpha^{*}}(n-l-1)!}\right]^{1/2}
\frac{1}{[n'-(l+1)]!}\frac{(-[n-l-1])_{n'-(l+1)}}{(2l+3-\alpha^{*})_{n'-(l+1)}}. \tag{4}
$$

Here, $(-[n-l-1])_{n'-(l+1)}$ and $(2l+3-\alpha^{*})_{n'-(l+1)}$ are the Pochhammer symbols. Therefore, ${}^{H}\psi^{(\alpha^{*})}$ and ${}^{M}\psi^{(\alpha^{*})}$ can also be expressed in terms of finite series of NISTO-HC and MNISTO-HC functions, respectively.

3. COMPUTATIONAL METHOD

We briefly describe the full procedure for the closed and open shell HFR calculations in our previous papers.[33–36] All the calculations were carried out employing a self-consistent field program constructed in Mathematica 7.0 international mathematical software. It should be noted that the Mathematica provides any degree of working precision in calculations. The coupling projection coefficients available in the literature for the first row atoms of the periodic table have been taken from Ref. 36

Using Eq. (3), all the atomic integrals needed over ${}^{H}\psi^{(\alpha^*)}$ and ${}^{M}\psi^{(\alpha^*)}$ functions in HFR calculations are reduced to the NISTO-HC and MNISTO-HC integrals, respectively, defined explicitly in Ref. 30,31. Then, integrals are analytically evaluated by the use of simple auxiliary functions. All the variational parameters in Eqs. (1b) and (2b) were variationally optimized with the help of Newton method. To avoid local minimums, different initial values for parameters have been used in optimization process. We also noted that the optimal value of parameter α^* is crucial for the iterative minimization process of the HFR energy. In all calculations, the virial ratios obtained for ${}^{H}\psi^{(\alpha^*)}$ and ${}^{M}\psi^{(\alpha^*)}$ basis sets do not deviate from the value of -2 by more than 1×10^{-6}.

4. NUMERICAL RESULTS AND DISCUSSION

By the use of the HFR method, the atomic self-consistent field calculations within the minimal basis set framework are performed for some open and closed shell atomic systems. The computations have been restricted to open shell of different symmetry. Some calculation difficulties arise in the case of open shell orbitals of the same symmetry because it is possible to eliminate the off-diagonal Lagrange multipliers between the closed shell orbitals, but it is not in general possible to eliminate rigorously the off-diagonal Lagrange multipliers coupling the open shell orbitals of the same symmetry.

In order to examine the new ETOs (${}^{H}\psi^{(\alpha^*)}$ and ${}^{M}\psi^{(\alpha^*)}$), the results of total energies for neutral atoms, their anions and cations are presented in Tables 8.1–8.3. The optimal values of the parameters α^* obtained from these basis sets are given in parentheses. In these tables, we have also compared them with each other and to the NISTO-HC, MNISTO-HC, conventional integer n-STO double-zeta (ISTO-CDZ) in which two principal quantum numbers are restricted to same integer values. As can be seen from these tables, the NISTO-HC results are better than corresponding ${}^{H}\psi^{(\alpha^*)}$ except Be, B, B^+, C^+, and N^+ atomic systems. Similarly, the ${}^{M}\psi^{(\alpha^*)}$ performance is poor when comparing to MNISTO-HC except Be, Li^-. We notice that the accuracy of ${}^{H}\psi^{(\alpha^*)}$ does not reach the accuracy of the ISTO-CDZ results. Therefore, in the sixth column in Table 8.1, the accuracy of ${}^{M}\psi^{(\alpha^*)}$ has been also compared with that of ISTO-CDZ. It can be seen that the total energy of ${}^{M}\psi^{(\alpha^*)}$ basis sets is slightly better

Table 8.1 The $\psi^{(\alpha^*)}$-ETOHC, $\psi^{(\alpha^*)}$-METOHC total energies (${}^H\psi^{(\alpha^*)}$ and ${}^M\psi^{(\alpha^*)}$ sign reversed), optimal α^* parameters given in parentheses, and total energy errors $\Delta E_{E_{H_{\psi}(\alpha^*)}-E_{NISTO-HC}}$, $\Delta E_{E_{M_{\psi}(\alpha^*)}-E_{MNISTO-HC}}$, $\Delta E_{E_{ISTO-CDZ}-E_{M_{\psi}(\alpha^*)}}$, and $\Delta E_{E_{H_{\psi}(\alpha^*)}-E_{M_{\psi}(\alpha^*)}}$ for the ground states of the neutral atoms (in Hartree)

Atoms	E^a	E^b	ΔE^c	ΔE^d	ΔE^e	ΔE^f
H	2.861433196	2.861672626	0.000045719	0.000000316	0.000000000	0.000239430
Be	14.57203367	14.57300317	−0.00080731	−0.00000798	0.00063256	0.00096950
	(2.034515137)	(2.152030704)				
B	24.52596394	24.52869013	−0.00054830	0.00025728	0.00076864	0.00272619
	(1.966899288)	(2.152030704)				
C	37.67900091	37.68754922	0.00231224	0.00077008	0.00079795	0.00854831
	(1.952825842)	(2.049191329)				
N	54.37758450	54.39872767	0.00918076	0.00157057	0.00077173	0.02114317
	(1.955482905)	(1.931762862)				
O	74.74598655	74.80507874	0.03136007	0.00301612	0.00074539	0.05909218
	(1.950535494)	(1.894914222)				
F	99.28484005	99.40200143	0.06443244	0.00497681	0.00068232	0.11716138
	(1.952932704)	(1.866301244)				
Ne	128.3386321	128.5357172	0.1092129	0.0074870	0.0005974	0.1970851
	(1.958609012)	(1.843481176)				
Mg	199.5067001	199.6000221	0.0487195	0.0119805	−0.0070124	0.0933220
	(1.637379782)	(2.480880139)				
Ar	526.6458362	526.753681	0.0613222	0.1249350	−0.1693087	0.1078448
	(1.529352521)	(2.229856029)				

[a] $E_{H_{\psi}(\alpha^*)}$. [b] $E_{M_{\psi}(\alpha^*)}$. [c] $\Delta E_{E_{H_{\psi}(\alpha^*)}} - E_{NISTO-HC}$. [d] $\Delta E_{E_{M_{\psi}(\alpha^*)}} - E_{MNISTO-HC}$. [e] $\Delta E_{E_{ISTO-CDZ}-E_{M_{\psi}(\alpha^*)}}$. [f] $\Delta E_{E_{H_{\psi}(\alpha^*)}-E_{M_{\psi}(\alpha^*)}}$.

NISTO-HC energy values are taken from Ref. 30, MNISTO-HC energy values are taken from Ref. 31 and ISTO-CDZ values are taken from Ref. 38

Table 8.2 The $\psi^{(\alpha^*)}$-ETOHC, $\psi^{(\alpha^*)}$-METOHC total energies (${}^{H}\psi^{(\alpha^*)}$ and ${}^{M}\psi^{(\alpha^*)}$ sign reversed), optimal α^* parameters given in parentheses, and total energy errors $\Delta E_{E_{H_{\psi^{(\alpha^*)}}} - E_{NISTO-HC}}$, $\Delta E_{E_{M_{\psi^{(\alpha^*)}}} - E_{MNISTO-HC}}$, and $\Delta E_{E_{H_{\psi^{(\alpha^*)}}} - E_{M_{\psi^{(\alpha^*)}}}}$ for the low-lying states of some anions (in Hartree)

Anions	E^a	E^b	ΔE^c	ΔE^d	ΔE^e
H^-	0.486985136	0.487824426	0.000052782	0.000022104	0.000839290
Li^-	7.425448933 (2.841732377)	7.428182610 (2.553300795)	0.000146611	−0.000082196	0.002733677
B^-	24.49448503 (2.256300766)	24.51684465 (2.181657794)	0.01165816	0.00166729	0.02235962
C^-	37.66076029 (2.233376202)	37.70460874 (2.103211681)	0.02422569	0.00295481	0.04384845
N^-	54.21238304 (2.188495829)	54.31333596 (2.044063900)	0.05768414	0.00578255	0.10095292
O^-	74.59489008 (2.165603963)	74.77560628 (1.998790697)	0.10327498	0.01312459	0.18071620
F^-	99.15370272 (2.152619814)	99.43864461 (1.962895254)	0.16197626	0.00497681	0.28494189
Na^-	161.6956168 (1.793955155)	161.8301463 (2.175030151)	0.0831871	0.0200203	0.1345295
Cl^-	459.3667868 (1.420027761)	459.4874143 (2.310907351)	0.2506557	0.0747155	0.1206275

[a] $E_{H_{\psi^{(\alpha^*)}}}$. [b] $E_{M_{\psi^{(\alpha^*)}}}$. [c] $\Delta E_{E_{H_{\psi^{(\alpha^*)}}}} - E_{NISTO-HC}$. [d] $\Delta E_{E_{M_{\psi^{(\alpha^*)}}}} - E_{MNISTO-HC}$. [e] $\Delta E_{E_{H_{\psi^{(\alpha^*)}}}} - E_{M_{\psi^{(\alpha^*)}}}$.
NISTO-HC energy values are taken from Ref. 33 and MNISTO-HC energy values are taken from Ref. 34.

Table 8.3 The $\psi^{(\alpha^*)}$-ETOHC, $\psi^{(\alpha^*)}$-METOHC total energies (${}^{H}\psi^{(\alpha^*)}$ and ${}^{M}\psi^{(\alpha^*)}$ sign reversed), optimal α^* parameters given in parentheses, and total energy errors $\Delta E_{E_{H_{\psi}(\alpha^*)}-E_{NISTO-HC}}$, $\Delta E_{E_{M_{\psi}(\alpha^*)}-E_{MNISTO-HC}}$ and $\Delta E_{E_{H_{\psi}(\alpha^*)}-E_{M_{\psi}(\alpha^*)}}$ for the low-lying states of some cations (in Hartree)

Cations	E^a	E^b	ΔE^c	ΔE^d	ΔE^e
Li^+	7.235740430	7.236412449	0.000171297	0.000000054	0.000672019
B^+	24.23673285	24.23754626	−0.00074258	0.00000969	0.00081342
	(1.874401317)	(1.964535284)			
C^+	37.29096440	37.29193723	−0.00175061	0.00019303	0.00097283
	(1.746285206)	(1.882687113)			
N^+	53.88433601	53.88722930	−0.00174619	0.00054245	0.00289329
	(1.726661591)	(1.832520915)			
O^+	74.36190003	74.37103923	0.00074134	0.00109379	0.00913919
	(1.742855003)	(1.798732860)			
F^+	98.79584197	98.82884542	0.01374009	0.00199346	0.03300345
	(1.758393132)	(1.774479192)			
Ne^+	127.7390220	127.8129948	0.0361105	0.0032664	0.0739728
	(1.776881332)	(1.756362606)			
Na^+	161.5355083	161.6694468	0.0687017	0.0049670	0.1339386
	(1.795639581)	(1.742220313)			
Al^+	241.6030641	241.6643523	0.0254458	0.0065749	0.0612882
	(1.509321654)	(1.608007598)			
K^+	598.8948332	598.9692603	0.9123878	0.0462248	0.0744271
	(1.411691938)	(2.145778628)			

[a] $E_{H_{\psi}(\alpha^*)}$. [b] $E_{M_{\psi}(\alpha^*)}$. [c] $\Delta E_{E_{H_{\psi}(\alpha^*)}} - E_{NISTO-HC}$. [d] $\Delta E_{E_{M_{\psi}(\alpha^*)}} - E_{MNISTO-HC}$. [e] $\Delta E_{E_{H_{\psi}(\alpha^*)}} - E_{M_{\psi}(\alpha^*)}$.
NISTO-HC energy values are taken from Ref. 33 and MNISTO-HC energy values are taken from Ref. 34.

than ISTO-CDZ ones for light neutral atoms. As far as we know, there have been no values reported in the literature for ISTO-CDZ results of anions and cations. For this reason, we could not compare the performance of ${}^{M}\psi^{(\alpha^*)}$ basis sets with ISTO-CDZ for ionic systems examined in this work. In the last column of tables, the total energies of ${}^{H}\psi^{(\alpha^*)}$ and ${}^{M}\psi^{(\alpha^*)}$ are compared. As we expect, the ${}^{M}\psi^{(\alpha^*)}$ energies are notably lower than the ${}^{H}\psi^{(\alpha^*)}$ energies. From the values of energy errors, we can conclude that the efficiency of ${}^{M}\psi^{(\alpha^*)}$ increases rapidly with the increasing atomic number Z with respect to the ${}^{H}\psi^{(\alpha^*)}$. We observed that the total energy errors for $\Delta E_{E_{{}^{H}\psi^{(\alpha^*)}}-E_{NISTO-HC}}$, $\Delta E_{E_{{}^{M}\psi^{(\alpha^*)}}-E_{MNISTO-HC}}$, $\Delta E_{E_{ISTO-CDZ}-E_{{}^{M}\psi^{(\alpha^*)}}}$, and $\Delta E_{E_{{}^{H}\psi^{(\alpha^*)}}-E_{{}^{M}\psi^{(\alpha^*)}}}$ increase with increasing atomic number. The energy errors ΔE between ${}^{H}\psi^{(\alpha^*)}$ and ${}^{M}\psi^{(\alpha^*)}$ do not exceed 0.1970851, 0.28494189, 0.1339386 Hartrees which are observed for the neutral Ne, anion F^-, and cation Na^+ systems. It should be noted that the total energy errors do not increase smoothly with an increasing atomic number for all atomic systems examined in this work.

The variation of the optimum parameters α^* shows that the optimal α^* values are generally close to 2 for all the species examined. These results are consistent with the results found in the case of optimal integer α^* values in Ref. 35 We also search the relationship between α^* and atomic number Z. However, we did not find any relationship between them. We found that the use of simple approximated linear equation $\alpha^*(Z) = aZ + b$ for α^* is not suitable in the energies of ${}^{H}\psi^{(\alpha^*)}$ and ${}^{M}\psi^{(\alpha^*)}$. It should be noted that the structures of ${}^{H}\psi^{(\alpha^*)}$ and ${}^{M}\psi^{(\alpha^*)}$ in the case of 1s orbital yield to NISTO-HC and MNISTO-HC functions, respectively. Therefore, the parameter α^* does not affect the results of calculation, for He, H^-, and Li^+, in the case of single atomic orbital. Other parameters in ${}^{H}\psi^{(\alpha^*)}$ and ${}^{M}\psi^{(\alpha^*)}$ basis sets with respect to the atomic number are found to change more or less similarly as in the case of NISTO-HC[30] and MNISTO-HC[31] basis sets.

To clearly demonstrate how the accuracy is increased when one moves from $\psi^{(\alpha^*)}$-ETOs to ${}^{H}\psi^{(\alpha^*)}$ and ${}^{M}\psi^{(\alpha^*)}$ by the use of hyperbolic functions, we prefer to compare our results with numerical Hartree–Fock (NHF) values. The accuracy of ${}^{H}\psi^{(\alpha^*)}$ and ${}^{M}\psi^{(\alpha^*)}$ basis sets can be easily measured by comparison with NHF results which are available for atoms and ions in the literature.[37] The energy differences of ${}^{H}\psi^{(\alpha^*)}$ and ${}^{M}\psi^{(\alpha^*)}$ are shown in Figures 8.1–8.3 as a function of the atomic number Z for neutral atoms, their anions and cations. As can be seen from these figures, the ${}^{H}\psi^{(\alpha^*)}$ basis sets total energy errors are considerably larger than those

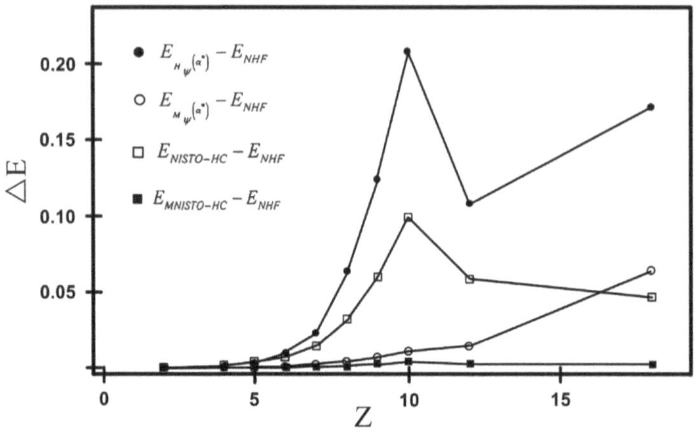

Figure 8.1 The $E_{H_{\psi}(\alpha^*)}$, $E_{M_{\psi}(\alpha^*)}$, $E_{NISTO-HC}$ and $E_{MNISTO-HC}$ energy differences ΔE relative to the E_{NHF} as a function of atomic number Z for the ground states of neutral atoms (in a.u.).

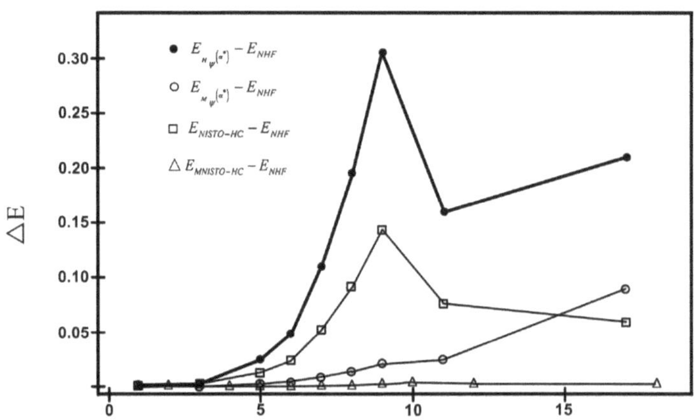

Figure 8.2 The $E_{H_{\psi}(\alpha^*)}$, $E_{M_{\psi}(\alpha^*)}$, $E_{NISTO-HC}$ and $E_{MNISTO-HC}$ energy differences ΔE relative to the E_{NHF} as a function of Atomic number Z for the low-lying states of anions (in a.u.).

of other basis sets. We also observed that the ${}^{M}\psi^{(\alpha^*)}$ total energy error increases consistently with an increasing atomic number for all atomic species examined. We notice that the performance of basis sets of which are poor for anions.

In our opinion, the main reason for the poor performances of the ${}^{H}\psi^{(\alpha^*)}$ and ${}^{M}\psi^{(\alpha^*)}$ with respect to the NISTO-HC and MNISTO-HC basis sets is due to the integer principal quantum number n used in this work. Based on this study, it can be concluded that the better performances

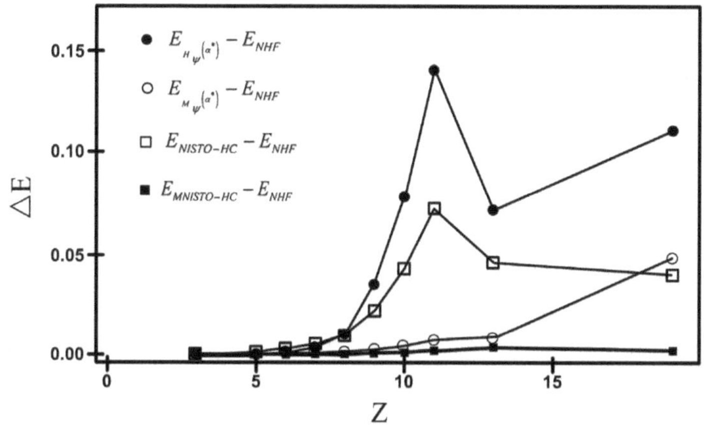

Figure 8.3 The $E_{^H\psi^{(\alpha^*)}}$, $E_{^M\psi^{(\alpha^*)}}$, $E_{NISTO-HC}$ and $E_{MNISTO-HC}$ energy differences ΔE relative to the E_{NHF} as a function of Atomic number Z for the low-lying states of cations (in a.u.).

of NISTO-HC and MNISTO-HC basis sets could be explained by their dependence of variational optimal noninteger principal quantum number n. Thus, variational noninteger principal quantum number n significantly improves the results as shown in the literature. It should also be noted that the parameters α^* are restricted to be common for all of the ${}^H\psi^{(\alpha^*)}$ and ${}^M\psi^{(\alpha^*)}$ basis sets. The quality of these basis sets and results of calculation can be easily improved by removal of this restriction.

We notice that the total number of nonlinear parameters in ${}^H\psi^{(\alpha^*)}$ and ${}^M\psi^{(\alpha^*)}$ functions is smaller than corresponding NISTO-HC and MNISTO-HC functions, respectively. Therefore, in the case of NISTO-HC and MNISTO-HC, the total number of nonlinear parameters increases more rapidly than ${}^H\psi^{(\alpha^*)}$ and ${}^M\psi^{(\alpha^*)}$ as the number of occupied orbitals increases. This is the reason why ${}^H\psi^{(\alpha^*)}$ and ${}^M\psi^{(\alpha^*)}$ total energies lie above the NISTO-HC and MNISTO-HC total energies.

In future, in order to test the quality of basis sets we are going to widely analyze the energies and cusp condition for many electron atoms and molecules. Then, the results of calculation obtained from the ${}^H\psi^{(\alpha^*)}$ and ${}^M\psi^{(\alpha^*)}$ basis sets could be compared with the other calculated data in the literature.

5. CONCLUSION

In this work, the ${}^H\psi^{(\alpha^*)}$ and ${}^M\psi^{(\alpha^*)}$ basis sets have been constructed and applied to atomic systems by performing the HFR calculations for the ground states of some atoms, anions, and cations within the minimal basis sets approximation.

The ${}^{M}\psi^{(\alpha^*)}$ total energies obtained in this study for the ground states of atoms and their ions are more accurate than those obtained previously using different type of ETOs such as Coulomb-Sturmian, Lambda, and BTOs and substantially more accurate than the ordinary $\psi^{(\alpha^*)}$-ETOs. The suitability of these basis sets for molecular calculations will be examined in future studies. The basis functions examined in this work play an important role in prediction of molecular properties but also for atomic processes such as electron-ion collision problem.

The resulting wave functions and parameters are available by request through merturk@comu.edu.tr

ACKNOWLEDGMENT

The authors would like to thank to Prof. T. Koga for providing his data.

REFERENCES

1. Roothaan, C. C. J. Self-Consistent Field Theory for Open Shells of Electronic Systems. *Rev. Mod. Phys.* **1960,** *32*, 179–185.
2. Roothaan, C. C. J.; Bagus, P. S. Methods in Computational Physics; Academic Press: New York, 1963.
3. Kato, T. On the Eigenfunctions of Many-Particle Systems in Quantum Mechanics. *Commun. Pure Appl. Math.* **1957,** *10*, 151–177.
4. Agmon, S. *Lectures on Exponential Decay of Solutions of Second Order Elliptic Equations: Bound on Eigenfunctions of N-Body Schrodinger Operators*; Princeton University Press: Princeton, NJ, 1982.
5. Hylleraas, E. A. Über den Grundzustand des Helium Atoms. *Z. Phys.* **1928,** *48* (7–8), 469–494.
6. Hylleraas, E. A. Neue Berechnung der Energie des Heliums im Grundzustande, sowie des tiefsten Terms von Ortho-Helium. *Z. Phys.* **1929,** *54* (5–6), 347–366.
7. Shull, H.; Lowdin, P. O. Role of the Continuum in Superposition of Configurations. *J. Chem. Phys.* **1955,** *23*, 1362.
8. Lowdin, P. O.; Shull, H. Natural Orbitals in the Quantum Theory of Two-Electron Systems. *Phys. Rev.* **1956,** *101*, 1730–1739.
9. Filter, E.; Steinborn, E. O. Extremely Compact Formulas for Molecular Two-Center One-Electron Integrals and Coulomb Integrals over Slater-Type Atomic Orbitals. *Phys. Rev. A* **1978,** *18*, 1–11.
10. Guseinov, I. I. New Complete Orthonormal Sets Of Exponential-Type Orbitals and Their Application to Translation of Slater Orbitals. *Int. J. Quant. Chem.* **2002,** *90*, 114–118.
11. Guseinov, I. I. New Complete Orthonormal Sets of Exponential-Type Orbitals in Standard Convention and Their Origin Bull. *Chem. Soc. Jpn.* **2012,** *85*, 1306–1309.
12. Bouferguene, A.; Fares, M.; Hoggan, P. E. STOP: A Slater-Type Orbital Package for Molecular Electronic Structure Determination. *Int. J. Quant. Chem.* **1996,** *57*, 801–810.
13. Avery, J. Many-Particle Sturmians. *J. Math. Chem.* **1997,** *21* (1), 285–304.

14. Rico, J. F.; López, R.; Ema, I.; Ramirez, G. Calculation of Many-Centre Two-Electron Molecular Integrals with STO. *Comp. Phys. Commun.* **1997,** *105* (2–3), 216–224.
15. Rico, J. F.; López, R.; Aguado, A.; Ema, I.; Ramírez, G. Reference Program for Molecular Calculations with Slater-Type Orbitals. *J. Comput. Chem.* **1998,** *19* (11), 1284–1293.
16. Ema, I.; Garcia de la Vega, J. M.; Miguel, B.; Dotterweich, J.; Meiβner, H.; Steinborn, E. O. Exponential-Type Basis Functions: Single- and Double-Zeta B Function Basis Sets for the Ground States of Neutral Atoms from Z = 2 to Z = 36. *Atomic Data and Nuclear Data Tables* **1999,** *72* (1), 57–99.
17. Steinborn, E. O.; Homeier, H. H. H.; Ema, I.; Lopez, R.; Ramírez, G. Molecular Calculations with B Functions. *Int. J. Quant. Chem.* **2000,** *76* (2), 244–251.
18. Steinborn, E. O.; Homeier, H. H. H.; Rico, J. F.; Ema, I.; Lopez, R.; Ramirez, G. An Improved Program for Molecular Calculations with B Functions. *J. Mol. Struct. (Theochem)* **1999,** *490* (1–3), 201–217.
19. Barnett, M. P. Symbolic Calculation of Auxiliary Functions for Molecular Integrals over Slater Orbitals. *Int. J. Quant. Chem.* **2000,** *76* (3), 464–472.
20. Rico, J. F.; Fernández, J. J.; López, R.; Ramírez, G. Molecular Integrals for Gaussian and Exponential-Type Functions: Shift Operators. *Int. J. Quant. Chem.* **2000,** *78* (3), 137–145.
21. Rico, J. F.; Fernández, J. J.; Ema, I.; López, R.; Ramírez, G. Four-Center Integrals for Gaussian and Exponential Functions. *Int. J. Quant. Chem.* **2001,** *81* (1), 16–28.
22. Rico, J. F.; López, R.; Aguado, A.; Ema, I.; Ramírez, G. New Program for Molecular Calculations with Slater-Type Orbitals. *Int. J. Quant. Chem.* **2001,** *81* (2), 148–153.
23. Safouhi, H.; Hoggan, P. New Methods for Accelerating the Convergence of Molecular Electronic Integrals over Exponential Type Orbitals. *Mol. Phys.* **2003,** *101* (1–2), 19–31.
24. Novosadov, B. K. Besselian: Calculation of Multicenter Matrix Elements. *J. Mol. Struct. (Theochem)* **2003,** *664–665*, 55–81.
25. Berlu, L.; Safouhi, H.; Hoggan, P. Fast and Accurate Evaluation of Three-Center, Two-Electron Coulomb, Hybrid, and Three-Center Nuclear Attraction Integrals Over Slater-Type Orbitals Using the SD Transformation. *Int. J. Quant. Chem.* **2004**, *99* (4), 221–235.
26. Absi, N.; Hoggan, P. E. Analytical Evaluation of Molecular Electronic Integrals Using Poisson's Equation: Exponential-Type Orbitals and Atom Pairs. *Int. J. Quant. Chem.* **2006,** *106* (14), 2881–2888.
27. Avery, J.; Avery, J. Can Coulomb Sturmians be Used as a Basis for N-Electron Molecular Calculations? *J. Phys. Chem. A* **2009,** *113* (52), 14565–14572.
28. Mitnik, D. M.; Colavecchia, Flavio D.; Gasaneo, G.; Randazzo, J. M. Computational Methods for Generalized Sturmians Basis. *Comp. Phys. Commun.* **2011,** *182* (5), 1145–1155.
29. Hoggan, P. E.; Bouferguene, A. Advances in the Theory of Quantum Systems in Chemistry and Physics. In *Progress in Theoretical Chemistry and Physics* Hoggan, P. E., Brandas, E. J., Maruani, J., Piecuch, P., Barrio, G. D., Eds.; **2012**; Vol. *22*, 83–102.
30. Koga, T. Hyperbolic Cosine Functions Applied to Atomic Roothaan–Hartree–Fock Wave Functions. *J. Mol. Struct. (Theochem)* **1998,** *451* (1–2), 1–10.
31. Koga, T. Hyperbolic Cosine Functions Applied to Atomic Roothaan–Hartree–Fock Wave Functions: Further Improvements. *J. Phys. B: At. Mol. Phys.* **1998,** *31* (8), 1637–1647.
32. Koga, T.; Kuroki, Y.; Tada, S. Hyperbolic Cosine Functions Applied to Atomic Calculations: Singly-Charged Cations and Anions. *Chem. Phys. lett.* **1999,** *304* (5–6), 445–449.
33. Guseinov, I. I.; Ertürk, M. Use of Noninteger n-Generalized Exponential Type Orbitals with Hyperbolic Cosine in Atomic Calculations Int. *J. Quant. Chem.* **2012,** *112* (6), 1559–1565.

34. Erturk, M. An Improvement on Noninteger n-Generalized Exponential Type Orbitals with Hyperbolic Cosine in Atomic Calculations Bull. *Chem. Soc. Jpn.* **2012,** *85* (7), 753–757.
35. Guseinov, I. I.; Sahin, E.; Erturk, M. An Improvement on $\psi(\alpha^{\star})$-Exponential Type Orbitals for Atoms in Standard Convention. *Mol. Phys.* **2013**. Doi:10.1080/00268976.2013.795665.
36. Guseinov, I. I.; Ertürk, M. Use of Noninteger n-Slater type Orbitals in Combined Hartree–Fock–Roothaan Theory for Calculation of Isoelectronic Series of Atoms Be to Ne. *Int. J. Quant. Chem.* **2009,** *109* (2), 176–184.
37. Koga, T.; Watanabe, S.; Kanayama, K.; Yasuda, R.; Thakkar, A. J. Improved Roothaan–Hartree–Fock Wave Functions for Atoms and Ions with $N \leq 54$. *J. Chem. Phys.* **1995,** *103* (8), 3000–3005.
38. Koga, T.; Seki, Y.; Thakkar, A. J. Improved Double-Zeta Description for the Atoms Li Through Xe. *Bull. Chem. Soc. Jpn.* **1993,** *66*, 3135–3141.

CHAPTER NINE

Reducing and Solving Electric Multipole Moment Integrals

N. Yükçü[a] and E. Öztekin[b]
[a]Department of Technical Programs, Vocational School, Sinop University, Sinop, Turkey
[b]Department of Physics, Faculty of Arts and Sciences, Ondokuz Mayıs University, Samsun, Turkey

Contents

Abstract

Electric multipole moment integrals with the same screening parameters are calculated using Slater type orbitals (STOs) translation formulae and the Fourier transform method. First, multipole moment operators which appear in the three-center electric multipole moment integrals are translated from the 0-center to the *b*-center. After the translation procedure and using Fourier transform convolution theorem, three-center electric multipole moment integrals were reduced to two-center molecular integrals. Then, the analytical expressions obtained were written in terms of overlap integrals. The numerical values for electric multiple moment integrals are useful especially for high quantum numbers and atomic parameters.

1. INTRODUCTION

Calculation of the electronic structure of atomic and molecular systems involves the determination of a number of integrals over the electronic coordinates. In molecular calculations the wave function is usually expanded in terms of atomic orbitals (AOs) centered on the nuclei of the molecule. The calculation of the electronic energy, electric dipole and other multipole moments, transition probabilities, fine and hyperfine coupling constants, and a number of other quantities reduces to the evaluation of electronic integrals involving the AOs used to construct the molecular wave function. In a general case one thus deals with one- and multicenter integrals. The purpose of this paper is to give a complete analysis of two-center

Advances in Quantum Chemistry, Volume 67
ISSN 0065-3276, http://dx.doi.org/10.1016/B978-0-12-411544-6.00009-1

integrals useful in calculations on diatomic molecules. It is well known that one-electron two-center electric multipole moment integrals (EMMIs) are one type of these integrals. Electric multiple moments have been of both experimental and theoretical interest recently. While direct measurements are usually limited to dipole and quadrupole moments,[1] molecular electronic structure theory has lately made investigations on electric moments exceeding tenth order.[2,3] Due to experimental limitations, studies of higher moments are not aimed at confirmation of measured result; rather they are driven by the needs of recent advances in molecular dynamics,[4] computation of high order polarizabilities,[5] molecular optical and magnetic properties,[6] intermolecular interactions in terms of the London series,[7] and quantum mechanical solvent effect in the formalism of the self-consistent reaction field theory[8] and some other analytical evaluations.[9,10] Quadrupole moment integrals were investigated by Wahl et al.,[11] and implemented for linear molecules by Bagus et al.[12] A specific and detailed evaluation of dipole moment integrals was studied by Rein et al.[13] and octopole moment integrals by Swissler and Rein.[14] Different methods of calculation of this integral are generally outlined by Harris and Michels,[15] Steinborn and Ruedenberg,[16] Guseinov and Sadichov,[17] Özdoğan and Orbay,[18] and Özdoğan,[19] and Yukcu et al.[20]

In this study, we expanded EMMIs over Slater type orbitals (STOs) by using partial-fraction decomposition and Taylor expansions of rational functions and Fourier transform method (FTM). It is well known that FTM is of considerable importance in the theory of two-center molecular integrals. With the help of FTM some six-dimensional integrals in coordinate space with nonseparated integration variables can be transformed into three-dimensional integrals in momentum space. In Section 3, we derive the mathematical expressions and present some values of EMMIs with the same screening parameters. The derivation of EMMIs over STOs with help of overlap integrals and FTM is discussed.

2. DEFINITIONS

General expression for one-electron two-center EMMIs with the same screening parameters is given[21]

$$M_{nlm,\nu\sigma}^{n'l'm'}\left(\alpha,\alpha;\vec{R}_{0b},\vec{R}_{ab}\right)=\int\chi_{nlm}^{*}\left(\alpha,\vec{r}_{a}\right)\hat{M}_{\nu\sigma}\left(\vec{r}_{0}\right)\chi_{n'l'm'}\left(\alpha,\vec{r}_{b}\right)\mathrm{d}^{3}r,\quad(1)$$

where $\chi_{n,l,m}(\alpha, \vec{r})$ is the STOs in the real spherical coordinates and α is screening parameter. In Eq. (1), the $\hat{M}_{\nu\sigma}(\vec{r}_0)$ is one-electron operator with the σ components of the 2^{ν}-pole of electric moment operator in spherical-polar coordinates[21] and it is defined as;

$$\hat{M}_{\nu\sigma}(\vec{r}) = \left(\frac{4\pi}{2\nu+1}\right)^{1/2} S_{\nu}^{\sigma}(\vec{r}), \tag{2}$$

where

$$S_{\nu}^{\sigma}(\vec{r}) = r^{\nu} Y_{\nu}^{\sigma}(\theta, \phi) \tag{3}$$

are regular solid harmonics and $Y_{\nu}^{\sigma}(\theta, \phi)$ is a spherical harmonic.[22] Any system of electric charges has associated with it set of electric multipole moments, in other words, with the 2^{ν}-pole (for $\nu = 0, 1, 2, 3, \ldots$; named monopole, dipole, quadrupole, octupole,..., etc.) moments.

In the literature, normalized STOs in real space and its Fourier transform (FT) form are defined by following forms[23]:

$$\chi_{n,l}^{m}(\alpha, \vec{r}) = \frac{2\alpha^{n+1/2}}{\sqrt{(2n)!}} r^{n-1} \exp(-\alpha r)\, Y_l^m(\theta, \phi), \tag{4}$$

$$\begin{aligned} U_{nl}^{m}(\alpha, \vec{p}) &= (2\pi)^{-3/2} \int e^{-i\vec{p}\cdot\vec{r}} \chi_{nl}^{m}(\alpha, \vec{r})\, d^3 r \\ &= \frac{\alpha^{2n-l-1}(n+l+1)!}{(2\pi)^{1/2}(1/2)_{l+1}} \frac{S_l^m\left(-\frac{i}{2}\vec{p}\right)}{\left[\alpha^2 + p^2\right]^{n+1}} \\ &\quad {}_2F_1\left(\frac{l-n}{2}, \frac{l-n+1}{2}; l+3/2; -\frac{p^2}{\alpha^2}\right), \end{aligned} \tag{5}$$

where $(a)_n = (a+n-1)!/(a-1)!$ is the Pochhammer symbol and ${}_2F_1(a, b; c; z)$ is a hypergeometric function. If $n-1$ is even, FT of an STO[23]:

$$\begin{aligned} U_{nl}^{m}(\alpha, \vec{p}) = {} & \frac{(-1)^{n/2+l}(n+l+1)!(n-l)!2^{n-2l}}{\alpha^{l+3}(2\pi)^{1/2}\left(\frac{n-l}{2}\right)!(1/2)_{(n+l+2)/2}} S_l^m(\vec{p}) \\ & \left[\frac{\alpha^2}{\alpha^2+p^2}\right]^{(n+l+2)/2} {}_2F_1\left(\frac{l-n}{2}, \frac{n+l+2}{2}; \frac{1}{2}; \frac{p^2}{p^2+\alpha^2}\right) \end{aligned} \tag{6}$$

and also if $n-1$ is odd[23]:

$$U_{nl}^{m}(\alpha,\vec{p}) = \frac{(-1)^{(n+3)/2+l}(n+l+1)!(n-l+1)!2^{n-2l+2}}{\alpha^{l+3}(2\pi)^{1/2}\left(\frac{n-l+1}{2}\right)!(1/2)_{(n+l+2)/2}}S_l^m(\vec{p})$$
$$\left[\frac{\alpha^2}{\alpha^2+p^2}\right]^{(n+l+3)/2}{}_2F_1\left(\frac{l-n+1}{2},\frac{n+l+3}{2};\frac{3}{2};\frac{\alpha^2}{\alpha^2+p^2}\right). \tag{7}$$

Now, we present multipole moment operators which appear in the EMMIs translated to the nuclear b-center from the origin 0-center. When multicenter molecular integrals are calculated, translation formulas provide a great advantage and have facilitated mathematical calculations. For $\phi_a = \phi_b = \phi$, this is done exactly using the finite formula[11,24]:

$$r_b{}^{\nu} Y_{\nu}^{\sigma}\left(\theta_b,\phi\right) = (-1)^{\sigma}\sum_{\mu=|\sigma|}^{\nu}\frac{(-1)^{\mu}}{(\nu-\mu)!}\sqrt{\frac{(2\nu+1)\,(\nu-|\sigma|)!\,(\nu+|\sigma|)!}{(2\mu+1)\,(\mu-|\sigma|)!\,(\mu+|\sigma|)!}}$$
$$R_{ba}^{\nu-\mu} r_a^{\mu} Y_{\mu}^{\sigma}\left(\theta_a,\phi\right). \tag{8}$$

By using Eqs. (2) and (8), the electric multipole moment operator is easily written in following form:

$$\hat{M}_{\nu\sigma}\left(\vec{r}_0\right) = \sum_{\mu=|\sigma|}^{\nu}\frac{(-1)^{\mu+\sigma}}{(\nu-\mu)!}\sqrt{\frac{(\nu+|\sigma|)!\,(\nu-|\sigma|)!}{(\mu+|\sigma|)!\,(\mu-|\sigma|)!}}R_{0b}{}^{\nu-\mu}\hat{M}_{\mu\sigma}\left(\vec{r}_b\right). \tag{9}$$

3. CALCULATION OF EMM INTEGRALS WITH THE SAME SCREENING PARAMETERS

In this section, we have calculated EMMIs with the same screening parameters by using FTM. For that purpose in this paper, firstly, we used translation formulae given by Eqs. (2), (8), (9) for an electric multipole moment operator. And then, we obtained a two-center integral expression of the EMMIs over STOs via the FTM evaluated two different methods.

In the first method, two-center EMMIs with the same screening parameters have been derived via the FTM and then integrant has been expressed as the product of two Gegenbauer polynomials and rational functions. Lately, the integral which appears in EMMIs with the same screening parameters has been obtained in terms of overlap integrals with the same screening parameters over STOs by using Eq. (15) of Ref. 25. Consequently, EMMIs are written as linear combination of overlap integrals with the same screening parameters.

If an operator form translates to STOs, one-electron two-center molecular integrals reduce to overlap integrals over STOs. To reduce EMMIs in Eq. (1) to two-center, by using Eqs. (2), (4), and (8), we can produce a function such as

$$\begin{aligned} f_{n'l'm',\nu\sigma}\left(\alpha,\vec{r}_0,\vec{r}_b\right) &= \hat{M}_{\nu\sigma}\left(\vec{r}_0\right)\chi_{n'l'm'}\left(\alpha,\vec{r}_b\right). \\ &= \sum_{\mu=|\sigma|}^{\nu}\frac{(-1)^{\mu+\sigma}}{(\nu-\mu)!}\sqrt{\frac{(\nu+|\sigma|)!\,(\nu-|\sigma|)!}{(\mu+|\sigma|)!\,(\mu-|\sigma|)!}} \\ &\quad {R_{0b}}^{\nu-\mu}\hat{M}_{\mu\sigma}\left(\vec{r}_b\right)\chi_{n'l'm'}\left(\alpha,\vec{r}_b\right). \end{aligned} \tag{10}$$

Equation (10) can be easily rewritten as[20]

$$f_{n'l'm',\nu\sigma}\left(\alpha,\vec{R}_{0b},\vec{r}_b\right) = \sum_{\mu=|\sigma|}^{\nu} G^{\sigma,\mu}_{n',\nu}\left(\alpha,R_{0b}\right)h^{1,l',m'}_{\mu+n',\mu,\sigma}\left(\alpha,\vec{r}_b\right), \tag{11}$$

where

$$\begin{aligned} h^{n_2,l_2,m_2}_{n_1,l_1,m_1}\left(\alpha,\vec{r}\right) &= \frac{2^{n_2-1/2}}{\sqrt{(2n_2)!}}r^{n_1+n_2-2}e^{-\alpha r} \\ &\quad \sum_{l=l_{\min}}^{l_{\max}}{}^{(2)}\langle lm_1+m_2\,|l_1m_1\,|l_2m_2\rangle\, Y_l^{m_1+m_2}\left(\theta,\phi\right) \end{aligned} \tag{12}$$

and

$$\begin{aligned} G^{\sigma,\mu}_{n,\nu}\left(\alpha,R_{0b}\right) &= \left(\frac{4\pi}{(2\mu+1)\,(2n)!}\right)^{1/2}(2\alpha)^{n+1/2} \\ &\quad (-1)^{\mu+\sigma}\sqrt{F_{(\mu+|\sigma|)}\left(\nu+|\sigma|\right)F_{(\mu-|\sigma|)}\left(\nu-|\sigma|\right)}{R_{0b}}^{\nu-\mu}. \end{aligned} \tag{13}$$

Here $F_a\left(b\right)$ are the Binomial coefficients and $\langle lm_1+m_2\,|l_1m_1\,|l_2m_2\rangle$ is the so-called Gaunt coefficient.[26] By using Eq. (11), we can rewrite Eq. (1) in terms of two-center EMMIs

$$\begin{aligned} M^{n'l'm'}_{nlm,\nu\sigma}\left(\alpha,\alpha;\vec{R}_{0b},\vec{R}_{ab}\right) &= \int \chi^*_{nlm}\left(\alpha,\vec{r}_a\right)f_{n'l'm',\nu\sigma}\left(\alpha,\vec{R}_{0b},\vec{r}_a-\vec{R}_{ab}\right)d^3r \\ &= \sum_{\mu=|\sigma|}^{\nu} G^{\sigma,\mu}_{n',\nu}\left(\alpha,R_{0b}\right)M^{n'l'm'}_{nlm,\mu\sigma}\left(\alpha,\alpha;\vec{R}_{ab}\right), \end{aligned} \tag{14}$$

where

$$M_{nlm,\mu\sigma}^{n'l'm'}\left(\alpha,\alpha;\vec{R}_{ab}\right)=\int\chi_{nlm}^{*}\left(\alpha,\vec{r}_a\right)h_{\mu+n',\mu,\sigma}^{1,l',m'}\left(\alpha,\vec{r}_a-\vec{R}_{ab}\right)d^3r. \tag{15}$$

For the evaluation of one-electron two-center EMMIs, the FT convolution theorem can be used in following form:

$$M_{nlm,\mu\sigma}^{n'l'm'}\left(\alpha,\alpha;\vec{R}_{ab}\right)=\int e^{-i\vec{p}.\vec{R}_{ab}}\left[U_{nl}^{m}\left(\alpha,\vec{p}\right)\right]^{*}\left[g_{\mu+n',\mu,\sigma}^{1,l',m'}\left(\alpha,\vec{p}\right)\right]d^3p, \tag{16}$$

where $g_{\mu+n',\mu,\sigma}^{1,l',m'}\left(\alpha,\vec{p}\right)$ is FT of $h_{\mu+n',\mu,\sigma}^{1,l',m'}\left(\alpha,\vec{r}_b\right)$ and it is given[20]:

$$g_{\mu+n',\mu,\sigma}^{1,l',m'}\left(\alpha,\vec{p}\right)=\sqrt{\frac{2}{\pi}}\sum_{L=L_{\min}}^{L_{\max}}{}^{(2)}\left\langle L\sigma+m'\,|\mu\sigma\,\middle|\,l'm'\right\rangle L!\left(\mu+n'-L\right)!$$
$$\frac{S_{L}^{\sigma+m'}\left(-2i\vec{p}\right)}{\left(\alpha^2+p^2\right)^{(\mu+n'+L+2)/2}}C_{\mu+n'-L}^{L+1}\left(\frac{\alpha}{\sqrt{\alpha^2+p^2}}\right), \tag{17}$$

where $C_n^l(x)$ are the Gegenbauer polynomials.[27,28]

With the help of Eqs. (5), (16), and (17) in this study and the product of two spherical harmonics in Eq. (3.2) of Ref. 29, general expression for EMMIs in Eq. (14) is rewritten in following form:

$$M_{nlm,\nu\sigma}^{n'l'm'}\left(\alpha,\alpha;\vec{R}_{0b},\vec{R}_{ab}\right)$$
$$=\frac{2^{n+l+3/2}\alpha^{n+1/2}l!(n-l)!}{\pi\sqrt{(2n)!}}\sum_{\mu=|\sigma|}^{\nu}G_{n',\nu}^{\sigma,\mu}\left(\alpha,R_{0b}\right)$$
$$\sum_{L=L_{\min}}^{L_{\max}}{}^{(2)}(i)^{l+L}(-2)^{L}L!\left(\mu+n'-L\right)!\left\langle L\sigma+m'\,|\mu\sigma\,\middle|\,l'm'\right\rangle$$
$$\sum_{L'=L'_{\min}}^{L'_{\max}}{}^{(2)}\sum_{M=-L'}^{L'}\left\langle L\left(\sigma+m'\right)|lm\,\middle|\,L'(\sigma+m'-m)\right\rangle A_{m,\sigma+m'}^{M} \tag{18}$$

$$\int e^{-i\vec{p}\cdot\vec{R}} \frac{C_{n-l}^{l+1}\left(\frac{\alpha}{\sqrt{\alpha^2+p^2}}\right) C_{\mu+n'-L}^{L+1}\left(\frac{\alpha}{\sqrt{\alpha^2+p^2}}\right)}{\left(\alpha^2+p^2\right)^{\left(n+\mu+n'+l+L+4\right)/2}} p^{l+L-L'} S_{L'}^{\sigma+m'-m}(\vec{p})\, d^3p$$

here

$$A_{m_1,m_2}^{m} = \frac{1}{\pi^{3/2}\sqrt{(1+\delta_{m,0})(1+\delta_{m_1,0})(1+\delta_{m_2,0})}} \int_0^{2\pi} \begin{Bmatrix} \cos m\phi \\ \sin|m|\,\phi \end{Bmatrix} \begin{Bmatrix} \cos(m_1+m_2)\phi + \cos(m_1-m_2)\phi \\ \sin|m_1+m_2|\,\phi + \sin|m_1-m_2|\,\phi \end{Bmatrix} d\phi. \tag{19}$$

Equation (18) is the simplest expression for EMMIs. If we notice that overlap integral form in Eq. (15) of Ref. 25 for $n_2 = \mu + n'$, $l_2 = L$ and $m_2 = \sigma + m'$, finally, general expression of EMMIs with the same screening parameters could be given in terms of overlap integrals with the same screening parameters by

$$M_{nlm,\nu\sigma}^{n'l'm'}(\alpha,\alpha;\vec{R}_{0b},\vec{R}_{ab}) = \sum_{\mu=|\sigma|}^{\nu} \sum_{L=L_{\min}}^{L_{\max}}{}^{(2)} G_{\nu,\sigma,\mu,L}^{n',l',m'}(\alpha,R_{0b}) S_{n,l,m}^{(\mu+n'),L,(\sigma+m')}(\alpha,\alpha;\vec{R}_{ab}), \tag{20}$$

where we denote following definition:

$$G_{\nu,\sigma,\mu,L}^{n',l',m'}\left(\alpha,R_{0b}\right) = (-1)^{\mu+\sigma}\left(4\pi \frac{(\nu+|\sigma|)!\,(\nu-|\sigma|)!\,\left(2\mu+2n'\right)!}{(\mu+|\sigma|)!\,(\mu-|\sigma|)!\,(2\mu+1)\left(2n'\right)!}\right)^{1/2} \frac{R_{0b}{}^{\nu-\mu}}{(\nu-\mu)!(2\alpha)^{\mu}} \left\langle L\sigma + m' \,|\mu\sigma\,\middle|\, l'm'\right\rangle \tag{21}$$

In the second method, for evaluated EMMIs with the same screening parameters over STOs via the FTM, we have used not only FTM and Taylor expansion or partial-fraction decomposition of rational functions but also Rayleigh expansions of plane wave and properties of integrals which involve spherical Bessel functions. Rayleigh expansions of the plane wave[30] play a central role in the analytical calculation of multicenter molecular integrals by using FTM. Finally, we are in a position to give explicit integral representation for EMMIs, which we will be treating in this paper. After using translation formulae for an electric multipole moment operator and then application of FTM to obtain integral form, one can rewrite following expression for EMMIs by using Rayleigh expansions of plane wave:

$$M_{nlm,\nu\sigma}^{n'l'm'}\left(\alpha,\alpha;\vec{R}_{0b},\vec{R}_{ab}\right)=\sum_{\mu=|\sigma|}^{\nu}G_{n',\nu}^{\sigma,\mu}\left(\alpha,R_{0b}\right)C_{nlm,n'l'm'}^{\sigma\mu}\left(\alpha;R_{ab},\theta,\phi\right), \tag{22}$$

where

$$\begin{aligned}C_{nlm,n'l'm'}^{\sigma\mu}\left(\alpha;R_{ab},\theta,\phi\right)&=\pi\frac{2^{2n+5}\alpha^{2n+1/2}(n-l)!}{\sqrt{(2n)!}}\\&\sum_{L=L_{\min}}^{L_{\max}}{}^{(2)}(-1)^{L}\left(\mu+n'-L\right)!\left\langle L\sigma+m'\left|\mu\sigma\right|l'm'\right\rangle\\&\sum_{L'=L'_{\min}}^{L'_{\max}}{}^{(2)}\sum_{M=-L'}^{L'}\left\langle L\left(\sigma+m'\right)|lm\left|L'(\sigma+m'-m)\right.\right\rangle A_{m,\sigma+m'}^{M}\end{aligned} \tag{23}$$

$$\sum_{s=0}^{[(n-l)/2]}\frac{(n-s)!}{s!(n-l-2s)!}\sum_{r=0}^{\left[(\mu+n'-L)/2\right]}\frac{(\mu+n'-r)!}{r!(\mu+n'-L-2r)!}2^{\mu+n'-2(s+r)+2\lambda}$$
$$\sum_{t=0}^{(l+L-L')/2}\frac{(-1)^{s+r+t}}{\alpha^{2(t+\lambda+s+r)-\mu-n'+1/2}}\frac{((l+L-L')/2)!}{t!\left((l+L-L')/2-t\right)!}$$
$$\sum_{k=1}^{\lambda-L'}g_{\lambda,k}^{L'}\chi_{k+L',L'}^{\sigma+m'-m}\left(\alpha,\vec{R}_{ab}\right)$$

and $\lambda=n+\mu+n'-s-r-t+1$.

Table 9.1 The numerical values for overlap integrals

n_1	l_1	m_1	n_2	l_2	m_2	α	R	θ	ϕ	Our numerical results	Results of literature	Ref.
3	2	2	3	2	2	5.8	0.2	0	0	$9.09920806903332 \times 10^{-1}$	$9.09920806903332 \times 10^{-1}$	25
5	2	2	4	2	2	6	2	0	0	$1.94083486492337 \times 10^{-2}$	$1.94083486492337 \times 10^{-2}$	25
10	7	1	8	1	1	3	10	0	0	$2.34478355886267 \times 10^{-3}$	$2.34478355886268 \times 10^{-3}$	25
25	6	3	5	4	3	0.5	16	0	0	$8.24612683258671 \times 10^{-3}$	$8.24612683258671 \times 10^{-3}$	25
20	5	5	8	5	5	1	250	0	0	$4.51138375363701 \times 10^{-80}$	$4.51138897499248 \times 10^{-80}$	31
26	12	10	16	12	10	2	20	0	0	$2.32953509405079 \times 10^{-3}$	$2.32953664607847 \times 10^{-3}$	32
4	2	1	3	2	1	1	25	0	0	$-5.65554471492814 \times 10^{-6}$	$-5.65554471492813 \times 10^{-6}$	33
15	14	14	15	14	14	1	15	0	0	$0.03747224970381 \times 10^{-2}$	$0.03747224970381 \times 10^{-2}$	33
1	0	0	1	0	0	10	1.4	0	0	$6.67994737679866 \times 10^{-5}$	$6.67994737679866 \times 10^{-5}$	34
2	1	0	2	1	0	2	1.4	0	0	$-1.00740382147441 \times 10^{-1}$	$-1.00740382147740 \times 10^{-1}$	34
11	8	5	9	5	5	1	1	0	0	$2.45593364389480 \times 10^{-3}$	$2.45593364 \times 10^{-3}$	34
1	0	0	1	0	0	10	1.4	0	0	$6.67994737679866 \times 10^{-3}$	$6.67994737679866 \times 10^{-3}$	35
20	5	5	8	5	5	1	250	60	45	$-1.30261805673296 \times 10^{-72}$	$-1.30261805673295 \times 10^{-72}$	31
2	1	0	2	1	0	1	5	60	120	$4.32351599107984 \times 10^{-2}$	$4.32351599107984 \times 10^{-2}$	25
5	4	2	5	4	2	1	5	36	120	$-1.44686204869417 \times 10^{-1}$	$-1.44686204869417 \times 10^{-1}$	32
10	9	8	12	9	8	2	6	180	108	$-2.63733203794236 \times 10^{-1}$	$-2.63733203794236 \times 10^{-1}$	32
80	20	8	25	20	8	2	10	51.4	180	$-1.83178453358437 \times 10^{-6}$	$-1.18040862548839 \times 10^{-6}$	32

4. SUMMARY AND CONCLUSION

In order to calculate the EMMIs with the same screening parameters via the FTM, three-center EMMIs over STOs can be evaluated by translating the orbitals on one-center to the other and reducing the system to an expansion of two-center integrals. EMMIs are then evaluated by using FTM and partial-fraction decomposition and Taylor expansion of rational functions. The resulting expressions depend on overlap integrals with the same screening parameters, Gaunt coefficients, and STOs. It is well known that the overlap integrals with the same screening parameters can be determined easily in all molecular integrals. The overlap integrals with the same screening parameters over STOs involve finite sums. EMMIs with the same screening parameters have been calculated in terms of overlap integrals via Eq. (20). And also, with help of Eq. (22) of Ref. 25, in Table 9.1, we have obtained numerical results for overlap integrals which completely agree with the results in the literature.[25,31–35]

The comparative values of EMMIs over STOs are tabulated in Table 9.2 for various quantum numbers. All numerical results in Table 9.2 are calculated by using Eq. (20) for $\phi = 0$. In Figures 9.1 and 9.2, curves of electric dipole moment versus interatomic distance and θ angle are plotted by using Eq. (20), respectively, for various atomic parameters. If we consider curves in Figure 9.1, we see that all curves oscillate and approach to zero but are not periodic. Only one of the three curves begins near the origin, and this curve occurs on condition that orbital quantum numbers have $l_1 \neq l_2$ (see state c, in Fig. 9.1). Moreover, we have obtained similar curves as in Figure 9.1

Table 9.2 The numerical values for EMMIs

n_1	l_1	m_1	n_2	l_2	m_2	v	σ	θ	α	R_{0b}	R_{ab}	Numerical results
1	0	0	1	0	0	1	0	0	4	2.3	4.5	$-5.47537647764789 \times 10^{-6}$
2	0	0	2	0	0	1	0	0	4	2.3	4.5	$-1.35539280082122 \times 10^{-4}$
2	1	0	2	1	0	2	0	0	1	0.2	4.5	$5.39499136051475 \times 10^{-4}$
3	2	2	2	1	1	4	1	30	1	0.2	4.5	$-6.81295969102066 \times 10^{1}$
6	5	5	5	3	3	5	2	30	10	2	5	$-1.22790792408944 \times 10^{-12}$
7	5	5	6	4	2	6	3	45	6	0.5	0.2	$2.33267295134455 \times 10^{-1}$
8	5	5	8	5	−1	6	6	20	3	0.3	0.5	$-1.10001002135298 \times 10^{2}$
12	11	2	10	8	0	8	0	70	8	0.8	0.7	$3.71423114216012 \times 10^{2}$
10	3	3	8	2	−2	15	5	90	1	3.5	4.5	$-2.08851840929211 \times 10^{12}$
12	5	3	7	2	−2	19	4	120	3	6	4	$-2.13471577216319 \times 10^{14}$
12	8	4	9	3	−3	20	7	180	2	8	5	$-1.30663309649809 \times 10^{19}$

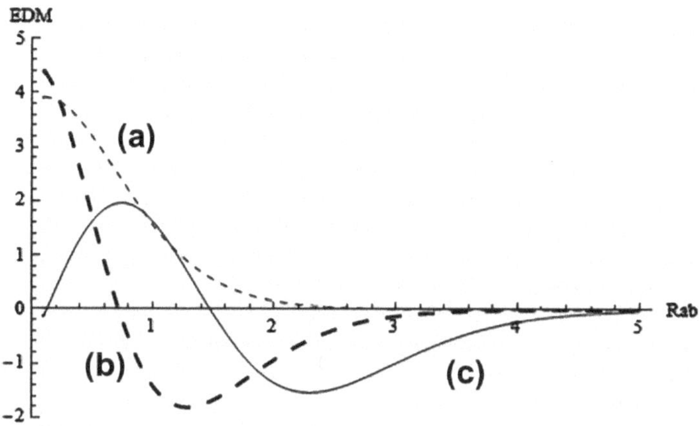

Figure 9.1 Electric dipole moment (EDM) versus R_{ab} for some atomic parameters such as: $v = 1$, $\sigma = 0$, $\alpha = 4$, $R_{Ob} = 4.5$, $\theta = 0°$, $\phi = 0°$. Curves in Figure 9.1 are represented by quantum numbers: (a) $n_1 = 1$, $l_1 = 0$, $m_1 = 0$ and $n_2 = 2$, $l_2 = 0$, $m_2 = 0$, (b) $n_1 = 3$, $l_1 = 2$, $m_1 = 1$ and $n_2 = 3$, $l_2 = 2$, $m_2 = 1$, (c) $n_1 = 5$, $l_1 = 3$, $m_1 = 1$ and $n_2 = 5$, $l_2 = 2$, $m_2 = 1$.

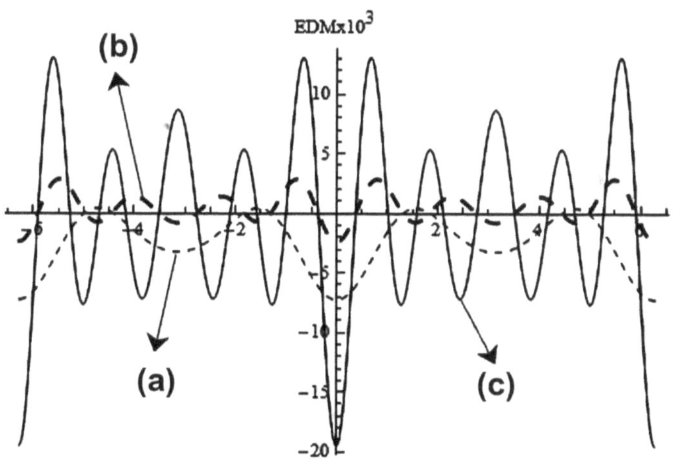

Figure 9.2 Electric dipole moment against θ for various atomic parameters such as: $v = 1$, $\sigma = 0$, $\alpha = 4$, $R_{ab} = 4.5$, $R_{Ob} = 4.5$, $\phi = 0°$; and in case of quantum numbers: (a) $n_1 = 4$, $l_1 = 1$, $m_1 = 0$ and $n_2 = 2$, $l_2 = 1$, $m_2 = 0$, (b) $n_1 = 3$, $l_1 = 2$, $m_1 = 1$ and $n_2 = 3$, $l_2 = 2$, $m_2 = 1$, (c) $n_1 = 5$, $l_1 = 3$, $m_1 = 1$ and $n_2 = 3$, $l_2 = 2$, $m_2 = 1$. Here, θ is in radians, and it ranges from -2π to 2π.

for quadrupole, octupole, and other multipoles. Curves in Figure 9.2 have symmetry for all atomic parameters and multipoles. Note that curves in Figures 9.1 and 9.2 have the behavior of changing amplitudes in the trigonometric function. Consequently, it can be seen from the curves in Figures 9.1 and 9.2 that the electric multipole moments highly depend on the geometry of the molecules.

The numerical values in tables and curves in figures are computed by using Mathematica 7.0 programming language.[36] For these computations, we do not use any package program, in other words, we created and used our own algorithms which are written in Mathematica 7.0. Furthermore, to compare our and literature results, we prefer to use 15 digits for our numerical results. However, it is well known that Mathematica can give any number of digits required. Atomic units are used in all numerical calculations.

REFERENCES

1. Buckingham, A. D. *J. Chem. Phys.* **1959,** *30*, 1580–1585.
2. Bounds, D. G.; Wilson, S. *Mol. Phys.* **1985,** *54*, 445–453.
3. Sundholm, D.; Pyykko, P. *Mol. Phys.* **1985,** *56*, 1411–1418.
4. Allen, M. P.; Tildesley, D. J. *Computer Simulation of Liquids;* Clarendon: Oxford, 1987.
5. Buckingham, A. D.; Clarke, K. L. *Chem. Phys. Lett.* **1978,** *57*, 321–325.
6. Stevens, R. M.; Karplus, M. *J. Chem. Phys.* **1968,** *49*, 1094–1101.
7. Sokalski, W. A.; Shibata, M.; Ornstein, R. L.; Rein, R. *J. Comput. Chem.* **1992,** *13*, 883–887.
8. Tapia, O.; Goscinski, O. *Mol. Phys.* **1975,** *29*, 1653–1661.
9. Zheng, X.; Zerner, M. C. *Int. J. Quantum Chem. Symp.* **1993,** *27*, 431–450.
10. Beck, M. E.; Hohlneicher, G. *Theor. Chem. Acc.* **1999,** *101*, 297–308.
11. Wahl, A. C.; Cade, P. E.; Roothaan, C. C. J. *J. Chem. Phys.* **1964,** *41* (9), 2578–2599.
12. Bagus, P. S.; Liu, B.; McLean, A. D.; Yoshimine, M. *ALCHEMY;* IBM Research Lab.: San Jose, CA, 1972.
13. Rein, R.; Clarke, G. A.; Harris, F. E. *Quantum Aspects of Heterocyclic Compounds in Chemistry and Biology;* The Israel Academy of Science and Humanities: Jerusalem, 1970.
14. Swissler, T. J.; Rein, R. *Chem. Phys. Lett.* **1972,** *15*, 617–620.
15. Harris, F. E.; Michels, H. H. *Adv. Chem. Phys.* **1967,** *13*, 205–266.
16. Steinborn, E. O.; Ruedenberg, K. *Int. J. Quantum Chem.* **1972,** *6*, 413–438.
17. Guseinov, I. I.; Sadichov, F. S. *J. Phys. B* **1977,** *10*, L261–L264.
18. Özdoğan, T.; Orbay, M. *Int. J. Quantum Chem.* **2002,** *87*, 15–22.
19. Özdoğan, T. *J. Math. Chem.* **2007,** *42* (2), 201–214.
20. Yükçü, N.; Şenlik, İ.; Öztekin, E. *Int. J. Quantum Chem.* **2012,** *112* (12), 414–425.
21. Hirschfelder, J. O.; Curtis, C. F.; Bird, R. B. *Molecular Theory of Gases and Liquids;* Wiley: New York, 1954.
22. Condon, E. U.; Shortly, G. H. *The Theory of Atomic Spectra;* Cambridge University Press: Cambridge, England, 1970.
23. Weniger, E. J.; Steinborn, E. O. *J. Chem. Phys.* **1983,** *78* (10), 6121–6132.
24. Barnett, M. P.; Coulson, C. A. *Philos. Trans. R. Soc.* **1951,** *A243*, 221–233.
25. Yavuz, M.; Yükçü, N.; Öztekin, E.; Döndür, S.; Yılmaz, H. *Commun. Theor. Phys.* **2005,** *43*, 151–158.
26. Gaunt, J. A. *Philos. Trans. R. Soc. A* **1929,** *228*, 151–196.
27. Abramowitz, M., Stegun, I. A., Eds.; *Handbook of Mathematical Functions with Formulas, Graphs, and Mathematical Tables*; Dover Publications: New York, 1965.
28. Gradshteyn, I. S.; Ryzhik, I. M. *Tables of Integrals Sums Series and Products,* 6th ed.; Academic Press: New York, 2000.

29. Weniger, E. J.; Steinborn, E. O. *Comput. Phys. Commun.* **1982,** *25*, 149–157.
30. Weissbluth, M. *Atoms and Molecules;* Academic Press: New York, 1978.
31. Guseinov, I. I.; Öztekin, E.; Hüseyin, S. *J. Mol. Struct. (Theochem)* **2001,** *536*, 59–63.
32. Guseinov, I. I.; Mamedov, B. A. *J. Math. Chem.* **2002,** *32*, 309–322.
33. Barnett, M. P. *Theor. Chem. Acc.* **2002,** *107* (4), 241–245.
34. Rico, J. F.; Lopez, R.; Ramirez, G. *J. Comput. Chem.* **1988,** *9* (7), 790–797.
35. Mekelleche, S. M.; Ahmed, A. B. *Int. J. Quantum Chem.* **1997,** *63* (4), 843–852.
36. Wolfram, S. *Mathematica a System for Doing Mathematics by Computer;* Addison-Wesley Pub. Comp., Inc.: Boston, MA, 1988.

CHAPTER TEN

Recurrence Relations for Radial Parts of STOs and Evaluation of Overlap Integrals via the Fourier Transform Methods

S. Özay and E. Öztekin
Department of Physics, Faculty of Science and Arts, University of Ondokuz Mayis, Atakum, Samsun 55139, Turkey

Contents

Abstract

In this paper, new analytical and numerical methods to evaluate overlap integrals of Slater type orbitals (STOs) with the same and different screening parameters and the recurrence relations for STO radial parts via the Fourier transform methods are presented. Our analytical method on the evaluation for radial parts of the STOs was based on the general recurrence relations of Gegenbauer polynomials. We analyzed the properties of STOs with respect to the Fourier transformation which is one of the most important methods for the evaluation of the two-center overlap integrals.

1. INTRODUCTION

Overlap integrals don't involve particle interaction but usual orbital overlap. They are some of the most important integrals for the evaluation of multicenter molecular systems and generally the first integrals to evaluate. We consider non-zero overlap integrals containing real atomic orbitals in configuration and momentum space

$$S_{n_1 l_1 m_1}^{n_2 l_2 m_2}(\alpha_1, \alpha_2; \vec{R}) = \int_{\text{All space}} [\chi_{n_1 l_1}^{m_1}(\alpha_1, \vec{r}_1)]^* \chi_{n_2 l_2}^{m_2}(\alpha_2, \vec{r}_2)\, \mathrm{d}V \tag{1}$$

Advances in Quantum Chemistry, Volume 67
ISSN 0065-3276, http://dx.doi.org/10.1016/B978-0-12-411544-6.00010-8

$$= \int e^{-iR\cdot p}[U_{n_1 l_1}^{m_1}(\alpha_1, \vec{p})]^* U_{n_2 l_2}^{m_2}(\alpha_2, \vec{p}) d^3\vec{p}, \quad (2)$$

where α and R are defined screening parameters and interatomic distances, respectively. The other atomic parameters in Eq. (1) have to be defined in the following paragraphs and sections. In that approach, the choice of a suitable basis set for the expansion of the atomic orbitals is of utmost importance. The radial parts of a good basis set for the evaluation of molecular calculations should satisfy two basic requirements: Firstly, short expansions of atomic orbitals in terms of the basis functions yield a good accuracy[1]; and secondly, the molecular multicenter integrals which occur inevitably and in very large numbers in that approach can be computed efficiently.[2]

STOs were introduced by Slater in his article on screening parameters with the intention of approximating numerical Hartree–Fock orbitals.[3] STOs have been used quite frequently and successfully as basis functions in atomic and molecular calculations in Hartree–Fock Roothaan methods. It was found that only short expansions are needed for an appropriate descriptions of molecular integrals. Among the ETOs, STOs have assumed a dominant position for this reason, having the simplest analytical structure of all exponentially decreasing functions. Other commonly occurring functions of ETOs can be expressed quite easily as linear combinations of STOs.[4] This implies that multicenter molecular integrals over the other ETOs can be expressed in terms of the basic multicenter molecular integrals over STOs.

Normalized STOs are given in terms of real spherical harmonics as the following form:

$$\chi_{nl}^{m}(\alpha, \vec{r}) = \frac{(2\alpha)^{n+1/2}}{\sqrt{(2n)!}} r^{n-1} e^{\alpha r} Y_l^m(\theta, \phi). \quad (3)$$

Real spherical harmonics are defined in terms of associated Legendre functions by Arfken and Weber[5]

$$Y_l^m(\theta, \phi) = P_l^m(\cos\theta) \begin{Bmatrix} \cos(m\phi) \\ \sin(|m|\phi) \end{Bmatrix}.$$

It is well known that the Fourier transform method is one of the most important methods for the evaluation of complicated multicenter molecular integrals.[6] There is extensive literature on the use of the Fourier transform method for ETOs in the analytical calculation of multicenter molecular integrals. In numerous papers it was demonstrated that the Fourier transform of an STO (FTSTO) are fairly complicated mathematical objects.[4,6,7] Therefore, in this paper, we prefer to use STOs in the calculations of overlap integrals with the same and different screening parameters by using Taylor expansions for the denominators and recurrence relations for the radial parts of FTSTOs. We think that the problems associated with the overlap integrals are now much better understood and that considerable progress could be achieved.

In the following sections of this paper, we will present definitions and basic formulas for the evaluation of two-center overlap integrals with the same and different screening parameters via the Fourier transform method. Then, we shall suggest another method of evaluating the radial parts of FTSTOs. This method is easy to handle and leads to simple results in many cases. It is based on some properties of the recurrence relations for Gegenbauer polynomials and Taylor expansions. It will be shown in this paper that the convolution product of the two Gegenbauer functions is expressed in terms of one function. New recurrence relations and master formulas for radial parts of the FTSTOs for the evaluation of overlap integrals are derived and tabulated. In the subsequent section, in order to derivation of analytical expressions for the overlap integrals with the same and different screening parameters are presented in terms of finite sums. We are devoted by using Fourier transform, Rayleigh plane wave, and Taylor expansions for the denominators. Consequently, it is relatively easy to express the Fourier transform method of two-center overlap integrals with the same and different screening parameters mentioned above as finite sums i.e., linear combinations of Gaunt coefficients, factorial functions, and STOs. Many functions are involved in the product of two denominators with different parameters and are expressed in terms of the sums of two series with one screening parameter by using Taylor expansion. Therefore, also for these specific expansions of general functions the convolution integrals can be obtained analytically in terms of STOs, making of further integration superfluous.

In Barnett's study on the symbolic calculation of molecular integrals over Slater orbitals, Barnett obtained some results in substantial disagreement with Öztekin and Guseinov who describe numerical schemes. Barnett believes that these schemes suffer from digital erosion, possibly because recurrence formulas were used outside their regions of stability.[8] In this study, a new analytical expression for overlap integrals is given where digital erosions have disappeared and therefore the numerical results are in best agreement with Barnett.

2. EVALUATION OF RECURRENCE RELATIONS FOR RADIAL PART FOR FTSTOS

In this paper, we shall use the symmetric version of the Fourier transformation, i.e., a given function and its Fourier transform are connected by the following relations:

$$\begin{aligned} U_{nl}^{m}(\alpha,\vec{p}) &= (2\pi)^{-3/2}\int e^{-ir\cdot p}\chi_{nl}^{m}(\alpha,\vec{r})\mathrm{d}^3r, \\ \chi_{nl}^{m}(\alpha,\vec{r}) &= (2\pi)^{-3/2}\int e^{ip\cdot r}U_{nl}^{m}(\alpha,\vec{p})\mathrm{d}^3p. \end{aligned} \tag{4}$$

For the evaluation of the FTSTOs, we only have to insert the well-known Rayleigh expansion of a plane wave in terms of spherical Bessel functions and spherical harmonics[9]

$$e^{\pm ir\cdot p} = 4\pi\sum_{l=0}^{\infty}\sum_{m=-l}^{l}(\pm i)^l j_l(pr)\left[Y_l^m(\theta_r,\phi_r)\right]^* Y_l^m(\theta_p,\phi_p). \tag{5}$$

The spherical Bessel functions $j_l(pr)$ are defined in terms of the Bessel functions of the first kind[10]

$$j_l(pr) = \sqrt{\frac{\pi}{2pr}}J_{l+1/2}(pr). \tag{6}$$

Substituting the Rayleigh plane wave expansion into the FTSTOs and using the orthogonality of spherical harmonics, the FTSTOs can be written in terms of Gegenbauer polynomials[11]

$$\begin{aligned} U_{n,l}^{m}(\alpha,\vec{p}) &= \frac{(2\alpha)^{n+1/2}}{[(2n)!]^{1/2}\sqrt{p}}(-i)^l Y_l^m(\theta_p,\phi_p)\int e^{-\alpha r}r^{n+1/2}J_{l+1/2}(pr)\mathrm{d}r \\ &= (-i)^l Q_{nl}(\alpha,p)Y_l^m(\theta_p,\phi_p), \end{aligned} \tag{7}$$

where Q functions constitute the radial part of the integral and are expressed in terms of Gegenbauer polynomials[11]

$$Q_{nl}(\alpha,p) = \frac{2^{n+l+1}l!(n-l)!}{\sqrt{\pi(2n)!}\alpha^{3/2}}x^{n+2}(1-x^2)^{l/2}C_{n-l}^{l+1}(x), \tag{8}$$

where

$$x = \frac{\alpha}{\sqrt{\alpha^2 + p^2}}.$$

By using the recurrence relationships for Gegenbauer polynomials,[10] we obtain new recurrence expressions for the radial parts of FTSTOs as follows

$$\begin{aligned} Q_{nl}(\alpha, p) = & \frac{2x^2}{\sqrt{n(2n-1)}}\Bigg\{\sqrt{2}nQ_{n-1,l}(\alpha, p) \\ & -\frac{(n-l-1)(n-l)}{\sqrt{(n-1)(2n-3)}}Q_{n-2,l}(\alpha, p)\Bigg\}, \end{aligned} \tag{9}$$

$$\begin{aligned} Q_{n,l+1}(\alpha, p) = & (n-l-1)\sqrt{\frac{2}{n(2n-1)}}Q_{n-1,l+1}(\alpha, p) \\ & +\frac{\sqrt{1-x^2}}{x}Q_{n,l}(\alpha, p), \end{aligned} \tag{10}$$

$$\begin{aligned} Q_{n+1,l+1}(\alpha, p) = & x\sqrt{\frac{2}{(n+1)(2n+1)}}\Bigg\{(n+l+2)\sqrt{1-x^2}Q_{n,l}(\alpha, p) \\ & + x(n-l)Q_{n,l+1}(\alpha, p)\Bigg\}, \end{aligned} \tag{11}$$

$$\begin{aligned} Q_{n,l}(\alpha, p) = & \frac{2}{\sqrt{2n(2n-1)}}\Bigg\{(n+l+1)x^2Q_{n-1,l}(\alpha, p) \\ & - x\sqrt{1-x^2}(n-l-1)Q_{n-1,l+1}(\alpha, p)\Bigg\}. \end{aligned} \tag{12}$$

To calculate and tabulate the radial part of FTSTOs, we use the following initial values

$$\begin{aligned} Q_{10} &= \frac{4}{\alpha^{3/2}}\sqrt{\frac{2}{\pi}}x^4, \\ Q_{20} &= \frac{4}{\alpha^{3/2}}\sqrt{\frac{2}{3\pi}}x^4(4x^2-1), \\ Q_{21} &= \frac{16}{\alpha^{3/2}}\sqrt{\frac{2}{3\pi}}x^5\sqrt{1-x^2}. \end{aligned} \tag{13}$$

Table 10.1 The master formulas for the radial parts of FTSTOs are tabulated for various quantum numbers by using Eqs. (9)–(12)

n	*l*	Q_{nl}	*n*	*l*	Q_{nl}
1	0	$\frac{4\sqrt{\frac{2}{\pi}}x^4}{\alpha^{3/2}}$	6	1	$\frac{512x^9\sqrt{1-x^2}(3-20x^2+24x^4)}{3\sqrt{231\pi}\alpha^{3/2}}$
2	0	$\frac{4\sqrt{\frac{2}{3\pi}}x^4(-1+4x^2)}{\alpha^{3/2}}$		3	$\frac{1024x^9(1-x^2)^{3/2}(-1+4x^2)}{\sqrt{231\pi}\alpha^{3/2}}$
	1	$\frac{16\sqrt{\frac{2}{3\pi}}x^5\sqrt{1-x^2}}{\alpha^{3/2}}$		5	$\frac{4096x^9(1-x^2)^{5/2}}{\sqrt{231\pi}\alpha^{3/2}}$
3	0	$\frac{32x^6(-1+2x^2)}{\sqrt{5\pi}\alpha^{3/2}}$	8	5	$\frac{4096\sqrt{\frac{2}{715\pi}}x^{11}(1-x^2)^{5/2}(-3+16x^2)}{3\alpha^{3/2}}$
	1	$\frac{32x^5\sqrt{5-5x^2}(-1+6x^2)}{15\sqrt{\pi}\alpha^{3/2}}$			
	2	$-\frac{64x^6(-1+x^2)}{\sqrt{5\pi}\alpha^{3/2}}$		7	$\frac{65536\sqrt{\frac{2}{715\pi}}x^{11}(1-x^2)^{7/2}}{3\alpha^{3/2}}$
4	3	$\frac{256\sqrt{\frac{2}{35\pi}}x^7(1-x^2)^{3/2}}{\alpha^{3/2}}$	11	6	$-\frac{262144\sqrt{\frac{2}{88179\pi}}x^{14}(-1+x^2)^3(3-40x^2+88x^4)}{11\alpha^{3/2}}$
				9	$\frac{1048576\sqrt{\frac{2}{88179\pi}}x^{13}(1-x^2)^{9/2}(-1+22x^2)}{11\alpha^{3/2}}$

Master formulas for radial parts of FTSTOs are given in Table 10.1. During the numerical calculations for the radial parts of FTSTOs, we consider the memory by the relation

$$k_n^l = \frac{n(n-1)}{2} + l + 1. \tag{14}$$

We shall show that the radial parts of FTSTOs become meaningful if we apply a suitable recursion relationship. And then, we want to compute the overlap integrals with the same and different screening parameters with respect to FTSTOs.

With the help of Eqs. (2), (4), (5), and (6), we are in a position to give an explicit expression for the overlap integral via the Fourier transform method which will be examined in this paper. If we combine Eqs. (5) and (6) in Eq. (2), we obtain the following representations for overlap integrals

$$
\begin{aligned}
S_{n_1 l_1 m_1}^{n_2 l_2 m_2}(\alpha, \beta; \vec{R}) = & 4\pi \sum_{l=0}^{\infty} \sum_{m=-l}^{l} (-i)^l [Y_l^m(\theta, \phi)]^* \\
& (-1)^{l_2} (i)^{l_1 + l_2} \int_0^{\infty} j_l(pR) [Q_{n_1 l_1}(\alpha, p)]^* Q_{n_2 l_2}(\beta, p) p^2 \mathrm{d}p \\
& \int_0^{\pi} P_{l_1}^{m_1}(\cos\theta) P_{l_2}^{m_2}(\cos\theta) P_l^m(\cos\theta) \sin\theta \mathrm{d}\theta \\
& \int_0^{2\pi} \Phi_{m_1}^*(\phi) \Phi_{m_2}(\phi) \Phi_m(\phi) \mathrm{d}\phi
\end{aligned} \tag{15}
$$

As can be seen from Eq. (15), the possibility of evaluating overlap integrals with the same and different screening parameters in closed forms over STOs via the Fourier transform methods depends crucially upon the functional form of the radial part of the FTSTOs involved. In Eqs. (8)–(15) we want to analyze the mathematical properties of radial parts for FTSTOs and evaluate the overlap integrals with respect to Fourier transform methods over STOs. Our emphasis will be on applications in multicenter molecular integrals over STOs via the Fourier transform methods. The remaining integrals of radial parts with the same and different screening parameters can be expressed in terms of the following integral representations of overlap integrals over FTSTOs in terms of Bessel functions[4]

$$
\int_0^{\infty} \frac{J_\nu(pR) p^{\nu+1}}{(\alpha^2 + p^2)^{\mu+1}} \mathrm{d}p = \frac{R^\mu \alpha^{\nu-\mu}}{2^\mu \mu!} K_{\nu-\mu}(\alpha R). \tag{16}
$$

In Eq. (15), the product of three associated Legendre polynomials which is called as Gaunt coefficients is expressed as follows[12]

$$
\begin{aligned}
C^{l|m|}(l_1 m_1, l_2 m_2) &= \int_0^{\pi} P_{l_1}^{m_1}(\cos\theta) P_{l_2}^{m_2}(\cos\theta) P_l^m(\cos\theta) \sin\theta \mathrm{d}\theta \\
&= \langle l_1 m_1 | l_2 m_2 | l m \rangle \\
&= \begin{cases} C^l(l_1 m_1, l_2 m_2), & \text{for } |m| = |m_1 - m_2| \\ C^l(l_1 m_1, l_2 - m_2), & \text{for } |m| = |m_1 + m_2| \end{cases}
\end{aligned} \tag{17}
$$

and the other angular integrals, which contain a combination of the product of the three trigonometric functions groups, are given[13]

$$
\begin{aligned}
f_{m_1,m_2}^{m} &= \int_0^{2\pi} \Phi_{m_1}(\phi)\Phi_{m_2}(\phi)\Phi_m(\phi)\mathrm{d}\phi \\
&= \frac{1}{\pi^{3/2}\sqrt{(1+\delta_{m,0})(1+\delta_{m_1,0})(1+\delta_{m_2,0})}} \\
&\cdot \int_0^{2\pi} \begin{Bmatrix} \cos m\phi \\ \sin|m|\phi \end{Bmatrix} \begin{Bmatrix} \cos m_1\phi \\ \sin|m_1|\phi \end{Bmatrix} \begin{Bmatrix} \cos m_2\phi \\ \sin|m_2|\phi \end{Bmatrix} \mathrm{d}\phi
\end{aligned} \tag{18}
$$

These integrals can be easily evaluated with the help of the following integral representations and trigonometric expansions[5]

$$
\begin{aligned}
&\int_0^{2\pi} \sin mx \sin nx \mathrm{d}x = \begin{cases} \pi\delta_{m,n}, & m \neq 0 \\ 0, & m = 0 \end{cases} \\
&\int_0^{2\pi} \cos mx \cos nx \mathrm{d}x = \begin{cases} \pi\delta_{m,n}, & m \neq 0 \\ 2\pi, & m = n = 0 \end{cases} \\
&\int_0^{2\pi} \sin mx \cos nx \mathrm{d}x = 0.
\end{aligned} \tag{19}
$$

By writing Legendre polynomials as real spherical harmonics and using the definition of Gaunt coefficients and integrals of radial part as given by Eqs. (16)–(19), we can immediately establish the following formula for overlap integrals:

$$
\begin{aligned}
S_{n_1 l_1 m_1}^{n_2 l_2 m_2}(\alpha,\beta;\vec{R}) &= 4\pi \sum_{l=0}^{\infty}\sum_{m=-l}^{l} (-i)^l Y_l^m(\theta_R,\phi_R) C^{l|m|}(l_1 m_1, l_2 m_2) f_{m_1 m_2}^{m} \\
&\quad (-1)^{l_2}(i)^{l_1+l_2}\int_0^{\infty} j_l(pR)[Q_{n_1 l_1}(\alpha,p)]^* Q_{n_2 l_2}(\beta,p)p^2\mathrm{d}p.
\end{aligned} \tag{20}
$$

In momentum space, the radial integrals like the integrals over STOs via the Fourier transform methods given by Eq. (20) had already been investigated quite frequently by Öztekin.[14] However, the numerical results for overlap integrals given in Refs. 14,15 were not able to give perfect numerical results for higher atomic parameters due to digital erosions (see Ref. 8a).

If the screening parameters change only slightly, all the representations of overlap integrals involving ETOs based upon the partial fraction decomposition become numerically unstable because of the occurrence of the canceling singularities. As an alternative, one has to use the multiplication theorem of STOs given by Eq. (20). In connection with the use of the FTSTOs, Eq. (20), one immediately obtains a finite series expansion of an overlap integral with different parameters in terms of the overlap integrals with the same screening parameters.

In the following sections of this article, we present new analytical expressions for overlap integrals with the same and different screening parameters over STOs via the Fourier transform method by using Taylor expansions.

2.1 Overlap integrals with the same screening parameters

To obtain a new expression for overlap integrals with the same screening parameters by using Fourier transform method, we must get $\alpha = \beta$ in the integrant of Eq. (20). If we substitute Eq. (8) in Eq. (20), we derive the following expression for the radial integral of overlap integral with the same screening parameters

$$\begin{aligned}
&\int_0^\infty j_l(pR)[Q_{n_1 l_1}(\alpha,p)]^* Q_{n_2 l_2}(\alpha,p)p^2 \mathrm{d}p \\
&= \frac{2^{n_1+n_2+l_1+l_2+2} l_1! l_2! (n_1-l_1)!(n_2-l_2)! \alpha^{n_1+n_2+1}}{\pi\sqrt{(2n_1)!(2n_2)!}} \\
&\int_0^\infty \frac{j_l(pR) p^{l_1+l_2+2} C_{n_1-l_1}^{l_1+1}\left(\frac{\alpha}{\sqrt{\alpha^2+p^2}}\right) C_{n_2-l_2}^{l_2+1}\left(\frac{\alpha}{\sqrt{\alpha^2+p^2}}\right)}{(\alpha^2+p^2)^{(n_1+l_1+n_2+l_2+4)/2}} \mathrm{d}p.
\end{aligned} \tag{21}$$

With the help of Taylor expansion given by Eq. (5.1) of Ref. 16, an integrable form can be written;

$$p^{2L} = (-1)^L \alpha^{2L} \sum_{t=0}^{L} (-1)^t \frac{L!}{t!(L-t)!} \left(\frac{\alpha^2+p^2}{\alpha^2}\right)^t, \tag{22}$$

where $l_1 + l_2 - l = 2L$ is always an even positive integer or zero because of some selection rules satisfied by the Gaunt coefficients. The radial integral can be calculated with the help of the relations by Eqs. (4.1)–(4.6) of Ref. 16. If we rearrange the product of two Gegenbauer polynomials and denominator in Eq. (21), we obtain an expression for overlap integrals with equal parameters in terms of reduced Bessel functions

$$
\begin{aligned}
&\int_0^\infty \frac{j_l(pR)p^{2L+l+2}C_a^b\left(\frac{\alpha}{\sqrt{\alpha^2+p^2}}\right)C_c^d\left(\frac{\alpha}{\sqrt{\alpha^2+p^2}}\right)}{(\alpha^2+p^2)^{(a+2b+c+2d)/2}}\mathrm{d}p \\
&\quad=(-1)^L\frac{2^{a+c}\alpha^{2L+a+c}}{(b-1)!(d-1)!} \\
&\quad\quad\sum_{s=0}^{[\frac{a}{2}]}\sum_{r=0}^{[\frac{c}{2}]}\frac{(-1)^{s+r}}{(2\alpha)^{2(s+r)}}\frac{(a+b-s-1)!(c+d-r-1)!}{s!r!(a-2s)!(c-2r)!} \\
&\quad\quad\sum_{t=0}^{\left(\frac{l_1+l_2-l}{2}\right)}(-1)^t\frac{\left(\frac{l_1+l_2-l}{2}\right)!}{\alpha^{2t}t!\left(\frac{l_1+l_2-l}{2}-t\right)!} \\
&\quad\quad\int_0^\infty \frac{p^{l+2}j_l(pR)}{(\alpha^2+p^2)^{a+b+c+d-s-r-t}}\mathrm{d}p.
\end{aligned}
\tag{23}
$$

With the help of Eqs. (3), (6), and (16) and after some mathematical manipulations, we obtain the following expression for the product of the spherical harmonics and radial integral in overlap integral with the same screening parameters

$$
\begin{aligned}
&Y_l^m(\theta,\phi)\int_0^\infty \frac{j_l(pR)p^{2L+l+2}C_a^b\left(\frac{\alpha}{\sqrt{\alpha^2+p^2}}\right)C_c^d\left(\frac{\alpha}{\sqrt{\alpha^2+p^2}}\right)}{(\alpha^2+p^2)^{(a+2b+c+2d)/2}}\mathrm{d}p \\
&\quad=(-1)^L\pi\frac{(2\alpha)^{-a-b-c-d-1/2}2^{-b-d}}{(b-1)!(d-1)!} \\
&\quad\quad\sum_{s=0}^{[\frac{a}{2}]}\sum_{r=0}^{[\frac{c}{2}]}\frac{(-1)^{s+r}}{(2\alpha)^{2(s+r)}}\frac{(a+b-s-1)!(c+d-r-1)!}{s!r!(a-2s)!(c-2r)!} \\
&\quad\quad\sum_{t=0}^{\left(\frac{l_1+l_2-l}{2}\right)}(-1)^t2^{2(t+1)}\begin{pmatrix}(l_1+l_2-l)/2\\ t\end{pmatrix} \\
&\quad\quad\sum_{q=1}^{\mu-l}\frac{(2\mu-2l-q-1)!}{(q-1)!\mu!(\mu-l-q)!}\sqrt{[2(l+q)]!}\chi_{q+l,l}^m(\alpha,\vec{R}),
\end{aligned}
\tag{24}
$$

where $\mu=n_1+n_2-s-r-t+1$ and $\binom{a}{b}$ are the binomial coefficients. As a result, the overlap integrals with the same screening parameters over

FTSTOs are written by in terms of linear combinations of STOs over finite series expansions as follows:

$$S_{n_1 l_1 m_1}^{n_2 l_2 m_2}(\alpha, \alpha; \vec{R}) = (-1)^{l_1} \pi \frac{(n_1 - l_1)!(n_2 - l_2)!}{\alpha^{3/2}\sqrt{2(2n_1)!(2n_2)!}} \sum_{l=|l_1-l_2|}^{l_1+l_2}$$

$$^{(2)} \sum_{m=-l}^{l} (-1)^l C^{l|m|}(l_1 m_1, l_2 m_2) f_{m_1, m_2}^{m}$$

$$\sum_{s=0}^{\left[\frac{n_1-l_1}{2}\right]} \sum_{r=0}^{\left[\frac{n_2-l_2}{2}\right]} (-1)^{s+r} \frac{(n_1 - s)!(n_2 - r)!}{s!r!(n_1 - l_1 - 2s)!(n_2 - l_2 - 2r)!}$$

$$\sum_{t=0}^{\left(\frac{l_1+l_2-l}{2}\right)} (-1)^t 2^{2(t+1)} \frac{\left(\frac{l_1+l_2-l}{2}\right)!}{t!\left(\frac{l_1+l_2-l}{2} - t\right)!}$$

$$\sum_{q=1}^{\mu-l} \frac{(2\mu - 2l - q - 1)!}{(q - 1)!\mu!(\mu - l - q)!} \sqrt{[2(l + q)]!}\, \chi_{q+l,l}^{m}(\alpha, \vec{R}). \tag{25}$$

2.2 Overlap integrals with the different screening parameters

Now, we shall show how the denominators and product of two Gegenbauer polynomials and Bessel functions, which occur in the integrals the representations given by Eq. (20), can be expressed in terms of functions such as $(\alpha^2 + p^2)^k$ using partial fraction decomposition or Taylor expansion. We shall start the following expressions for the overlap integrals with different screening parameters

$$S_{n_1 l_1 m_1}^{n_2 l_2 m_2}(\alpha, \beta; \vec{R})$$

$$= (-1)^{l_2} (i)^{l_1+l_2} \frac{2^{n_1+n_2+l_1+l_2+4} l_1! l_2! (n_1 - l_1)!(n_2 - l_2)! \alpha^{n_1+1/2} \beta^{n_2+1/2}}{\sqrt{(2n_1)!(2n_2)!}}$$

$$\sum_{l=0}^{\infty} \sum_{m=-l}^{l} (-i)^l C^{l|m|}(l_1 m_1, l_2 m_2) Y_l^m(\theta, \phi) f_{m_1, m_2}^{m}$$

$$\int_0^\infty \frac{j_l(pR) p^{l_1+l_2+2} C_{n_1-l_1}^{l_1+1}\left(\frac{\alpha}{\sqrt{\alpha^2+p^2}}\right) C_{n_2-l_2}^{l_2+1}\left(\frac{\beta}{\sqrt{\beta^2+p^2}}\right)}{(\alpha^2 + p^2)^{(n_1+l_1+2)/2} (\beta^2 + p^2)^{(n_2+l_2+2)/2}} \mathrm{d}p. \tag{26}$$

This integral representation by a finite numbers of terms can be derived if one uses a decomposition of the denominator into fraction by using Eq. (4.1) of Ref. 16. If we insert this partial fraction decomposition into the integral (26), we obtain a linear combination of integrals of the general type which contains the product of two Gegenbauer polynomials and reduced Bessel functions

$$
\int_0^\infty \frac{j_l(pR)p^{2L+l+2} C_a^b\left(\frac{\alpha}{\sqrt{\alpha^2+p^2}}\right) C_c^d\left(\frac{\beta}{\sqrt{\beta^2+p^2}}\right)}{(\alpha^2+p^2)^{(a+2b)/2}(\beta^2+p^2)^{(c+2d)/2}} \mathrm{d}p
= \frac{(2\alpha)^a(2\beta)^c}{(b-1)!(d-1)!}
\sum_{s=0}^{[\frac{a}{2}]}\sum_{r=0}^{[\frac{c}{2}]} \frac{(-1)^{s+r}}{(2\alpha)^{2s}(2\beta)^{2r}} \frac{(a+b-s-1)!(c+d-r-1)!}{s!r!(a-2s)!(c-2r)!}
\int_0^\infty \frac{j_l(pR)p^{2L+l+2}}{(\alpha^2+p^2)^{a+b-s}(\beta^2+p^2)^{c+d-r}} \mathrm{d}p
$$

$$
= (-1)^L\sqrt{\frac{\pi}{2}} \frac{(2\alpha)^a(2\beta)^c}{(b-1)!(d-1)!}
\sum_{s=0}^{[\frac{a}{2}]}\sum_{r=0}^{[\frac{c}{2}]} \frac{(-1)^{s+r}}{(2\alpha)^{2s}(2\beta)^{2r}} \frac{(a+b-s-1)!(c+d-r-1)!}{s!r!(a-2s)!(c-2r)!}
\sum_{t=0}^{\left(\frac{l_1+l_2-l}{2}\right)} (-1)^t 2^t \frac{\left(\frac{l_1+l_2-l}{2}\right)!}{t!\left(\frac{l_1+l_2-l}{2}-t\right)!R^{t+1/2}} (\alpha^2-\beta^2)^{-a-b-c-d+s+r+1}
$$

$$
(-1)^{c+d-r}\left\{\frac{\alpha^{2L+l-t+1/2}}{(c+d-r-1)!}\sum_{q=0}^{a+b-s-1} \frac{(a+b+c+d-s-r-q-2)!}{(a+b-s-q-1)!}
\left(\frac{R}{2\alpha}\right)^q \frac{(\alpha^2-\beta^2)^q}{(q-t)!} K_{l+t-q+1/2}(\alpha R)
- \frac{\beta^{2L+l-t+1/2}}{(a+b-s-1)!}\sum_{q=0}^{c+d-r-1} \frac{(a+b+c+d-s-r-q-2)!}{(c+d-r-q-1)!}
\left(-\frac{R}{2\alpha}\right)^q \frac{(\alpha^2-\beta^2)^q}{(q-t)!} K_{l+t-q+1/2}(\beta R)\right\}, \quad (27)
$$

where modified Bessel functions can be represented by an exponential multiplied by a polynomial[10]

$$K_{n+1/2}(x) = \sqrt{\frac{\pi}{2x}} e^{-x} \sum_{k=0}^{n} \frac{(n+k)!}{k!(n-k)!(2x)^k}. \tag{28}$$

The product of modified Bessel functions and spherical harmonics is well defined in terms of STOs by using Eq. (3)

$$Y_l^m(\theta,\phi) K_{l+t-q+1/2}(\alpha R) = \sqrt{\pi}(2\alpha)^{-3/2} \sum_{k=0}^{l+t-q} \frac{(l+t-q+k)!\sqrt{[2(q-t-k)]!}}{k!(l+t-q-k)!(2\alpha R)^{q-t-1/2}} \chi_{q-t-k,l}^{m}(\alpha, \vec{R}). \tag{29}$$

With the help of Eq. (29), we obtain a new analytical expression for the overlap integral with the different screening parameters as follows

$$\begin{aligned}
S_{n_1 l_1 m_1}^{n_2 l_2 m_2}(\alpha,\beta;\vec{R}) &= (-1)^{n_2+l_1} \pi \frac{(n_1-l_1)!(n_2-l_2)!}{\sqrt{2(2n_1)!(2n_2)!}} 2^{l_1+l_2+2} \\
&\sum_{l=|l_1-l_2|}^{l_1+l_2} \sum_{m=-l}^{l} (-1)^l \langle l_1 m_1 | l_2 m_2 | lm \rangle f_{m_1,m_2}^{m} \\
&\sum_{s=0}^{\left[\frac{n_1-l_1}{2}\right]} \sum_{r=0}^{\left[\frac{n_2-l_2}{2}\right]} (-1)^s \frac{(n_1-s)!(n_2-r)!(\alpha^2-\beta^2)^{s+r-n_1-n_2-1}}{s!r!(n_1-l_1-2s)!(n_2-l_2-2r)!} \\
&(2\alpha)^{2(n_1-s)-l_1+1/2}(2\beta)^{2(n_2-r)-l_2+1/2} \sum_{t=0}^{\left(\frac{l_1+l_2-l}{2}\right)} (-2)^t \frac{\left(\frac{l_1+l_2-l}{2}\right)!}{t!\left(\frac{l_1+l_2-l}{2}-t\right)!} \\
&\left\{-\frac{\alpha^{l_1+l_2-t-1/2}}{(n_2-r)!} \sum_{q=0}^{n_1-s} \frac{(n_1+n_2-s-r-q)!}{(n_1-s-q)!(q-t)!} \left(\frac{\alpha^2-\beta^2}{2\alpha}\right)^q \right. \\
&\sum_{k=0}^{l+t-q} \frac{(l+t-q+k)!}{k!(l+t-q-k)!} \frac{\sqrt{(2(q-t-k))!}}{(2\alpha)^{q-t}} \chi_{q-t-k,l}^{m}(\alpha,\vec{R}) \\
&+ \frac{\beta^{l_1+l_2-t-1/2}}{(n_1-s)!} \sum_{q=0}^{n_2-r} \frac{(n_1+n_2-s-r-q)!}{(n_2-r-q)!(q-t)!} \left(\frac{\beta^2-\alpha^2}{2\beta}\right)^q \\
&\left. \sum_{k=0}^{l+t-q} \frac{(l+t-q+k)!}{k!(l+t-q-k)!} \frac{\sqrt{(2(q-t-k))!}}{(2\beta)^{q-t}} \chi_{q-t-k,l}^{m}(\beta,\vec{R}) \right\}.
\end{aligned} \tag{30}$$

3. RESULTS AND DISCUSSIONS

In this paper, we present two new analytical expressions for overlap integrals with the same and different screening parameters of real STOs via the Fourier transform method. As can be seen from Eqs. (25) and (30), final results for the overlap integrals are expressed in terms of linear combinations of STOs, Gaunt coefficients, and factorial functions. With respect to Eqs. (25) and (30), the representations of overlap integrals with the same and different screening parameters contain only a finite sum of terms. In the literature, especially, analytical expressions for overlap integrals with different screening parameters are given in terms of infinite series. For that reason, Eqs. (25) and (30) given in this study for the evaluation for overlap integrals over STOs with respect to Fourier transform method show good rate of convergence and great numerical stability under a wide range of quantum numbers and atomic parameters.

The master formulas for the radial parts of FTSTOs and numerical results for overlap integrals with the same and different screening parameters are given in Tables 10.1–10.4, respectively. The computer program for calculating numerical results is written in Mathematica 7 programming language.[17] The computer time required for the calculation of overlap integrals is not given in Tables 10.2–10.4 due to the fact that the comparison cannot be made for the different computers used in the literature. However, our CPU times are satisfactory.

As can be seen in Tables 10.2–10.4, our computer results for overlap integrals with the same and different screening parameters in the line- and nonline-up coordinate systems over STOs via the Fourier transform methods are in best agreement with Refs. 6,14,15 and 8c,18–24. In spite of the fact that agreements are best with the earlier literature, striking results have been obtained with Refs. 14,15 and 8c,18–24. Whereas the best agreement is obtained with the values of Barnett (see Tables 10.1 and 10.2 of Ref. 8a) for all atomic parameters in the lined-up coordinate systems, some discrepancies occurred compared to the values of Özdoğan and Guseinov and co-workers for very high quantum numbers and very small interatomic distances in the lined- and nonlined-up coordinate systems. We observed that our computer results agree well in all digits with Barnettt's values given by three overlap integrals labeled 8, 17, and 20 in Tabel II of Özdoğan. In Table 10.2, our results by given bold values are completely disagreement with the results of Ref. 18

In early studies, the digital erosions that occurred in the calculations for overlap integrals are not due to the recurrence relations.[14,15] Now, we believe that these schemes suffer from digital erosion, possibly because

Table 10.2 The numerical results for overlap integrals with the same screening parameters in the lined-up coordinate systems

	n_1	l_1	n_2	l_2	m	*alphaR*	In this study (Eq. (25))	Ref. 18
1	4	2	3	2	1	25	$-5.65554471492813643477769 8\times 10^{-6}$	$-5.65554471492813644\times 10^{-6}$
2	15	4	8	4	4	25	$2.5432413255990502534619 81\times 10^{-3}$	$2.54324132559904823\times 10^{-3}$
3	18	10	18	17	9	20	$-1.1490869466029531184165 18\times 10^{-2}$	$-1.14908694110728558\times 10^{-2}$
4	25	12	20	17	11	30	$4.8812974127096336729119 02\times 10^{-2}$	$4.88129736080919383\times 10^{-2}$
5	30	15	29	10	8	50	$7.3662718636871584286511 63\times 10^{-2}$	$7.36627874148956217\times 10^{-2}$
6	34	14	34	12	10	75	$1.3527999157490239442408 59\times 10^{-4}$	$1.35616683543570101\times 10^{-4}$
7	**40**	**29**	**40**	**15**	**15**	**15**	$\mathbf{-1.1980892914947998085132 08\times 10^{-7}}$	$\mathbf{-1.12831889959425283\times 10^{-7}}$
8	**45**	**29**	**44**	**19**	**17**	**0.05**	$\mathbf{1.47153434333818712078\times 10^{-28}}$	$\mathbf{1.97965762959554458\times 10^{-8}}$
9	50	15	15	13	11	10	$2.9010070720058874505317 91\times 10^{-4}$	$2.9010070720063166\times 10^{-4}$
10	55	20	52	23	20	35	$-2.58332568976225152663\times 10^{-1}$	$-2.58332568946277651\times 10^{-1}$
11	60	14	52	17	12	35	$-2.62097248347309231335\times 10^{-2}$	$-2.62097248368272357\times 10^{-2}$
12	**65**	**24**	**65**	**20**	**18**	$\mathbf{10^{-2}}$	$\mathbf{2.95052488660206327282\times 10^{-14}}$	$\mathbf{2.40037259979039436\times 10^{-11}}$
13	70	15	65	13	10	10^{-4}	$3.34993306902369502366\times 10^{-11}$	$3.34752062574253062\times 10^{-11}$
14	70	25	70	15	14	25	$-3.46058313664739518445\times 10^{-6}$	$-3.46017432399857001\times 10^{-6}$
15	**75**	**30**	**75**	**20**	**18**	$\mathbf{10^{-6}}$	$\mathbf{-8.19297549621687882025\times 10^{-78}}$	$\mathbf{-1.03225259163782463\times 10^{-32}}$

Table 10.3 The numerical results for overlap integrals with the different screening parameters in the lined-up coordinate systems

	n_1	l_1	n_2	l_2	m	α	β	R	In this study (Eq. (30))	Literature	Ref.
1	1	0	1	0	0	1.3	6.2	0.2	$4.125849742489160894439706 \times 10^{-1}$	$4.12584974248916089 \times 10^{-1}$	13
2	1	0	2	1	0	10	2	1.4	$-1.17413789686628284854907 3 \times 10^{-1}$	$-1.174137896866282 \times 10^{-1}$	20
3	2	1	2	1	0	0.6	2.3	2.5	$-1.765767418748205865689800 \times 10^{-2}$	$-1.76576741875 \times 10^{-2}$	19
4	3	2	3	2	1	6.5	2.0	1.4	$-9.595775344088398154310522 \times 10^{-2}$	$-9.59577534408839815 \times 10^{-2}$	18
5	3	2	3	2	2	5.8	4.5	8.7	$7.934310549472991320856153 \times 10^{-15}$	$7.93431054948 \times 10^{-15}$	19
6	4	0	4	1	0	0.5	0.4	1.4	$-1.230350868942323080523488 \times 10^{-1}$	$-1.23035086894232308 \times 10^{-1}$	18
7	5	0	5	0	0	0.1	0.1	1.4	$9.996371894103858561392756 \times 10^{-1}$	$9.996371894103862 \times 10^{-1}$	20
8	5	2	5	2	2	1.5	1.4	1.0	$9.389786330449060778836973 \times 10^{-1}$	$9.38978633044906079 \times 10^{-1}$	18
9	5	4	6	3	2	7.0	3.0	1.2	$2.769070225696625582474567 \times 10^{-2}$	$2.76907022569662673 \times 10^{-2}$	18

(Continued)

Table 10.3 Continued

	n_1	l_1	n_2	l_2	m	α	β	R	In this study (Eq. (30))	Literature	Ref.
10	7	4	7	4	4	3.0	1.0	1.0	$1.101104777649702174335891 \times 10^{-1}$	$1.101104777649702 \times 10^{-1}$	20
11	8	7	8	7	7	0.06	0.04	0.5	$7.068020766365193821836148 \times 10^{-1}$	$7.06802076636519367 \times 10^{-1}$	18
12	10	0	10	9	0	2.5	1.0	3.0	$1.829925377782952911503784 \times 10^{-5}$	$1.82992537778180995 \times 10^{-5}$	18
13	12	9	11	10	8	9.5	3.2	0.6	$-5.42197898043609735681753 7 \times 10^{-2}$	$-5.42197898043610 \times 10^{-2}$	21
14	18	17	18	17	17	5.0	5.0	2.4	$1.569738129172988237506223 \times 10^{-1}$	$1.56973812917299342 \times 10^{-1}$	18
15	21	6	20	7	4	0.02	0.02	0.14	$-3.60848109154608201008325 3 \times 10^{-4}$	$-3.60848109154608202 \times 10^{-4}$	18

Table 10.4 The numerical results for overlap integrals with the same and different screening parameters in the nonlined-up coordinate systems. In table, t and p are denoted, respectively, $t = \frac{\alpha-\beta}{\alpha+\beta}$, $p = R(\alpha+\beta)/2$.

	n_1	l_1	m_1	n_2	l_2	m_2	p	T	θ	ϕ	In this study	Literature	Ref.
1	2	1	0	2	1	0	25	0.5	60	120	$-2.01912763169782351044068 6\times10^{-5}$	$-2.01912763169777\times10^{-5}$	22
2	3	2	0	3	2	0	25	0.5	60	120	$-6.80340033602071132889581 8\times10^{-5}$	$-6.80340033108473\times10^{-5}$	24
3	3	2	1	2	1	−1	15	0.7	30	60	$-1.97894655854632920845042 2\times10^{-3}$	$-1.97894655862444\times10^{-3}$	8c
4	5	4	2	5	4	0	150	0.7	45	210	$-2.27511176683283043597258 5\times10^{-17}$	$-2.27511175896203\times10^{-17}$	8c
5	7	3	2	4	3	2	150	0.7	0	0	$-1.76861050692264859080888 4\times10^{-18}$	$-1.76861050692265\times10^{-18}$	23
6	7	4	1	5	3	1	200	0.7	45	210	$-2.00183868910395189767894 5\times10^{-23}$	$-2.00183867476671\times10^{-23}$	8c
7	14	11	7	13	10	6	12	−0.6	15	10	$4.48032190106317531036390 1\times10^{-5}$	$4.48032207619290\times10^{-5}$	8c
8	19	9	8	14	13	7	15	0.8	18	300	$2.18261954539413971517851 1\times10^{-5}$	$2.18261954784081\times10^{-5}$	8c

(Continued)

Table 10.4 Continued

	n_1	l_1	m_1	n_2	l_2	m_2	p	T	θ	ϕ	In this study	Literature	Ref.
9	20	13	5	20	9	5	40	−0.5	120	225	$1.70386203431244488855756 \times 10^{-3}$	$1.70386168880515 \times 10^{-3}$	8c
10	20	8	2	10	8	1	25	0	60	45	$7.93865478066131311456698 5 \times 10^{-4}$	$7.93865462132564 \times 10^{-4}$	22
11	28	7	3	18	6	5	10	0.5	12	150	$-1.305079596042023615129968 \times 10^{-4}$	$-1.31318185950514 \times 10^{-4}$	8c
12	28	14	10	28	16	16	20	0.7	30	60	$-4.735045067864533159531955 \times 10^{-10}$	$-4.73486792914621 \times 10^{-10}$	8c
13	35	8	8	25	5	−4	25	0.5	60	120	$-2.408212608290414665303749 \times 10^{-4}$	$-2.30030196064108 \times 10^{-4}$	8c
14	46	2	1	14	1	1	10	0.5	15	24	$3.16442030563724283564008 7 \times 10^{-1}$	$3.16465568973146 \times 10^{-1}$	8c
15	50	4	4	50	4	4	22	0.7	45	210	$1.418051730221017060998600 \times 10^{-11}$	$1.41805162040407 \times 10^{-11}$	8c

Mathematica 7 and early versions calculated binomial coefficients and factorial functions which contained negative integers erroneously. For example, Mathematica 7 and early versions give the following result in the case of factorials for negative integers and binomial coefficients given in Ref. 25,

$$\text{In}[1] := \frac{(-3)!}{(-5)!}$$

- ∞::indet: Indeterminate expression 0 ComplexInfinity encountered.
- Out[1] = Indeterminate
- IM[1]:Binomial[−2,5] = −6

REFERENCES

1. Ahlrichs, R. Asymptotic Behaviour of Atomic Bound State Wave Functions. *Chem. Phys. Lett.* **1972,** *15*, 609. Asymptotic Behaviour of Molecular Bound State Wave Functions **1973,** 18, 521
2. Kato, T. On the Eigenfunction of Many-Particle System Quantum Mechanics. *Commun. Pure Appl. Math.* **1957,** *10*, 151.
3. Slater, J. C. Atomic Shielding Constants. *Phys. Rev.* **1930,** *36*, 57.
4. Weniger, E. J.; Steinborn, E. O. The Fourier Transform of Some Exponential Type Basic Functions and Their Relevance to Multicenter Problems. *J. Chem. Phys.* **1983,** *78* (10), 6121.
5. Arfken, G. B.; Weber, H. J. *Mathematical Methods for Physicists*; Academic Press: London, 2001.
6. Prosser, F. P.; Blanchard, C. H. On the Evaluation of Two-Center Integrals. *J. Chem. Phys.* **1962,** *36* (4), 1112.
7. Geller, M. Two-Center, Nonintegral, Slater-Orbital Calculations, Integral Formulation and Application to the Hydrogen Molecule-Iyon. *J. Chem. Phys.* **1962,** *36*, 2424. Two Electrons One and Two-Center Integrals **1963,** *39*, 84.Silverstone, H. J. On the Evaluation of the-Center Overlap and Coulomb Integrals with Noninteger-n Slater Type Orbitals. *J. Chem. Phys.* **1966,** *45*, 4337. Series Expansion for Two-Center Noninteger-n Overlap Integrals **1967,** 46, 4368 Weniger, E. J.; Steinborn, E. O. Numerical Properties of the Convolution Theorems of B-Functions. *Phys. Rev. A* **1983,** *28* (4), 2026. The Fourier-Transform of Some Exponential Type Basic Functions and their Relevance to Multicenter Problems. *J. Chem. Phys.* **1983,** *78*, 6121. Overlap Integrals of B Functions-A Numerical Study of Infinitive Series Representations and Integral-Representations. *Theor. Chem. Acta* **1988,** *73*, 323.Filter, E.; Steinborn, E. O. The Three Dimensional Convolution of Reduced Bessel Functions and Other Functions of Physical Interest. *J. Math. Phys.* **1978,** *19* (1), 79. Extremely Compact Formulas for Molecular Two-Center One-Electron Integrals and Coulomb Integrals Over Slater-Type Atomic Orbitals. *Phys. Rev. A* **1978,** *18* (1), 1.Weniger, E. J. Weakly Convergent Expansions of a Plane Wave and their Use in Fourier Integrals. *J. Math. Phys.* **1985**, *26* (2), 276.Edwards, S. A.; Gottlieb, H. P. W.; Doddrell, D. M. A General Analytic Procedure for Calculating Two-Center Integrals Involving the One-Electron Dipolar Coupling Operator and Slater Atomic Orbitals. *Mol. Phys.* **1979,** *38*, 1147.Kaijser, P.; Smith, V. H. Evaluation of Momentum Distribution and Compton Profiles for Atomic and Molecular Systems. *Adv. Quant. Chem.* **1977,** *10*, 37. Antolovic, D.; Delhalle, J. Multipole and Overlap Integrals Over Reduced Bessel-Function in Molecular Quantum Mechanics. *Phys. Rev. A* **1980,** *21*, 1815.Guseinov, I. I. Use of Expansions Formulas for Slater-Type Orbitals in Evaluation of Overlap Integrals. *J. Mol. Sci. (Wuhan China)* **1987,** *5*, 169. Guseinov, I. I.; Öztekin, E.; Hüseyin, S.; Calculation of Overlap Integrals with the Same Screening

Parameters Using Gegenbauer Coefficients, "Computation of Molecular Integrals Over Slater-Type Orbitals. Part VI. Computation of Molecular Integrals Over Slater-Type Orbitals. Part VI. Calculation of Overlap Integrals with the Same Screening Parameters Using Gegenbauer Coefficients. *J. Mol. Struct. (THEOCHEM)* **2001,** *536,* 59. Yavuz, M.; Yükçü, N.; Öztekin, E.; Yılmaz, H.; Döndür, S. On the Evaluation Overlap Integrals with the Some and Different Screening Parameters. Over Slater Type Orbitals via the Fourier Transform Method. *Commun. Theor. Phys.* **2005,** *43,* 151. Öztekin, E.; Özcan, S.; Orbay, M.; Yavuz, M. Calculation of Nuclear-Attraction and Modified Overlap Integrals Using Gegenbauer Coefficients. *Int. J. Quant. Chem.* **2002,** *90,* 136.

8. (a)Barnett, M. P. Digital Erosion in the Evaluation or Molecular Integrals. *Theor. Chem. Acc.* **2002,** *107,* 241.(b)Öztekin, E.; Yavuz, M.; Atalay, Ş. Formulas and Numerical Table for the Radial Part of Overlap Integrals with the Same Screening Parameters of Slater Type Orbitals. *Theor. Chem. Acta.* **2001,** *106,* 264.(c)Guseinov, I. I.; Mamedov, B. A. Computation of Molecular Integrals Over Slater Type Orbitals I. Calculations of Overlap Integrals Using Recurrence Relations. *J. Mol. Struct. (THEOCHEM)* **1999,** *465,* 1.
9. Weissbluth, M. *Atoms and Molecules*; Academic Press: New York, 1978.
10. Gradshteyn, I. S.; Ryzhik, I. M. *Tables of Integrals, Series and Product*; Academic Press: New York, 1995.
11. Guseinov, I. I.; İlik, A.; Allahverdiyev, S. I. On the Evaluation of Overlap Integrals with the Same screening Parameters of Slater-Type Prbitals Using Binomial Coefficients. *Int. J. Quant. Chem.* **1996,** *60,* 637.
12. Gaunt, J. A. *Philos. Trans. R. Soc. London, Ser. A* **1929,** *228,* 192.
13. Guseinov, I. I. Analytical Evaluation of Two-Center Coulomb, Hybrid and One-Electron Integrals for Slater-Type Orbitals. *J. Phys. B* **1970,** *3,* 1399.
14. Öztekin, E. Overlap Integrals with Respect to Quantum Numbers over Slater Type Orbitals via the Fourier Transform Method. *Int. J. Quant. Chem.* **2004,** *100,* 236.
15. Öztekin, E.; Yavuz, M.; Atalay, Ş. Calculation of Overlap Integrals using Recurrence-based Algorithms. *J. Mol. Struct. (THEOCHEM)* **2001,** *544,* 69.
16. Weniger, E. J.; Grotendorst, J.; Steinborn, E. O. Efficient Evaluation of Infinite-Series Representations for Overlap, Two-Center Nuclear Attraction, and Coulomb Integrals Using Nonlinear Convergence Eccelerators. *Phys. Rev. A* **1986,** *33* (6), 3688.
17. Wolfram, S. *Mathematica, A System for Doing Mathematics by Computer*; Addision Wesley: New York, 1998.
18. Özdoğan, T. Fast and Stable Algorithm for Analytical Evaluation of Two-Center Overlap Integrals Over Slater-Type Orbitals With Integer and Noninteger Principal Quantum Numbers. *Int. J. Quant. Chem.* **2004,** *100,* 69.
19. Magnasco, V.; Rapallo, A.; Casanova, M. New Translation Method for STOs and Application to Calculation of Overlap Integrals. *Int. J. Quant. Chem.* **1999,** *73,* 333.
20. Jones, H. W. Comprehensive Strategy for the Calculation of Overlap Integrals With Slater-Type Orbitals. *Int. J. Quant. Chem.* **1997,** *61,* 881.
21. Özdoğan, T.; Orbay, M. Evaluation of Two-Center Overlap and Nuclear Attraction Integrals Over Slater-Type Orbitals with Integer and Noninteger Principal Quantum Numbers. *Int. J. Quant. Chem.* **2002,** *87,* 15.
22. Özdoğan, T.; Orbay, M.; Değirmenci, S. Evaluation of Two-Center Overlap Integrals Using Slater Type Orbitals in Terms of Bessel Type Orbitals. *J. Math. Chem.* **2005,** *37,* 1.
23. Özdoğan, T. Evaluation of Two-Center Overlap Integrals Over Slater Type Orbitals Using Fourier Transform Convolution Theorem. *Collect. Czech. Chem. Commun.* **2004,** *69,* 279.
24. Guseinov, I. I.; Mamedov, B. A.; Orbay, M. Calculation of Orbitals Using Basic Overlap Integrals. *Commun. Theor. Phys.* **2000,** *33,* 161.
25. Yükçü, N.; Öztekin, E. Strategies on the Evaluation of Binomial Coefficients for All Integers. *Comp. Math. Math. Phys.* **2013,** *53* (1), 1. <http://link.springer.com/article/10.1134/S0965542513010119>.

CHAPTER ELEVEN

On the β^--Decay in the ^{8}Li and ^{9}Li Atoms

M. B. Ruiz and A. M. Frolov
[a]Department of Theoretical Chemistry, Friedrich-Alexander-University Erlangen-Nürnberg, Egerlandstraße 3, Erlangen 91058, Germany
[b]Department of Chemistry, University of Western Ontario, Ontario, London, Canada, N6H 5B7

Contents

Abstract

The nuclear β^--decay from the ground and some excited states of the three-electron ^{8}Li and ^{9}Li atoms is considered. The final state probabilities for product Be^+ ion are determined numerically using highly accurate bound state wave functions of the Li atom and Be^+ ion. The probability of electron ionization during the nuclear β^--decay of the Li atom is evaluated numerically. We also discuss the possibility of observing double β^--decay using known values of the final state probabilities for the regular nuclear β^--decay.

1. INTRODUCTION

In our earlier study[1] we considered the atomic excitations during the nuclear β^--decay in light atoms and ions. In[1] we determined the final state probabilities for a number of bound states in the final atoms and ions. Recently, all these probabilities have been re-calculated to much better accuracy. They are presented in this work. Also, here we discuss two problems related with the atomic ionization during the nuclear β^--decay: (a) excitation of the internal electron shells, and (b) evaluating the probability of "additional" electron ionization. In reality these two problems are very complex and below we have made a few preliminary steps to the final

Advances in Quantum Chemistry, Volume 67
ISSN 0065-3276, http://dx.doi.org/10.1016/B978-0-12-411544-6.00011-X

solutions. In addition to this we restrict ourselves by the analysis of the three-electron Li atom.

The focus below is given to the ^{8}Li and ^{9}Li atoms which are of interest in some industrial applications. In general, the β^--decay of the Li atom(s) can be written in the following form

$$\text{Li} \rightarrow \text{Be}^+ + e^- + \overline{\nu}, \tag{1}$$

where the notation e^- stands for the fast electron emitted during the nuclear β^--decay, while $\overline{\nu}$ designates the electron anti-neutrino. In general, the nuclear β^--decay of the Li atom leads to the following re-distribution of the bound atomic electrons. As a result of this re-distributed incident electron density the final Be^+ ion can be found in a variety of bound states, or even in a number of unbound states. Briefly, this means the formation of the two-electron Be^{2+} ion during the nuclear β^--decay of the Li atom. In actual applications it is important to predict the probabilities to form the final Be^+ ion in different final states. Note that there are a few selection rules which are applied to the $\beta^{\pm}$-decays in atomic systems (see, e.g., Ref. 2). These rules can be concisely formulated as the conservation laws for the angular momenta **L** and the total electron spin **S**. The wave functions of the incident and final system must also have the same spatial parity. For instance, if the incident Li atom was in its $3^2\text{P}(L=1)$-state, then the final Be^+ can be found only in one of its $n^2\text{P}(L=1)$-states. In other words, after the nuclear β-decay of the Li atom in the $3^2\text{P}(L=1)$-state it is impossible to detect the final Be^+ ion, e.g., in the $3^2\text{S}(L=0)$-state, or in the $3^2\text{D}(L=2)$-state.

The advantage of considering three-electron atoms and ions is obvious: the wave functions of such systems can be approximated to very high numerical accuracy. For simplicity, throughout this study we shall assume that the original Li atom was in its ground $^2\text{S}(\text{L} = 0)$-state. The choice of the ground state of the incident Li atom is not a fundamental restriction for our method. Formally, such a state can be arbitrary, e.g., either ground state, or "vibrationally," or "rotationally" excited atomic state with the given angular momentum L (see below) and the total electron spin S.

By analyzing the properties of known Li-isotopes one finds that there are two β^--decaying isotopes of lithium: ^{8}Li ($\tau_\beta \approx 0.84$ s) and ^{9}Li ($\tau_\beta \approx 0.17$ s). These two isotopes are formed in the (n;γ)-reactions during thermonuclear explosions in which light thermonuclear fuel (^{6}LiD) is compressed to very high densities $\rho \geq 100$ g sm^{-3} by extremely intense flux of soft X-ray radiation from the primary. Larger compressions mean, in general, greater output of these two lithium isotopes. In the laboratory, the ^{8}Li isotope is produced with the use of the (n; γ)-reaction at ^{7}Li. In contrast with this, the ^{9}Li isotope is produced by using either (d; 2p)- and (n; p)-reactions with the ^{9}Be nuclei, or (t; p)-reaction with nuclei of ^{7}Li.

The lifetimes of these two lithium isotopes are relatively short from the chemical point of view. Therefore, it is hard to study the regular chemical properties of these isotopes. An alternative approach is based on detailed analysis of the optical radiation emitted by the final Be^+ ions which are formed after the nuclear β^--decay of these two Li isotopes. This can be achieved, if we know the corresponding final state probabilities, i.e., the probabilities to form the final Be ions in certain bound states. The first goal of this study is to evaluate the final state probabilities of formation of various final states in product Be^+ ions. Note that all evaluations of the final state probabilities during the nuclear β^--decay in atoms and molecules are based on the sudden approximation[3,4] which applies to both atomic systems (original atom and final ion) involved in the process. The sudden approximation is appropriate for all $\beta^\pm$-decaying atoms, since the velocities of the $\beta^\pm$ electrons are significantly greater than those of regular atomic electrons.

The final state probabilities, i.e. probabilities to form different atomic states during nuclear β^--decay in various light atoms, have been evaluated numerically in a number of earlier papers (see, e.g.,[1,2]). All such evaluations, however, have been based on the assumption that the total number of bound electrons is constant in nuclear β^--decay. In reality, the nuclear β-decay in light atoms often leads to an "additional" electron ionization. For the Li atom this process can be written in the form

$$\mathrm{Li} \rightarrow \mathrm{Be}^{2+} + \mathrm{e}^- + \beta^- + \overline{\nu}, \tag{2}$$

where e^- designates the secondary atomic electron which becomes free during atomic β^--decay. It is interesting to evaluate the probability of this process and obtain the actual energy spectra of the emitted secondary electrons. Formally, all secondary electrons emitted during atomic β-decay must be considered as δ-electrons. On the other hand, the original definition of δ-electrons means that these electrons are fast and their total energies significantly exceed the usual energies of "regular" atomic electrons. The energy of the free electron from reaction, Eq. (2), is comparable with atomic energies. Therefore, here we deal with the regular atomic ionization during β^--decay. In earlier works the process of additional ionization only from the atomic K-shell was considered (see discussion and references in Ref. 5).

The main goal of this study is to determine the final state probabilities to form various bound states in the Be^+ ion. These calculations are discussed in the fourth Section. Another aim of our study is to evaluate the probability of "additional" ionization during the nuclear β^--decay and investigate the energy spectrum of secondary electrons emitted during the nuclear β^--decay. This problem is considered in the third Section. We also briefly investigate the long-standing problem of the double nuclear β-decay. Concluding remarks are in the last Section.

2. EVALUATION OF THE FINAL STATE PROBABILITIES FOR THE BOUND STATES

As follows from the general theory of perturbations in Quantum Mechanics (see, e.g., Ref. 5) in sudden approximation the final state probabilities are determined as overlap integrals between the wave function of the incident atomic system (i.e., the wave function of the Li atom in our case) and the wave function of the final atomic system (i.e., the wave function of the Be^+ ion). To compute such a three-electron integral we need to assume that the total numbers of bound electrons in the incident and final atomic systems are the same. In the sudden approximation the general formula for the transition probability $w_{\rm if}$ for the transitions from the incident i-state into the final f-state takes the form (see, e.g., Ref. 5)

$$w_{\rm if} = \frac{1}{\hbar^2} \mid \int_0^{+\infty} V_{\rm if} \exp(\imath \omega_{\rm if} t) {\rm d}t \mid^2 \approx \frac{1}{\hbar^2} \mid V_{\rm if} \mid^2, \tag{3}$$

where $V_{\rm if}$ is the overlap integral computed with the use of time-independent incident and final atomic wave functions, i.e.,

$$V_{\rm if} = \langle \psi_{{\rm Be}^+}(\mathbf{x}_1, \mathbf{x}_2, \mathbf{x}_3) \mid \Psi_{\rm Li}(\mathbf{x}_1, \mathbf{x}_2, \mathbf{x}_3) \rangle, \tag{4}$$

where $\Psi_{\rm Li}(\mathbf{x}_1, \mathbf{x}_2, \mathbf{x}_3)$ and $\psi_{{\rm Be}^+}(\mathbf{x}_1, \mathbf{x}_2, \mathbf{x}_3)$ are the wave functions of the Li atom and Be^+ ion, respectively. The derivation of the formula, Eq. (3), is based on the facts that: (1) the velocity of the β-electron is substantially larger than the velocities of atomic electrons, and (2) the final ion does not move during the nuclear $\beta^{\pm}$-decay. In atomic units we have $\hbar = 1$, $m_e = 1$, and $e = 1$ and, therefore, $w_{\rm if} = \mid V_{\rm if} \mid^2$. The notation $\mathbf{x}_i$ in Eq. (4) designates the spin-spatial coordinates of the ith electron, i.e., $\mathbf{x}_i = (\mathbf{r}_i, \mathbf{s}_i)$. Note that in some works the integral $V_{\rm fi}$, Eq. (4), or the ratio $\frac{V_{\rm if}}{\hbar}$) is called the probability amplitude. The two wave functions in Eq. (4) depend only upon spatial and spin coordinates of three electrons and do not depend upon the time . All wave functions used in Eq. (4) are assumed to be properly symmetrized in respect to all spin-spatial permutations of identical particles (electrons).

As follows from Eq. (4) the final state probability for β^--decay in the ^{8}Li and ^{9}Li atoms can be determined, if the wave functions of the incident and final atomic systems (bound states) are known. The construction of highly accurate variational wave functions for three-electron atoms and ions is considered in the fourth Section. The final state probabilities determined using such wave functions can be found in Tables 11.1 and 11.2. Here we assume that the incident Li atom was in its ground $^2S(L = 0)$-state (2^2S-state). It should be mentioned that in reality the incident Li atoms are formed in the (n; γ)-, (n; p)-, and some other nuclear reactions with neutrons of different energies (see above). In such cases it is hard to expect that all incident Li atoms will always be in the ground $2^2S(L = 0)$-state. In fact, these β^--decaying Li atoms can be found in a variety of the rotationally and/or vibrationally excited states. After reactions with neutrons the incident Li atom before nuclear β^--decay will be likely to have non-zero speed in some direction. Therefore, some other (excited) bound states in the Li atom must also be considered as the incident atomic states before the nuclear β^--decay.

Numerical computation of the overlap integrals, Eq. (4), is reduced to calculations of some separated integrals, which include different spin components of the incident and final atomic wave functions.

Discussing e.g. the construction of three-electron variational wave function for the Li atom. Without loss of generality, below we restrict ourselves to the consideration of the ground $^2S(L = 0)$-state of the Li atom. As is well known (see, e.g.,[6,7]) the accurate variational wave function of the ground (doublet) $^2S(L = 0)$-state of the Li atom is written in the following general form

$$\begin{aligned}\Psi(\mathrm{Li})_{L=0} &= \psi_{L=0}(A; \{r_{ij}\})(\alpha\beta\alpha - \beta\alpha\alpha) \\ &\quad + \phi_{L=0}(B; \{r_{ij}\})(2\alpha\alpha\beta - \beta\alpha\alpha - \alpha\beta\alpha),\end{aligned} \tag{5}$$

where $\psi_{L=0}(A; \{r_{ij}\})$ and $\phi_{L=0}(B; \{r_{ij}\})$ are the two independent radial parts (= spatial parts) of the total wave function. Everywhere below in this study, we shall assume that all mentioned wave functions have unit norms. The notation α and β in Eq. (5) denotes one-electron spin-up and spin-down functions, respectively (see, e.g., Ref. 8). The notations A and B in Eq. (5) mean that the two sets of non-linear parameters associated with the radial functions ψ and ϕ can be optimized independently. In general, each

Table 11.1 Convergency of the β^--decay transition amplitudes, final-state probabilities, and total energies for the ground state of Li atom and different n^2S states of Be^+ ion. Energy in a.u., energy differences in μh

State	Basis[a]	N[b]	Amplitude	Probability	Virial	Energy	N_{Ref}	Ref. Ener.	Ref.	Diff.
Li 2^2S	$n=4$	308			2.000004	−7.478053222				7.1
Li 2^2S	$n=5$	517			2.000001	−7.478057825				2.5
Li 2^2S	$n=6$	620			2.000000	−7.478058892	13,944	−7.478060323909560	15	1.4
Be^+ 2^2S	$n=4$	308	0.759681281	0.577116	2.000002	−14.324757377				
Be^+ 2^2S	$n=5$	612	0.759686424	0.577123	2.000001	−14.324760412				2.8
Be^+ 2^2S	$n=6$	637	0.759683487	0.577119	2.000000	−14.324761723	13,944	−14.32476317685542	15	1.4
Be^+ 3^2S	$n=4$	307	0.514947878	0.265171	2.000000	−13.922759980				
Be^+ 3^2S	$n=5$	459	0.514892996	0.265115	2.000009	−13.922781623				7.6
Be^+ 3^2S	$n=6$	637	0.514929058	0.265152	2.000005	−13.922784968	10,000	−13.922789268542	16	4.3
Be^+ 4^2S	$n=4$	252	0.075214427	0.005657	2.001593	−13.798520453				
Be^+ 4^2S	$n=5$	372	0.074214825	0.005508	2.000484	−13.798704722				16.3
Be^+ 4^2S	$n=6$	451	0.073790160	0.005445	2.000154	−13.798706849	8000	−13.7987166092	25	9.8
Be^+ 5^2S	$n=5$	502	0.039839395	0.001587	2.004399	−13.744513336				
Be^+ 5^2S	$n=6$	698	0.043113179	0.001859	2.001584	−13.744589135	1940	−13.74463182	26	42.7
Be^+ 6^2S	$n=6$	618	0.029411301	0.000865	2.003166	−13.716152057	2058	−13.71628624	26	134.2
Be^+ 7^2S	$n=7$	619	0.021688396		2.003570	−13.699131127				

[a] Basis set, i.e., $n = 4$ stays for [4s3p2d1f] or [1s2s3s4s2p3p4p3d4d4f].
[b] N is the number of Hy-CI symmetry adapted configurations. N_{Ref} is the number of configurations employed in the calculation of Ref. Ener.

Table 11.2 Transition probabilities for the nuclear β^--decay from the ground 2^2S-state of the Li atom into the ground and various excited S-states of the Be^+ ion. The probability of ionization from Be^+ ion to Be^{2+} ion is calculated as $P_{ion} = 1 - \sum_{i=1}^{\infty} P_i$

State of Be^+	Amplitude	Probability (P_i)	P_i in %
2^2S	0.759683487	0.577119	57.71
3^2S	0.514929058	0.265152	26.52
4^2S	0.073790160	0.005445	0.54
5^2S	0.043113179	0.001859	0.19
6^2S	0.029411301	0.000865	0.09
7^2S	0.021688396	0.000470	0.05
Total		0.850910	85.09
P_{ion}		0.149090	14.91

of the radial basis functions in Eq. (5) explicitly depends upon all six interparticle (or relative) coordinates $r_{12}, r_{13}, r_{23}, r_{14}, r_{24}, r_{34}$, where the indices 1, 2, 3 stand for the three electrons, while index 4 means the nucleus.

In atomic systems, the wave function must be completely antisymmetric with respect to all electron spin-spatial variables. For three-electron wave function this requirement is written in the form $\hat{\mathcal{A}}_e\Psi(1,2,3) = \hat{\mathcal{A}}_{123}\Psi(1,2,3) = -\Psi(1,2,3)$, where the wave function Ψ is given by Eq. (5) and $\hat{\mathcal{A}}_e$ is the electron antisymmetrizer. In our case $\hat{\mathcal{A}}_e$ is the three-electron (or three-particle) antisymmetrizer, i.e., $\hat{\mathcal{A}}_e = \hat{e} - \hat{P}_{12} - \hat{P}_{13} - \hat{P}_{23} + \hat{P}_{123} + \hat{P}_{132}$. Here $\hat{e}$ is the identity permutation, while $\hat{P}_{ij}$ is the permutation of the ith and jth particles. Analogously, the operator $\hat{P}_{ijk}$ is the permutation of the ith, jth, and kth particles.

Suppose that the incident three-electron wave function of the Li atom has been chosen in the form of Eq. (5). Applying the three-electron antisymmetrizer $\hat{\mathcal{A}}_{123}$ to the first part of the total wave function, Eq. (5), one finds

$$\begin{aligned}
&\hat{\mathcal{A}}_{123}\left[\psi_{L=0}(A;\{r_{ij}\})(\alpha\beta\alpha - \beta\alpha\alpha)\right] \\
&\quad= (\hat{e}\psi)(\alpha\beta\alpha - \beta\alpha\alpha) + (\hat{P}_{12}\psi)(\alpha\beta\alpha - \beta\alpha\alpha) \\
&\quad\quad -(\hat{P}_{13}\psi)(\alpha\beta\alpha - \alpha\alpha\beta) - (\hat{P}_{23}\psi)(\alpha\alpha\beta - \beta\alpha\alpha) \\
&\quad\quad +(\hat{P}_{123}\psi)(\alpha\alpha\beta - \alpha\beta\alpha) + (\hat{P}_{132}\psi)(\beta\alpha\alpha - \alpha\alpha\beta),
\end{aligned} \tag{6}$$

$$\begin{aligned}
&\hat{\mathcal{A}}_{123}\left[\phi_{L=0}(B;\{r_{ij}\})(2\alpha\alpha\beta - \beta\alpha\alpha - \alpha\beta\alpha)\right] \\
&\quad= (\hat{e}\phi)(2\alpha\alpha\beta - \beta\alpha\alpha - \alpha\beta\alpha) - (\hat{P}_{12}\phi)(2\alpha\alpha\beta - \beta\alpha\alpha - \alpha\beta\alpha) \\
&\quad\quad -(\hat{P}_{13}\phi)(2\beta\alpha\alpha - \alpha\alpha\beta - \alpha\beta\alpha) - (\hat{P}_{23}\phi)(2\alpha\beta\alpha - \alpha\alpha\beta - \beta\alpha\alpha) \\
&\quad\quad +(\hat{P}_{123}\phi)(2\beta\alpha\alpha - \alpha\beta\alpha - \alpha\alpha\beta) + (\hat{P}_{132}\phi)(2\alpha\beta\alpha - \alpha\alpha\beta - \beta\alpha\alpha)
\end{aligned} \tag{7}$$

where the notations $(\hat{P}_{ij}\phi)$ and $(\hat{P}_{ijk}\phi)$ mean permutations of spatial coordinates in $\phi_{L=0}(B; \{r_{ij}\})$ radial function, Eq. (5).

Now, by using the expressions, Eqs. (6) and (7), we can obtain the formulae used in computations of the final state probabilities in the case of nuclear β^--decay, Eq. (1), in the three-electron Li atom. For instance, if the final wave function has the same spin-symmetry, i.e., it is written in the form

$$\Psi_{\rm fi} = \psi_{\rm fi}(\mathbf{r}_1, \mathbf{r}_2, \mathbf{r}_3)(\alpha\beta\alpha - \beta\alpha\alpha) + \phi_{\rm fi}(\mathbf{r}_1, \mathbf{r}_2, \mathbf{r}_3)(2\alpha\alpha\beta - \beta\alpha\alpha - \alpha\beta\alpha) \tag{8}$$

then the final state probabilities are determined using the following formulae

$$P_{\psi\psi} = \langle \psi_{\rm fi}(\mathbf{r}_1, \mathbf{r}_2, \mathbf{r}_3) \mid \frac{1}{2\sqrt{3}} \left(2\hat{e} + 2\hat{P}_{12} - \hat{P}_{13} - \hat{P}_{23} - \hat{P}_{123} - \hat{P}_{132}\right) \psi_{\rm Li}(A; \{r_{ij}\})\rangle, \tag{9}$$

$$P_{\phi\psi} = \langle \phi_{\rm fi}(\mathbf{r}_1, \mathbf{r}_2, \mathbf{r}_3) \mid \frac{1}{2} \left(\hat{P}_{13} - \hat{P}_{23} + \hat{P}_{123} - \hat{P}_{132}\right) \psi_{\rm Li}(A; \{r_{ij}\})\rangle, \tag{10}$$

$$P_{\psi\phi} = \langle \psi_{\rm fi}(\mathbf{r}_1, \mathbf{r}_2, \mathbf{r}_3) \mid \frac{1}{2} \left(\hat{P}_{13} - \hat{P}_{23} + \hat{P}_{123} - \hat{P}_{132}\right) \phi_{\rm Li}(B; \{r_{ij}\})\rangle, \tag{11}$$

$$P_{\phi\phi} = \langle \phi_{\rm fi}(\mathbf{r}_1, \mathbf{r}_2, \mathbf{r}_3) \mid \frac{1}{2\sqrt{3}} \left(2\hat{e} - 2\hat{P}_{12} + \hat{P}_{13} + \hat{P}_{23} - \hat{P}_{123} - \hat{P}_{132}\right) \phi_{\rm Li}(B; \{r_{ij}\})\rangle. \tag{12}$$

Note they coincide with the known formulae[9] which correspond to the case when both incident and final atomic states contain three electrons in the doublet spin configuration (the total electron spin equals $\frac{1}{2}$). This is the "classical" $\beta^{\pm}$-decay in few-electron atoms, when the incident and final electron configurations have conserved L and S quantum numbers.

In reality, another process is also possible in few-electron atoms during the nuclear $\beta^{\pm}$-decay in few- and many-electron atoms. This process leads to the formation of the final ion/atom in some excited states. For instance, consider the case when the three final electrons form the doublet configuration with the spin function $\alpha\alpha\beta$. It is clear that such a wave function cannot represent the ground state of the Be^+ ion. However, some excited states (with internal electron shell vacancies) can have this spin function, e.g., 1s2s3p-, 1s2s4d- and 1s3s3p-states of the Be^+ ion. Another example is discussed in the next section. It represents an additional electron ionization during the nuclear β^- -decay in three-electron atoms. If this free electron moves away in the β-spin state, then the final Be^{2+} ion can be only in its triplet spin state (not singlet state). Formally this means formation of the final ion in an excited state (with some vacancies on its internal electron shells). In this case the formulae for the final state probabilities become

$$P_{tr\psi} = \langle \psi_{\rm fi}(\mathbf{r}_1, \mathbf{r}_2, \mathbf{r}_3) \mid \frac{1}{2\sqrt{3}} \left(\hat{P}_{13} - \hat{P}_{23} + \hat{P}_{123} - \hat{P}_{132}\right) \psi_{\rm Li}(A; \{r_{ij}\})\rangle, \tag{13}$$

$$P_{tr\phi} = \langle \psi_{\rm fi}(\mathbf{r}_1, \mathbf{r}_2, \mathbf{r}_3) \mid \frac{1}{2\sqrt{3}} \left(2\hat{e} - 2\hat{P}_{12} + \hat{P}_{13} + \hat{P}_{23} - \hat{P}_{123} - \hat{P}_{132}\right) \phi_{\rm Li}(B; \{r_{ij}\})\rangle, \tag{14}$$

where it is assumed that the incident electron wave function was written in the form of Eq. (5). These formulae indicate clearly that the probability of finding the final Be^{2+} ion in the excited triplet spin states is non zero. In all earlier studies the transitions to the final atomic states with different spin states were ignored. Moreover, any possibility to form the final few-electron ion/atom in excited states (with some vacancies on internal electron shells) during nuclear $\beta^{\pm}$-decay was rejected. Demonstrating the very existence of such transitions is a great achievement of this study.

In general, during nuclear β^--decay of the Li atom the final Be^+ ion can be formed in many different bound and/or unbound states. If such a state is unbound, then we deal with the additional ionization during atomic β^--decay. It is discussed in the next section. This process is of great interest, since it often leads to the formation of the final ion in an excited state(s) with various vacancies in the internal electron shells. For light atoms and ions this means a possibility to observe emission of the optical quanta after the nuclear $\beta^{\pm}$-decay in many-electron atoms with the total number of electrons ≥ 3.

3. ELECTRON IONIZATION DURING THE NUCLEAR β^--DECAY

The probability of ionization (or "additional" ionization) of the final Be^+ ion during the nuclear β-decay can also be evaluated using the sudden approximation. In this case the final wave function is constructed as the product of the bound state wave function of the two-electron Be^{2+} ion and the wave function of the unbound electron which moves in the central Coulomb field of this two-electron ion. To determine the corresponding final state probability one needs to compute the following overlap integral between the wave functions of the incident and final atomic systems

$$\mathcal{A}_{\rm if} = \langle \psi_{\rm Be^+}(\mathbf{x}_1, \mathbf{x}_2)\phi(\mathbf{x}_3) \mid \Psi_{\rm Li}(\mathbf{x}_1, \mathbf{x}_2, \mathbf{x}_3)\rangle, \quad (15)$$

where $\phi(\mathbf{x}_3)$ is the wave function of the unbound electron which moves in the Coulomb field of the Be^{2+} ion and $\mathbf{x}_i = (\mathbf{r}_i, s_i)$ is the set of the four spin-spatial coordinates of the particle i. This function must include the continuous parameter k which is the electron's wave number (see below). It should be mentioned that such an "additional" ionization is unrelated to the interaction between the emitted β-electron and atomic electron. In fact, the additional ionization is a direct consequence of the non-zero component $\simeq \phi(\mathbf{x}_3)$ in the incident atomic wave function.

The probability of additional ionization has been determined for a number of β^--decaying atoms in a number of earlier studies (see, e.g., Refs. 4 and 5). This work, however, is restricted to the analysis of electron ionization from the internal K-shells only. In this case the original problem was reduced to the solution of the model one-electron problem. Analogous reduction for few-electron atomic systems is much more difficult to perform, since all electron–nucleus and electron–electron coordinates are not truly independent. It complicates accurate computation of integrals which contain electron–electron coordinates explicitly. Nevertheless, numerical computations of the final state probabilities can be conducted even with highly accurate wave functions known for many few-electron atoms. In this section we discuss some details of such calculations.

In atomic units the explicit form of the one-electron wave function is $\phi(\mathbf{r}) = \phi_{kl}(r)Y_{lm}(\mathbf{n})$, where $\phi_{kl}(r)$ is the one-electron radial function, while $Y_{lm}(\mathbf{n})$ is the corresponding spherical harmonics and $\mathbf{n} = \frac{\mathbf{r}}{r}$ is the unit vector associated with $\mathbf{r}$. In this Section the parameter k is $k = \sqrt{\frac{2m_eE}{\hbar^2}} = \sqrt{2E}$ (in atomic units). The explicit formula for the radial function $\phi_{kl}(r)$ (in atomic units) is (see, e.g., Ref. 5)

$$\phi_{kl}(r) = \frac{C_{kl}}{(2l+1)!}(2Qkr)^l \cdot \exp(-\imath Qkr) \cdot {}_1F_1\left(\frac{\imath}{Qk} + l + 1, 2l + 2, 2\imath Qkr\right), \quad (16)$$

where ${}_1F_1(a, b; x)$ is the confluent hypergeometric function (see, e.g., Ref. 11), while C_{kl} is the following constant

$$C_{kl} = \left[\frac{8\pi Qk}{1 - \exp\left(-\frac{2\pi}{Qk}\right)} \right]^{\frac{1}{2}} \cdot \prod_{s=1}^{l} \sqrt{s^2 + \frac{1}{Q^2 k^2}}. \tag{17}$$

In these two equations the parameter Q is the electric charge of the remaining double-charged (positive) ion, i.e., $Q = 2$. In reality, this parameter must slightly be varied (around 2) to obtain better agreement with the experimental data. Such variations formally represent ionizations from different electronic shells of the incident Li atom.

Accurate numerical computations of the final state probabilities during the nuclear β^{-}-decay in few-electron atoms with additional electron ionization are very difficult to perform, since all highly accurate wave functions of the bound states explicitly include the electron–electron coordinates (see above). As a rule, the better accuracy of the bound state wave function means more complete and accurate involvement of the terms which describe various electron–electron correlations. On the other hand, the crucial step of the whole procedure is the numerical and/or analytical computation of the Fourier transforms of the one-electron wave function. This corresponds to the free motion of the final electron. During that step, it is better to consider all electrons as particles independent of each other, i.e., ignore all electron–electron correlations. In the general case, this two-fold problem has no simple solution which is accurate and relatively simple for Fourier transform at the same time.

In this study we have developed an approximate procedure which can be used to perform approximate numerical evaluations for the β^{-}-decaying isotopes of three-electron atoms. In this approach the trial wave function is constructed as the sum of many terms and each of these terms contains the products of the electron–nucleus functions. None of the three electron–electron coordinates r_{32}, r_{31}, r_{21} is included in such trial wave functions. For the ground state (the doublet ^{2}S(L=0)-state) of the Li atom the radial wave function $\psi_{L=0}(A; \{r_{ij}\})$ is chosen in the following form:

$$\begin{aligned} &\psi_{L=0}(r_{14}, r_{24}, r_{34}, 0, 0, 0) \\ &= \sum_{k=1}^{N_s} C_k r_{14}^{m_1(k)} r_{24}^{m_2(k)} r_{34}^{m_3(k)} exp(-\alpha_k r_{14} - \beta_k r_{24} - \gamma_k r_{34}) \\ &= \sum_{k=1}^{N_s} C_k r_1^{m_1(k)} r_2^{m_2(k)} r_3^{m_3(k)} exp(-\alpha_k r_1 - \beta_k r_2 - \gamma_k r_3) \end{aligned} \tag{18}$$

where C_k are the linear (or variational) coefficients, while $m_1(k)$, $m_2(k)$, and $m_3(k)$ are the three integer (non-negative) parameters, which are, in fact, the powers of the three electron–nucleus coordinates $r_{i4} = r_i$ $(i = 1, 2, 3)$. Below, we shall assume that the trial wave function Eq. (18) has a unit norm. Furthermore, in all calculations performed for this study only one spin function $\chi_1(\chi_1 = \alpha\beta\alpha - \beta\alpha\alpha)$ is used. It is clear that the wave function Eq. (18) contains only electron-nucleus coordinates and does not include any of the electron–electron coordinates. The real (and non-negative) parameters $\alpha_k, \beta_k, \gamma_k$ are the $3N_s$ varied parameters of the variational expansion, Eq. (18). The wave function, Eq. (18), must be properly symmetrized upon all three electron coordinates. The main question regarding the wave function, Eq. (18), is related to its overall accuracy. If (and only if) it is relatively accurate, then the trial wave function, Eq. (18), can be used in actual computations of the probability amplitudes. For this study we have constructed the 23-term variational wave function shown in Table 11.1 of Ref. 10. This wave function is represented in the form of Eq. (18) and contains no electron–electron coordinates. All sixty-nine $(69 = 3 \times 23)$ non-linear parameters in this trial wave function have been optimized carefully in a series of bound state computations performed for the ground $^2S(L=0)$-state of the Li atom. Finally, the total energy E of the ground ^2S-state of the $^{\infty}$Li atom obtained with this independent-electron wave function is -7.4485927661 a.u. Note that this energy value, E, is close to the exact total energy for the ground state $^{\infty}$Li atom. It indicates the good overall quality of our approximate wave function with 23 terms with no electron–electron coordinates (r_{12}, r_{13}, r_{23}). This wave function is used in computations of the final state probabilities (see below) for the nuclear β^--decay with additional electron ionization in the three-electron Li atom.

Note also that in atomic physics based on the Hartree-Fock and even hydrogenic approximations the ground state in the Li atom is designated as the 2^2S-state, while in the classification scheme developed in highly accurate computations the same state is often denoted 1^2S-state, where the symbol '1' stands for the first (lowest) eigenvalue of the Hamiltonian matrix. This classification scheme is very convenient to work with truly correlated few-electron wave functions. It is clear that no hydrogenic quantum numbers are good in such cases, and we have to use the more appropriate (and convenient) classification scheme. However, to avoid conflicts between these two schemes in this study we follow the traditional system of notation.

The wave function of the final two-electron Be^{2+} ion arising during nuclear β^--decay with the additional ionization can also be approximated by basis functions depending upon the electron–nucleus coordinates only and do not include the electron–electron coordinate r_{21}. For the bound S ($L=0$)-states of the Be^{2+} ion such an expansion takes the form

$$\psi_{L=0}(r_1, r_2, 0) = \sum_{k=1}^{N_s} C_k r_1^{m_1(k)} r_2^{m_2(k)} \exp(-\alpha_k r_1 - \beta_k r_2). \tag{19}$$

The use of the approximate wave functions Eqs. (18) and (19) with no explicit electron–electron coordinate dependence simplifies numerical computations of all integrals required for numerical evaluation of the final state probabilities during the nuclear β^--decay in the three-electron atoms and ions. The remaining part of the problem is the analytical computation of the integral between the product of the factor $r^m \exp(-\gamma r)$ and radial function from Eq. (16). Such an integral is computed with the use of the formula (see, e.g., Eq. (7.522.9) from Ref. 11):

$$\int_0^{+\infty} \exp(-\lambda x) x^{\nu} \cdot {}_1F_1(a, b; cx)\, dx = \frac{\Gamma(\nu+1)}{\lambda^{\nu+1}} \cdot {}_2F_1\left(a, \nu+1; b; \frac{c}{\lambda}\right), \tag{20}$$

where $\Gamma(x)$ is the usual γ-function (see, e.g., Section 8.31 in Ref. 11). Our results for numerical computation of the final states probabilities for the nuclear β^--decay with additional electron ionization will be published elsewhere.

4. BOUND STATE WAVE FUNCTIONS OF THE THREE-ELECTRON ATOMS AND IONS

To determine the final state probabilities during the nuclear β^--decay (see above), one needs to know the accurate wave functions of the incident and final atoms and ions. In sudden approximation the angular momentum L, electron spin S, and spatial parity π of the atomic wave function Ψ are conserved during nuclear β^--decay. Therefore, all approximate wave functions must be constructed as the eigenfunctions of angular momentum $\hat{L}^2$ and total electron spin $\hat{S}^2$. In this study we use the variational wave functions constructed by Hylleraas-configuration interaction (Hy-CI). In general, the wave functions in Hylleraas-type expansions rapidly converge to the exact wave functions. The Hylleraas-Configuration Interaction wave function[12,13] is a linear combination of symmetry adapted configurations Φ_p:

$$\Psi_{\text{Hy-CI}} = \sum_{p=1}^{N} C_p \Phi_p, \quad \Phi_p = \hat{O}(\hat{L}^2)\hat{\mathcal{A}}\psi_p \chi, \tag{21}$$

where the spatial part of the basis functions are Hartree products of Slater orbitals containing up to one inter-electronic distance r_{ij} per configuration:

$$\psi_p = r_{ij}^{\nu} \prod_{k=1}^{n} \phi_k(r_k, \theta_k, \varphi_k). \tag{22}$$

If $\nu = 1$, then the wave function, Eq. (21), corresponds to Hy-CI. In the case when $\nu = 0$, it is the usual configuration interaction (CI) wave function. The higher powers of the electron–electron distances r_{ij}^{ν} can effectively be reduced to the r_{ij} term (or $\nu = 1$). Indeed, all higher terms $\nu > 1$ can be expressed as a product of r_{ij}, a polynomial of r_i, r_j, and some angular functions. Also, in Eq. (21) N is the number of configurations used in computations. The coefficients C_p are determined variationally. The operator $\hat{O}(\hat{L}^2)$ in Eq. (21) projects over the appropriate space, so that every configuration is an eigenfunction of the square of the angular momentum operator $\hat{L}^2$. $\hat{\mathcal{A}}$ is the antisymmetrization operator upon all electron spin-spatial coordinates and χ is the electron-spin eigenfunction. For the lithium atom and three-electron ions one can choose the total spin function in the one-component form, i.e., $\chi = (\alpha\beta - \beta\alpha)\alpha$.

The basis functions ϕ_k in this work are the s-, p-, d-, and f-Slater orbitals. Since the convergence of Hy-CI wave functions is usually very fast, there is no need to use orbitals with higher angular momentum. The un-normalized complex Slater orbitals are defined as:

$$\phi(\mathbf{r}) = r^{n-1} e^{-\alpha r} Y_l^m(\theta, \varphi), \tag{23}$$

where the parameter α is the adjustable variable (for each orbital) and $Y_l^m(\theta, \varphi)$ are the complex spherical harmonics. The basis sets employed in this work are n=4, 5, 6, and 7, where the basis $n = 4$ means the orbital set [1s2s3s4s2p3p4p3d4d4f]). With all these orbitals from our basis set we have constructed the most important configurations of the S($L = 0,\ M = 0$)-, P($L = 1,\ M = 0$)-, and D($L = 2,\ M = 0$)-symmetries. All details of construction of the symmetry adapted configurations Φ_p of Eq. (21) can be found in Ref. 14.

The orbital exponents have been optimized for each atomic state of the Li atom and Be^+ ion. A set of two exponents were used, one for the *K*-shell and the other for the odd-electron of the *L*-shell. This is constant for all configurations, to accelerate numerical computations. The results obtained are sufficiently accurate for the purposes of our investigation. It is clear that to obtain highly accurate energies one needs to apply more flexibility in the exponents. It was shown in recent calculations on the lithium atom and beryllium ion.[6,15–20] The virial factor $\chi = -\frac{\langle V \rangle}{\langle T \rangle}$ is used to check the quality of the wave function and guide the numerical optimization of the exponents in the trial wave functions.

As for a given basis set the number of possible configurations would be too large, we have selected the "most important" configurations according to their contribution to the total energy. In our case the selection criteria is an energy contribution $> 1 \times 10^{-8}$ a.u. with respect to the previous configuration. For that, blocks containing all possible Hy-CI configurations of the same type have been filtered and the configurations with less energetic weight have been thrown out. More details can be found in.[14] Note that the length of the wave functions varies then for every state, and the selected configurations are for every state different. As a result, higher excited states must be approximated with the use of the longer trial functions.

For our calculations in this study we have written a three-electron Hy-CI computer program in the Fortran 90 language. The calculations have been done in quadruple precision. The program has been thoroughly checked by comparing results of our numerical calculations with the analogous results obtained earlier by Sims and Hagstrom[18] and King[19] for the lithium atom. Note that in such calculations we observe complete agreement.

Ground and excited S-state energy calculations of Li both the atom and Be^+ ion are shown in Table 11.1, together with their convergence with the basis set used. For Li atom and Be^+ ion ground states, an accuracy of $1.4 \cdot 10^{-6}$ a.u. has been achieved with the techniques described in this paper. For the first two excited states of both Li atom and Be^+ ion the accuracy is of about $(4 - 9) \cdot 10^{-6}$ a.u. In the third and higher excited states within a given symmetry of the S-,P-, and D-states the accuracy is of order $\pm 1 \cdot 10^{-4}$ a.u. For the accurate calculation of these higher excited states it would be necessary to introduce different sets of orbital exponents and to increase the orbital basis. However, numerical calculations on higher excited states are rare in the literature.

For numerical calculations of the amplitudes and transition probabilities during nuclear β^--decays in three-electron atomic systems we have developed a new computer program which calculates the overlap integrals, Eq. (4), between the wave functions of Li atom and Be^+ ion. The previous step is the calculation of the wave functions of the different states of the Li and Be^+ atoms using the Hy-CI method. The algorithms we have employed for the calculation of the kinetic and potential energy integrals can be found in Refs. 21–23. The energy values obtained for the ground and S-, P-, and D-excited states are given in Ref. 14 as more details on the calculation and the comparison with the best data of the bibliography. Conversely, in the calculation of the overlaps we need only the overlap integrals between the configurations, and the coefficients of the Hy-CI wave functions. Therefore in this program only integrals of the types $\langle r_{12}^n \rangle$, $\langle r_{12} r_{13} \rangle$ are needed, while the fully-linked three-electron integral $\left\langle \frac{r_{12} r_{13}}{r_{23}} \right\rangle$ is not needed, when the overlap between the wave functions containing the r_{ij} terms is calculated.

In this work we have improved our earlier method of calculation of the final state probabilities during the nuclear β^--decay.[1] Now, we calculate the overlap between the wave functions of different length. This overlap is the sum of the matrix elements of a rectangular overlap matrix. This method of calculation has an advantage, since there are several possible verifications. First, the permutation symmetry of the overlap matrix $\langle \Psi_1 \mid \Psi_2 \rangle = \langle \Psi_2 \mid \Psi_1 \rangle$ and its unit-norm condition, i.e. $\langle \Psi_1 \mid \Psi_1 \rangle = \langle \Psi_2 \mid \Psi_2 \rangle = 1$.

The convergence of the probability amplitudes and probabilities increases with the improvement of the total energies of the incident and final atomic systems. The final transition probabilities are calculated with an error $\leq 0.001\%$ (they are summarized in Table 11.2). We have obtained the transition probability from the ground ^{2}S-state of the Li atom to the ground state of the Be^+ ion $\approx 57.712\%$. The transition probability for the transition from the ground ^{2}S-state of the Li atom to the first excited ^{2}S-state of Be^+ ion is $\approx 26.515\%$ and to the second excited ^{2}S-state such a probability is $\approx 0.544\%$. The sum of probabilities is then $\approx 85.09\%$ and the ionization probability calculated as 1.0 minus this sum is $\approx 14.91\%$. In addition we have computed the transition probabilities from the lower-lying excited states of S-, P-, and D- symmetry of the Li atom to the states of the same symmetry states in the Be^+ ion. The probability distributions can be found in Tables 11.3–11.5, respectively. It is clear that the sum of the probabilities of transition from one state of the incident atom to the states of the final one must always be less (or equal) unity. In this work we have checked this condition everywhere. In general we have found that

Table 11.3 Transition probabilities for the nuclear β^- decay from the excited 3^2S-, 4^2S-, 5^2S-, 6^2S- and 7^2S-states of the Li atom[a] into the ground and various excited states of the Be^+ ion[b]

States Li → Be^+	Amplitude	Probability	P_i in %
$3^2S \rightarrow 2^2S$	0.239962786	0.057582	5.76
$3^2S \rightarrow 3^2S$	0.466529800	0.217650	21.76
$3^2S \rightarrow 4^2S$	0.757456066	0.573740	57.37
$3^2S \rightarrow 5^2S$	0.055586071	0.003090	0.31
$3^2S \rightarrow 6^2S$	0.012740357	0.000162	0.02
$3^2S \rightarrow 7^2S$	0.013723711	0.000188	0.02
Total		0.852412	85.24
P_{ion}		0.147588	14.76
$4^2S \rightarrow 2^2S$	0.132669559	0.017601	1.76
$4^2S \rightarrow 3^2S$	0.236587524	0.055974	5.60
$4^2S \rightarrow 4^2S$	0.122373066	0.014975	1.50
$4^2S \rightarrow 5^2S$	0.828124464	0.685790	68.58
$4^2S \rightarrow 6^2S$	0.277774076	0.077158	7.72
$4^2S \rightarrow 7^2S$	0.007347388	0.000054	0.01
Total		0.851552	85.16
P_{ion}		0.148447	14.84
$5^2S \rightarrow 2^2S$	0.087318854	0.007625	0.76
$5^2S \rightarrow 3^2S$	0.148984137	0.022196	2.22
$5^2S \rightarrow 4^2S$	0.109684232	0.012031	1.20
$5^2S \rightarrow 5^2S$	0.175864858	0.030928	3.09
$5^2S \rightarrow 6^2S$	0.698154162	0.487419	48.74
$5^2S \rightarrow 7^2S$	0.503067106	0.253076	25.31
Total		0.813276	81.33
P_{ion}		0.186724	18.67
$6^2S \rightarrow 2^2S$	0.063750613	0.004064	0.41
$6^2S \rightarrow 3^2S$	0.104007178	0.010817	1.08
$6^2S \rightarrow 4^2S$	0.079072301	0.006252	0.63
$6^2S \rightarrow 5^2S$	0.071415619	0.005100	0.51
$6^2S \rightarrow 6^2S$	0.350972033	0.123181	12.32
$6^2S \rightarrow 7^2S$	0.430715551	0.185516	18.55

[a]The calculated total energy of the first excited S-state of lithium atom is −7.354093706 a.u. (3^2S-state), while the total energies of the second and higher excited states are −7.318517759 a.u. (4^2S-state), −7.303458818 a.u. (5^2S-state), and −7.295734702 a.u. (6^2S-state), respectively.

[b]The total energies of the same *S*-states in the Be^+ can be found in Table 11.1.

Table 11.4 Transition probabilities between states of P-symmetry for the nuclear β^--decay of the Li[a] atom to the Be^+ ion[b]

States Li → Be^+	Amplitude	Probability	P_i in %
$2^2P \to 2^2P$	0.697549959	0.486576	48.66
$2^2P \to 3^2P$	0.603885572	0.364678	36.47
$2^2P \to 4^2P$	0.003979607	0.000016	0.00
$2^2P \to 5^2P$	0.020232690	0.000409	0.04
$2^2P \to 6^2P$	0.013143263	0.000173	0.02
$2^2P \to 7^2P$	0.013285358	0.000176	0.02
Total		0.852028	85.20
P_{ion}		0.147972	14.80
$3^2P \to 2^2P$	0.275908160	0.076125	7.61
$3^2P \to 3^2P$	0.319479925	0.102067	10.21
$3^2P \to 4^2P$	0.801261129	0.642019	64.20
$3^2P \to 5^2P$	0.166010974	0.027560	2.76
$3^2P \to 6^2P$	0.004047006	0.000016	0.00
$3^2P \to 7^2P$	0.004025567	0.000016	0.00
Total		0.847804	84.78
P_{ion}		0.152196	15.22
$4^2P \to 2^2P$	0.161045822	0.025936	2.59
$4^2P \to 3^2P$	0.195960248	0.038400	3.84
$4^2P \to 4^2P$	0.046100299	0.002125	0.21
$4^2P \to 5^2P$	0.724469360	0.524856	52.49
$4^2P \to 6^2P$	0.425779325	0.181288	18.13
$4^2P \to 7^2P$	0.425325535	0.180902	18.09
Total		0.953507	95.35
P_{ion}		0.046493	4.65
$5^2P \to 2^2P$	0.113441928	0.012869	1.29
$5^2P \to 3^2P$	0.135765197	0.018432	1.84
$5^2P \to 4^2P$	0.017086489	0.000292	0.03
$5^2P \to 5^2P$	0.328135053	0.107673	10.77
$5^2P \to 6^2P$	0.547505865	0.299763	29.98
$5^2P \to 7^2P$	0.550438371	0.302982	30.30
$6^2P \to 2^2P$	0.081224665	0.006597	0.66
$6^2P \to 3^2P$	0.099158755	0.009832	0.98
$6^2P \to 4^2P$	0.029284955	0.000858	0.09
$6^2P \to 5^2P$	0.177208892	0.031403	3.14
$6^2P \to 6^2P$	0.353201418	0.124751	12.48
$6^2P \to 7^2P$	0.348353690	0.121350	12.13

[a]The total energies of the incident P-states of Li atom are: −7.410149067 a.u. (2^2P-state), −7.337050609 a.u. (3^2P-state), −7.311770213 a.u. (4^2P-state), −7.299899542 a.u. (5^2P-state) and −7.293494640 a.u. (6^2P-state), respectively.

[b]The total energies of the same (final) P-states of the Be^+ ion are: −14.179326074 a.u., −13.885034739 a.u., −13.783519845 a.u., −13.733901878 a.u., −13.711935225 a.u., and −13.711378665 a.u., respectively.

Table 11.5 Transition probabilities between states of D-symmetry for the nuclear β^--decay of the Li atom[a] to the Be^+ ion[b]

States Li → Be^+	Amplitude	Probability	P_i in %
$3^2D \rightarrow 3^2D$	0.613 879 768	0.376 848	37.68
$3^2D \rightarrow 4^2D$	0.675 444 736	0.456 226	45.62
$3^2D \rightarrow 5^2D$	0.124 573 183	0.015 518	1.55
$3^2D \rightarrow 6^2D$	0.005 331 066	0.000 028	0.00
$3^2D \rightarrow 7^2D$	0.008 233 705	0.000 068	0.01
$3^2D \rightarrow 8^2D$	0.004 516 156	0.000 020	0.00
Total		0.848 709	84.87
P_{ion}		0.151 291	15.13
$4^2D \rightarrow 3^2D$	0.297 395 858	0.088 444	8.84
$4^2D \rightarrow 4^2D$	0.092 613 446	0.008 577	0.86
$4^2D \rightarrow 5^2D$	0.645 316 813	0.416 434	41.64
$4^2D \rightarrow 6^2D$	0.314 342 441	0.098 811	9.88
$4^2D \rightarrow 7^2D$	0.002237872	0.000005	0.00
$4^2D \rightarrow 8^2D$	0.000807365	0.000001	0.00
Total		0.612272	61.23
P_{ion}		0.387728	38.77
$5^2D \rightarrow 3^2D$	0.221060550	0.048868	4.89
$5^2D \rightarrow 4^2D$	0.131551368	0.017306	1.73
$5^2D \rightarrow 5^2D$	0.303521357	0.092125	9.21
$5^2D \rightarrow 6^2D$	0.662141399	0.438431	43.84
$5^2D \rightarrow 7^2D$	0.408116762	0.166559	16.66
$5^2D \rightarrow 8^2D$	0.051512344	0.002653	0.26
Total		0.765943	76.59
P_{ion}		0.234057	23.41
$6^2D \rightarrow 3^2D$	0.219012937	0.047966	4.80
$6^2D \rightarrow 4^2D$	0.109576684	0.012007	1.20
$6^2D \rightarrow 5^2D$	0.275765182	0.076046	7.60
$6^2D \rightarrow 6^2D$	0.172623479	0.029799	2.98
$6^2D \rightarrow 7^2D$	0.238431045	0.056849	5.68
$6^2D \rightarrow 8^2D$	0.469886268	0.220793	22.08
$7^2D \rightarrow 3^2D$	0.250505561	0.062753	6.28
$7^2D \rightarrow 4^2D$	0.224414920	0.050362	5.04
$7^2D \rightarrow 5^2D$	0.113445739	0.012870	1.29
$7^2D \rightarrow 6^2D$	0.227483595	0.051749	5.17
$7^2D \rightarrow 7^2D$	0.476006811	0.226582	22.66
$7^2D \rightarrow 8^2D$	0.429479022	0.184452	18.45

[a] The total energies of the incident D-states of Li atoms are: −7.335511694 a.u (3^2D-state), −7.311211047 a.u. (4^2D-state), −7.298 835 884 a.u. (5^2D-state), −7.288 077 393 a.u. (6^2D-state), and −7.268731551 a.u. (7^2D-state).

[b] The total energies of the same (final) states of the Be^+ ion are: −13.878 004 697 a.u., −13.778 986 828 a.u., −13.733832498 a.u., −13.705903173 a.u., −13.677 409085 a.u., and −13.660271947 a.u.

the highest transition probability within a group is between an initial state and its one order higher final state (i.e. 3^2P $\rightarrow$ 4^2P). This is consequently fulfilled in all groups of probability distributions. For low-lying states, it converges rapidly to zero. In these groups we calculate the probability of ionization to be around 15%. For higher transitions, the probability of ionization is not calculated here because the transitions to higher excited states like 8 ^{2}S, 8^2P, and 9^2D are expected to be important and these states are not considered here.

5. ON THE DOUBLE β DECAY

The idea of the double nuclear $\beta^{\pm}$-decay in some nuclei was proposed in 1935 by Maria Goeppert Mayer.[24] In this study we do not discuss either theoretical significance of the double β^{-}-decay for nuclear physics, or its possible applications. Instead, consider the difference in the final atomic probabilities which can be obtained in the two following cases: (a) the double nuclear β-decay, and (b) two consecutive (single) nuclear β^{-}-decays. From atomic point of view we need to compare the time $\tau_{2\beta}$ for which two β^{-}-particles leave the nucleus with the regular atomic time $\tau_e = \frac{\hbar^2}{m_e e^4 Q^2}$, where Q is the electric charge of the nucleus expressed in e, i.e., Q = Qe. The condition $\tau_{2\beta} \ll \tau_e$ means sudden emission of the two fast β^{-} particles. In this case the probability amplitude is determined as the overlap integral of the incident and final (atomic) wave functions, Eq. (4). If the equation of the double β^{-} decay is written in the form $X \rightarrow Z^{2+} + \beta_1^- + \beta_2^- + \overline{\nu}_1 + \overline{\nu}_2$, then for the final state probability one finds (in atomic units)

$$w_{\text{ii}} = | \langle \psi_{Z^{2+}}(\mathbf{x}_1, \mathbf{x}_2, \ldots, \mathbf{x}_n) \mid \Psi_X(\mathbf{x}_1, \mathbf{x}_2, \ldots, \mathbf{x}_n) \rangle |^2 . \tag{24}$$

In the opposite case, i.e., when $\tau_{2\beta} \gg \tau_e$, we deal with the two consecutive nuclear β^{-} decays. In this case we need to use the sudden approximation twice. The corresponding equations are $X \rightarrow Y^{+} + \beta_1^- + \overline{\nu}_1$ and $Y^{+} \rightarrow Z^{2+} + \beta_2^- + \overline{\nu}_2$. The amplitude of the final state probability takes the form

$$\begin{aligned} A_{\text{if}} = & \langle \Psi_X(\mathbf{x}_1, \mathbf{x}_2, \ldots, \mathbf{x}_n) \mid \psi_{Y^+}(\mathbf{x}_1, \mathbf{x}_2, \ldots, \mathbf{x}_n) \rangle \\ & \langle \Psi_{Y^+}(\mathbf{x}_1, \mathbf{x}_2, \ldots, \mathbf{x}_n) \mid \psi_{Z^{2+}}(\mathbf{x}_1, \mathbf{x}_2, \ldots, \mathbf{x}_n) \rangle \end{aligned} \tag{25}$$

and the final state probability is $w_{fi} = | A_{fi} |^2$. The sum over all states of the Y^+ ion will lead us back to Eq. (24). However, if $\tau_{2\beta} \gg \tau_e$, then the final state of the Y^+ ion is uniformly defined and Eq. (25) can be used in this case only for this unique state of the Y^+ ion. The sum over all "intermediate" states of the Y^+ ion is reduced to the one term only. From here one easily finds that

$$w_{\rm if}(X \rightarrow Z^{2+}) \geq w_{\rm fi}(X \rightarrow Y^{+}) w_{\rm if}(Y^{+} \rightarrow Z^{2+}). \tag{26}$$

Based on these formulas one can expect to observe substantial differences in the final state (atomic) probabilities of the double β^- decay and two consequtive β^- decays. Such differences can be found for the ground-ground and ground-excited transition amplitudes in the case of bound state transitions. The corresponding ionization probabilities can also be very different for any atom in which the central nucleus decays with the use of the double β^- decay, or by the two consequtive β^- decays. In an ideal case, we can observe and measure such differences in "traditional" atoms with the double beta-decaying nuclei, e.g., in the ^{76}Ge, ^{128}Te, ^{138}Xe and ^{238}U atoms.

In general, the study of the double nuclear β^--decay in atoms and molecules can be used as a natural and powerful tool to study electron–nucleus and electron–electron correlations at the femto- and attosecond time-scale. Unfortunately, at this moment no experimental group in the World performs similar research even for atoms and ions.

6. CONCLUSION

We have considered the nuclear β^--decays in the three-electron ^{8}Li and ^{9}Li atoms. The final state probabilities to form different bound states in the Be^+ ion have been determined to very good accuracy which is better than analogous accuracy obtained in our earlier study.[1] The Hylleraas-CI wave functions constructed for atoms/ions involved in the β^--decay are substantially more accurate than wave functions used in earlier studies. They provide a better description of the electron density near of the nucleus. For the first time, the wave functions of the excited states are determined to the same numerical accuracy as the wave functions of the ground states. We can determine the final state probabilities to very high accuracy using them.

We also discuss a possible observation of double nuclear β^--decay and nuclear β^--decay with the additional electron ionization. It is shown that the Be^{2+} ion formed during the last process can be detected not only in the singlet bound states, but also in the triplet bound states. It was never observed/predicted in earlier studies. It may lead to the fundamental re-structuring of the internal electron shells of the incident atom during β^--decay. We also derive some useful formulae which will be used in future studies to determine the probabilities of electron ionization (in different channels) during the nuclear β^--decay.

REFERENCES

1. Frolov, A. M.; Ruiz, M. B. Atomic Excitations During the Nuclear β^- Decay in Light Atoms. *Phys. Rev. A* **2010,** *82*, 042511.
2. Frolov, A. M.; Talman, J. D. Final-state Probabilities for β^--Decaying Light Atoms. *Phys. Rev. A* **2005,** *72*, 022511.
3. Migdal, A. B. Ionization of Atoms Accompanying α- and β-Decay. *J. Phys. Acad. Sci. U.S.S.R.* **1941,** *4*, 449–453.
4. Migdal, A. B.; Krainov, V. *Approximation Methods in Quantum Mechanics*; W.A. Benjamin: New York, 1969.
5. Landau, L. D.; Lifshitz, E. M. *Quantum Mechanics: Non-relativistic Theory*; Pergamon Press: New York, 1976 Chpt VI.
6. Frolov, A. M. Compact and Accurate Variational Wave Functions Of Three-Electron Atomic Systems Constructed From Semi-Exponential Radial Basis Functions. *Eur. Phys. J. D* **2011,** *61*, 571–577.
7. Larsson, S. Calculations on the 2S Ground State of the Lithium Atom Using Wave Functions of Hylleraas Type. *Phys. Rev.* **1968,** *169*, 49–54.
8. Dirac, P. A. M. *The Principles of Quantum Mechanics*; Clarendon Press: Oxford, 1958.
9. Frolov, A. M.; Wardlaw, D. M. On Bound State Computations in Three- and Four-Electron Atomic Systems. *JETP* **2010,** *138*, 5–15.
10. Frolov, A. M.; Wardlaw, D. M. On the Nuclear *(n;t)*-Reaction in the Three-Electron ^{6}Li atom. *Phys. Rev. A* **2013,** *87*, 042702.
11. Gradstein, I. S.; Ryzhik, I. M. *Tables of Integrals*, 6th revised ed.; Series and Products; Academic Press: New York, 2000.
12. Sims, J. S.; Hagstrom, S. A. One-Center Be^{2+} Integrals Over Slater Orbitals. *J. Chem. Phys.* **1971,** *55*, 4699–4710.
13. Sims, J. S.; Hagstrom, S. A. Combined Configuration-Interaction-Hylleraas-Type Wave-Function Study of the Ground State of the Beryllium Atom. *Phys. Rev. A* **1971,** *4*, 908–916.
14. Ruiz, M. B.; Margraf, J. T.; Frolov, A. M. Hylleraas-Configuration-Interaction Analysis of the Low-lying States in the Three-Electron Li atom and Be^+ ion. *Phys. Rev. A* **2013,** *88*, 012505.
15. Puchalski, M.; Kedziera, D.; Pachucki, K. Ground State of Li and Be^+ Using Explicitly Correlated Functions. *Phys. Rev. A* **2009,** *80*, 032521.
16. Puchalski, M.; Pachucki, K. Relativistic, QED, and Finite Nuclear Mass Corrections for Low-lying States of Li and r_{ij}. *Phys. Rev. A* **2008,** *78*, 052511.
17. Wang, L. M.; Yan, Z.-C.; Qiao, H. X.; Drake, G.W. F. Variational Energies and the Fermi Contact Term for the Low-lying States of Lithium: Basis-set Completeness. *Phys. Rev. A* **2012,** *85*, 052513.
18. Sims, J. S.; Hagstrom, S. A. Hylleraas-Configuration-Interaction Study of the 22S Ground State of Neutral Lithium and the First Five Excited 2S States. *Phys. Rev. A* **2009,** *80*, 052507.

19. King, F. W. High-Precision Calculations for the Ground and Excited States of the Lithium Atom. *Adv. At. Mol. Opt. Phys.* **1999,** *40*, 57–112.
20. Thakkar, A. J.; Koga, T.; Tanabe, T.; Teruya, H. *Chem. Phys. Lett.* **2002,** *366*, 95.
21. Ruiz, M. B. Evaluation of Hylleraas-CI Atomic Integrals. III. Two-Electron Kinetic Energy Integrals. *J. Math. Chem.* **2011,** *49*, 2457–2485.
22. Ruiz, M. B. Evaluation of Hylleraas-CI Atomic Integrals by Integration Over the Coordinates of One Electron I. Three-electron Integrals. *J. Math. Chem.* **2009,** *46*, 24–64.
23. Sims, J. S.; Hagstrom, S. A. Mathematical and Computational Science Issues in High Precision Hylleraas-Configuration Interaction Variational Calculations: I. Three-electron Integrals. *J. Phys. B: At. Mol. Opt. Phys.* **2004,** *37*, 1519–1540.
24. Giuliani, A.; Poves, A. Neutrinoless Double-Beta Decay. *Adv. High Energy Phys.* **2012,** 38, 857016.
25. Stanke, M.; Komasa, J.; Kedziera, D.; Bubin, S.; Adamowicz, L. Three Lowest S States of ^{9}Be$^+$ Calculated with Including Nuclear Motion and Relativistic and QED Corrections. *Phys. Rev. A* **2008,** 77, 062509.
26. King, F. W. High-precision Calculations of the Hyperfine Constants for the Low-lying Excited ^{2}S States of Be$^+$. *J. Phys. Chem. A* **2009,** *113*, 4110–4116.

Index

T

W

X

Z

Kolibri 360 GmbH
Ferdinand-Jühlke-Straße 7
99095 Erfurt, Deutschland
produktsicherheit@kolibri360.de